STRATIGRAPHY
TERMINOLOGY AND PRACTICE

Jacques REY and Simone GALEOTTI (Eds.)

STRATIGRAPHY
TERMINOLOGY AND PRACTICE

Work initiated by the French Committee of Stratigraphy

2008

25 rue de Ginoux, 75015 PARIS, FRANCE

FROM THE SAME PUBLISHER

Essentials of Reservoir Engineering
P. DONNEZ

A Geoscientist's Guide to Petrophysics
B. ZINSZNER, F.M. PELLERIN

Sedimentary Geology
B. BIJU-DUVAL

Basics of Reservoir Engineering
R. COSSÉ

Geophysics of Reservoir and Civil Engineering
J.L. MARI, G. ARENS, D. CHAPELLIER, P. GAUDIANI

Comprehensive Dictionary of Earth Science
M. MOUREAU, G. BRACE

Geophysics for Sedimentary Basins
G. HENRY

Dictionary of Seismic Prospecting
FRENCH OIL GAS INDUSTRY ASSOCIATION

All rights reserved.

No part of this publication may be reproduced or transmitted in any form or by any means, electronic or mechanical, including photocopy, recording, or any information storage and retrieval system, without the prior written permission of the publisher.

© Editions Technip, Paris, 2008.

Printed in France

ISBN 978-2-7108-0910-4

TABLE OF CONTENTS

Acknowledgments	VIII
Foreword	IX

Chapter 1. STRATIGRAPHY: FOUNDATIONS AND PERSPECTIVES
J. REY

1. – DEFINITION OF STRATIGRAPHY 1
2. – THE LANGUAGE OF STRATIGRAPHY 1
3. – THE NEED FOR A COMMON LANGUAGE 1
4. – STRATIGRAPHIC METHODS AND PROCESSES 2
   - 4.1. From geometric stratigraphy to chronometric stratigraphy 2
   - 4.2. Stratigraphic methods 2
   - 4.3. Stratigraphic markers 3
   - 4.4. Applications of stratigraphic methods 4
5. – STRATIGRAPHIC TRENDS 4
   - 5.1. A multidisciplinary system 4
   - 5.2. Universality 4
   - 5.3. Accuracy 4
6. – ORGANIZATION OF THE BOOK 5

Chapter 2. LITHOSTRATIGRAPHY from lithologic units to genetic stratigraphy
L. COUREL (COORD.), J. REY, P. COTILLON, J. DUMAY, P. MAURIAUD, P. RABILLER, J.F. RAYNAUD & G. RUSCIADELLI

1. – DEFINITION (L. COUREL & J. REY) 7
2. – LITHOSTRATIGRAPHIC UNITS 8
   - 2.1. Terminology 8
     - 2.1.1. Surface units (L. COUREL) 8
     - 2.1.2. Subsurface units (J.F. RAYNAUD, P. MAURIAUD & P. RABILLER) 11
   - 2.2. Practice 11
     - 2.2.1. Chronostratigraphic value of lithostratigraphic units (L. COUREL & J. REY) 11
     - 2.2.2. Specific issues in the definition of subsurface units (P. MAURIAUD & J. DUMAY) 12
3. – FROM SEQUENCES OF OBJECTS TO GENETIC SEQUENCES (J. REY) 17
   - 3.1. Terminology 17
     - 3.1.1. Sequences of objects 17
     - 3.1.2. Facies sequences 19
     - 3.1.3. Genetic sequences 19
   - 3.2. Practice 20
     - 3.2.1. Chronostratigraphic significance of sequences of objects 20
     - 3.2.2. Relations between lithostratigraphic units, sequences of objects and genetic sequences 20
4. – SEQUENCE STRATIGRAPHY (G. RUSCIADELLI) 20
   - 4.1. Definition 20
   - 4.2. Terminology and concepts 21
     - 4.2.1. Accomodation space 21
     - 4.2.2. The sedimentary signature of accomodation changes 22
     - 4.2.3. The sequence stratigraphic signature of the accomodation 24
   - 4.3. Practice 29
     - 4.3.1. The time-stratigraphic framework of sequences 29
     - 4.3.2. The hierarchy of relative sea level cycles 31
5. – LITHOLOGIC CYCLES (P. COTILLON) 32
   - 5.1. Definition 32
   - 5.2. Practice 33
     - 5.2.1. Cycles characterization 33
     - 5.2.2. Cycles, lithostratigraphic tools 34
6. – CONVENTIONS IN LITHOSTRATIGRAPHY 37
   - 6.1. Naming, definition, publication of lithostratigraphic units (L. COUREL) 37
   - 6.2. Procedure for codification of surface stratigraphic units (J. REY) 38
     - 6.2.1. The case of lithostratigraphic units 38
     - 6.2.2. The case of facies sequences 38
     - 6.2.3. The case of genetic sequences 38
   - 6.3. Procedure for codification of subsurface stratigraphic units (J.-F. RAYNAUD) 38
     - 6.3.1. Well logs and reference section of a lithostratigraphic unit 38
     - 6.3.2. Reference material, nature, preservation and availability 39
     - 6.3.3. Selection 39
     - 6.3.4. Definition 39

Chapter 3. CHEMOSTRATIGRAPHY
M. RENARD, J.C. CORBIN, V. DAUX, L. EMMANUEL, F. BAUDIN & F. TAMBURINI

1. – DEFINITION 41
2. – TERMINOLOGY 41
3. – PRACTICE 44
   - 3.1. $CaCO_3$ fluctuations in pelagic carbonates 44
   - 3.2. Oxygen stable isotope variations 44
   - 3.3. Carbon stable isotope variations 45
   - 3.4. Sulfur stable isotopes 46
   - 3.5. Strontium stable isotopes 47
   - 3.6. Trace elements in carbonates 47
   - 3.7. Iridium anomalies 49
   - 3.8. REE, Rare Earth Elements 50
   - 3.9. Organic carbon and geochemical biomarkers 50
   - 3.10. New Frontier in Chemostratigraphy 52

3.10.1 Oxygen isotopes 52
3.10.2 Impact of gas hydrates for the carbon isotopes... 52
3.10.3 Strontium isotopes and other isotopes (e.g., osmium) 52
4.- CONCLUSION 52

Chapter 4. MAGNETOSTRATIGRAPHY
B. Galbrun (Coord.), N.K. Belkaaloul & L. Lanci

1. – DEFINITION 53
   1.1. The earth's magnetic field 53
   1.2. The time-averaged geomagnetic field: the axial dipole hypothesis 54
   1.3. Reversals of the earth's magnetic field 54
2. – TERMINOLOGY 55
   2.1. Magnetostratigraphic polarity units 55
   2.2. Nomenclature 55
3. – PRACTICE 56
   3.1. Field sampling 56
   3.2. Sampling in sediments cores 58
   3.3. Measurement of remanent magnetization 58
   3.4. Samples demagnetization 58
      3.4.1. Alternating field (AF) demagnetization 59
      3.4.2. Thermal demagnetization 59
   3.5. Rock magnetism 59
      3.5.1 Identification of ferrimagnetic minerals in a rock 59
      3.5.2 Curie Temperature analysis 59
      3.5.3 Acquisition of thermal demagnetization of IRM... 59
   3.6. Analysis of magnetization components 59
      3.6.1. Analysis of vector diagrams 60
      3.6.2. Analysis of remagnetization circles 60
   3.7. Corrections of the measured direction of remanent magnetization 61
      3.7.1. Correction for orientation of sample (geographic correction) 61
      3.7.2. Correction for tilt of bedding (tectonic correction) 61
   3.8. Field tests of magnetization stability 61
      3.8.1. Reversal test 62
      3.8.2. Fold test 62
      3.8.3. Conglomerate test 62
   3.9. Statistical analysis of directions and poles 62
      3.9.1. The Fisher distribution and its use in paleomagnetism 62
      3.9.2. Calculation of virtual geomagnetic pole (VGP) position and its confidence limits 64
      3.9.3. VGP latitude 64
4. – CONCLUSIONS 64

Chapter 5. BIOSTRATIGRAPHY from taxon to biozones and biozonal schemes
J. Thierry & S. Galeotti

1. – DEFINITION AND AIM 65
   1.1. Introduction 65
   1.2. The biostratigraphic procedures 65
   1.3. Index fossils, isochrony, dating and correlations 66
   1.4. Evolution of concepts and methods 66
   1.5. The biozone, the basic unit of biostratigraphy 66
   1.6. Definition and identification of biozones 66
2. – CONCEPTS, METHODS AND TERMS IN BIOSTRATIGRAPHY 67

   2.1. Classical biostratigraphy and biostratigraphic units 68
      2.1.1. Preliminary remarks 68
      2.1.2. Kinds of biostratigraphic units 69
   2.2. The logical biostratigraphy or Unitary Associations method 74
      2.2.1. Definitions 74
      2.2.2. The construction of biochronozones 74
   2.3. "Statistical biostratigraphy" 74
      2.3.1. Graphical methods 75
      2.3.2. Semi-empirical methods 76
      2.3.3. Multivariate analysis 76
      2.3.4. Probabilistic methods 77
   2.4. Hierarchy and sub-categories of biostratigraphic units 77
      2.4.1. Biohorizon, zonule and marker-bed 77
      2.4.2. Subbiozones and superbiozones 78
      2.4.3. Unitary associations and biochronozones 78
3. – PRACTICE 78
   3.1. The steps of the procedure of biostratigraphy 78
   3.2. Results of the classical biostratigraphy 79
      3.2.1. Use of the classic biozones 79
      3.2.2. Possible "diachronism of bioevents" 82
      3.2.3. Significance, precision and reliability of the classical biozones 82
   3.3. Results of the logical biostratigraphy 83
   3.4. Results of the statistical biostratigraphy 83
   3.5. Relations between the classic biostratigraphic units, the units of the logical biostratigraphy and the units of the statistical biostratigraphy 85
   3.6. Relations between biochronology, geochronology and chronostratigraphy 85
      3.6.1. Biochronology and chronostratigraphy: biozone and stage boundary 85
      3.6.2. Biochronology and geochronology: the geochronologic calibration of biostratigraphic scales 86
   3.7. Recommendations in biostratigraphy 88
      3.7.1. Definition and denomination of biostratigraphic units 88
      3.7.2. Validation of biostratigraphic units 89
      3.7.3. Use of biozones and biostratigraphic scales 89

Chapter 6. ISOTOPE GEOCHRONOLOGY
N. Clauer & A. Cocherie

1. – INTRODUCTION 91
2. – DESCRIPTION OF METHODS 92
   2.1. Potassium-argon methods: K-Ar, $^{40}Ar/^{39}Ar$ 92
   2.2. Isochron type methods 93
   2.3. U-Th-Pb method on separated minerals 93
3. – TECHNICAL ASPECTS 94
   3.1. Which geochronometers? 94
   3.2. Specific characteristics of sedimentary geochronometers 94
   3.3. Initial isotopic homogenization 95
4. – DIRECT ISOTOPIC DATING OF SEDIMENTARY ROCKS 96
   4.1. K-Ar, Rb-Sr and Sm-Nd dating of Precambrian sediments 96
   4.2. Other methods of stratigraphic dating of sediments 97
5. – INDIRECT ISOTOPIC DATING OF SEDIMENTARY ROCKS 97
6. – CONCLUSIONS 98

Chapter 7. SPECIFIC STRATIGRAPHIES
P. Lebret, R. Capdevila, M. Campy,
M. Isambert, J.P. Lautridou, J.J. Macaire,
F. Menillet, R. Meyer & A. De Goër de Herve

1. – STRATIGRAPHY OF THE METAMORPHIC
   AND PLUTONIC TERRAINS
   (R. Capdevila) 101
   1.1. The study of lithology, space-relations
        and chronology of plutonic and metamorphic
        bodies is pertaining to stratigraphy 101
   1.2. How should plutonic and metamorphic bodies
        be considered? as lithostratigraphic units
        or as units of lithostratigraphic class or
        as different units of class? 101
   1.3. Stratigraphic classification of bodies of igneous
        and metamorphic rocks 102
        1.3.1. Bodies of volcanic rocks and low-grade
               metavolcanic and metasedimentary rocks ... 102
        1.3.2. Bodies of unstratified plutonic
               and metamorphic rocks 102
        1.3.3. Layered intrusions 103
2. – GEOCHRONOMETRY OF PRECAMBRIAN TIME
   (R. Capdevila) 103
   2.1. Principles of geochronometric subdivision
        of the Precambrian 104
   2.2. Subdivision method 104
   2.3. Geochronometric units of the Precambrian 104
        2.3.1. Archean 104
        2.3.2. Protérozoic 106
3. – STRATIGRAPHY OF SURFICIAL FORMATIONS
   (P. Lebret (coord.), M. Campy, M. Isambert,
   J.-P. Lautridou, J.-J. Macaire, F. Menillet & R. Meyer) 106
   3.1. Définition 106
   3.2. Vocabulary: surficial formations ; study concepts
        and units 107
   3.3. Study methods 107
4. – QUATERNARY STRATIGRAPHY
   (P. Lebret (coord.), M. Campy, M. Isambert,
   J.-P. Lautridou, J.-J. Macaire, F. Menillet & R. Meyer) 108
   4.1. Définition 108
   4.2. Study methods 108
        4.2.1. The Quaternary in an oceanic environment 108
        4.2.2. The Quaternary in a continental environment ... 108
   4.3. Practice: stratigraphic scales for the Quaternary ... 109
5. – STRATIGRAPHY OF VOLCANIC TERRAINS
   (A. de Goër de Herve) 109
   5.1. Presentation 109
   5.2. Vocabulary 109
        5.2.1. Volcanic and volcanogenic materials 109
        5.2.2. Facies associations and lithostratigraphic
               units 111
        5.2.3. Questionable vocabulary 112
   5.3. Practice 113
        5.3.1. Periodicity 113
        5.3.2. Dating 113

Chapter 8. CHRONOSTRATIGRAPHIC UNITS
AND CORRELATIONS
J. Rey (Coord.), L. Courel, J. Thierry,
J.-F. Raynaud & S. Galeotti

1. – DEFINITION 117

2. – TERMINOLOGY 117
   2.1. The chronostratigraphic units 117
        2.1.1. Chronostratigraphic, geochronologic
               and geochronometric units 118
        2.1.2. The Stage 118
        2.1.3. The System 118
        2.1.4. The Era 118
        2.1.5. The chronozone 118
   2.2. The chronostratigraphic scale 118
        2.2.1. The standard chronostratigraphic scale 118
        2.2.2. The regional chronostratigraphic scales 119
        2.2.3. The global chronostratigraphic scale
               and other stratigraphic scales 119
   2.3. The stratotype 119
        2.3.1. Definition 119
        2.3.2. The unit stratotype 119
        2.3.3. The boundary stratotype 119
   2.4. The chrono-correlations 120
3. – PRACTICE 121
   3.1. Duration of chronostratigraphic units 121
   3.2. Denomination of chronostratigraphic units 121
   3.3. A semi-quantitative approach:
        the composite standard reference section 121
   3.4. The procedure of integrated stratigraphy 122
   3.5. Geochronometry of the Precambrian 122
   3.6. Chronostratigraphy of the Plio-Quaternary interval .. 123
4. – CONCLUSIONS: THE STRATIGRAPHIC TOOL 123
5. – CONVENTIONS 123
   5.1. Main rules in the definition of chronostratigraphic units . 123
   5.2. Procedures for the ratification of chronostratigraphic
        units and their stratotypes 123

Chapter 9. THE GEOLOGICAL TIME SCALE
F.M. Gradstein, James Ogg & Gabi Ogg

1. – INTRODUCTION 125
2. – BOUNDARY STRATOTYPES 125
   2.1. Definition 125
   2.2. Klonk 126
   2.3. Progress with GSSP's 126
3. – RECONCILE PROTEROZOIC ROCK RECORD
   WITH ABSTRACT TIME 128
4. – UNITS OF TIME 128
5. – BUILDING THE GEOLOGICAL TIME SCALE 130
6. – SEDIMENTARY CYCLES 130
7. – DECAY OF ATOMS 130
8. – INTERPOLATION AND STATISTICS 132
9. – THE GEOLOGICAL TIME SCALE 132
   9.1. GTS 2004 132
   9.2. GTS 2010 133
10. – TIME SCALE CREATOR 134

Glossary 137

References 149

Editors, Authors and Translators 165

ACKNOWLEDGMENTS

The editors are grateful to Total, the BRGM, the Commission for the Geological Map of the World – and particularly P. MAURIAUD, J.F. RAYNAUD, D. VASLET and J.P. CADET – for their valuable helps that allowed the publishing of this work.

FOREWORD

« Stratigraphie. Terminologie française » is one of those educational books that take off where the International Stratigraphic Guide stops. What I mean is that the teaching of stratigraphy from the latter book would not particularly inspire students; of course that is not its function. Having said this, that is precisely what struck me when I first became familiar with this erudite stratigraphic text book in the French language, so ably put together by Professor JACQUES REY and his excellent team on behalf of the "Comité Français de Stratigraphie". In my enthusiasm, when I leafed through the attractively illustrated text, I immediately started thinking that the English language community also should have the advantage of using this rich text book. The eight chapters and the glossary make us familiar with stratigraphic principles and terms in a narrative format, interspersed with applications. Equally important the book dwells on theoritical and practical aspects of e.g. quantitative biostratigraphy, magnetostratigraphy and chemostratigraphy. The chapter on lithostratigraphy demonstrates how a seemingly boring subject can become alive, and attractively also covers sequence stratigraphy, without becoming sidetracked by some of its bad applications; this chapter also spends time on cycle stratigraphy.

Hence, it is with great pleasure that I commended Professors REY and GALEOTTI for assembling a team of experts to translate the book in English – and also to update it resulting in a second edition. This has now been accomplished. In the process of translation and updating, I suggested to add an introduction to theoretical and practical aspects of the Geologic Time Scale (GTS 2004). The new book edition is a very welcome addition to the limited literature on the combined fundamental and practical aspects of stratigraphy. Its content, attractive format and splendid illustrations make this book mandatory for all universitary libraries, and I would not be surprised if petroleum company libraries also will like to have it on quick standby!

The International Commission on Stratigraphy thanks Professors REY and GALEOTTI and their team for this splendid undertaking, and wishes the book a wide circulation.

Felix GRADSTEIN
Chairman of the International
Commission on Stratigraphy (I.C.S.)

The definitions of the terms highlighted in **bold italic** can be found in the glossary at the end of the work.

Chapter 1

STRATIGRAPHY: FOUNDATIONS AND PERSPECTIVES

J. Rey

CONTENTS

1. – DEFINITION OF STRATIGRAPHY 1
2. – THE LANGUAGE OF STRATIGRAPHY 1
3. – THE NEED FOR A COMMON LANGUAGE 1
4. – STRATIGRAPHIC METHODS AND PROCESSES 2
   4.1. From geometric stratigraphy to chronometric stratigraphy 2
   4.2. Stratigraphic methods 2
   4.3. Stratigraphic markers 3
   4.4. Applications of stratigraphic methods 4
5. – STRATIGRAPHIC TRENDS 4
   5.1. A multidisciplinary system 4
   5.2. Universality 4
   5.3. Accuracy 4
6. – ORGANIZATION OF THE BOOK 5

1. – DEFINITION OF STRATIGRAPHY

Stratigraphy, etymologically "description of strata" is the scientific discipline which studies the organization of geological formations and of the events which they produce in space and time, in order to reconstruct the history of the Earth and its evolution through time.

This definition in itself demonstrates that the field of investigation in the domain of stratigraphy is vast, given that it covers all types of terrain: sedimentary, metamorphic and magmatic. The aims of stratigraphy are also diverse: in the broadest sense of the term, this discipline comprises a temporal component (the establishing of a time frame) and a spatial component (organization of paleogeography and paleoenvironments). More strictly speaking – and this is a definition which has become accepted through usage – it is the chronological aspect which constitutes the essence of stratigraphy. Although the two aims are closely linked, we will focus principally in this book on the problems of stratigraphic terminology relating to time, in other words, those relating to chronostratigraphic connotations or effects.

Stratigraphy is a harmonizing discipline which unites the sciences and other, more varied disciplines: petrology, mineralogy, sedimentology, paleontology, structural analysis, physics, astronomy, chemistry, statistics... etc. It is also, however, a crucial discipline for geology, which, among the Earth Sciences, gains its unique character from its constant reference to the passage of time.

2. – THE LANGUAGE OF STRATIGRAPHY

As in every scientific discipline, stratigraphy has its own particular language. This terminology must lead to a *stratigraphic classification* which divides the various types of terrain into units which are classed according to their relative importance or duration. Within this language then, the following elements can be distinguished:

– **a vocabulary which is specific** to each type of approach, and which allows tools, markers, concepts, or processes to be named: (for example: "seismic reflectors", "species-index", "isotopic dating", "logician approach"...);

– **common nouns**, qualifying the various types of unit used in stratigraphy in order to divide up geological compositions (for example: "sequence", "association biozone", "magnetozone"...). These terms constitute **stratigraphic terminology**, and are specific to each one of the analytical processes used.

– **proper nouns** referring to the stratigraphic units (for example: "Albian", "Annot Formation", "Brunhes magnetic era", "Lamberti Zone"). These terms belong to **stratigraphic nomenclature**. These terms were chosen by national committees, or in the case of chronostratigraphic units, by the International Commission on Stratigraphy of the International Union of Geological Sciences.

Although stratigraphic terminology is the main aim of this volume, specific vocabulary has been used when we deemed it necessary for communication between stratigraphers or with geologists in other disciplines. Moreover, we considered it useful to provide a reminder of standard practices or procedures used when establishing stratigraphic nomenclature.

3. – THE NEED FOR A COMMON LANGUAGE

Stratigraphy is a discipline which is in a state of constant renewal. It tends to become more diverse, more universal, more quantitative. The increasing plethora of new techniques used is accompanied by a kind of explosion within the discipline and its vocabulary. Consequently, different terms can often relate to the same concept (for example: "genetic", "sequential"); or indeed inversely, the same term may be used with a different meaning for different approaches (for example: "zone", "sequence"). With the recent developments in the Earth Sciences, new terms have appeared which are often poorly defined or not yet in common use and there are certain specialists who would like to give a more modern sheen to a discipline which is nearly two centuries old, by introducing neologisms in order to refer to classic concepts which are still valid.

Developments within stratigraphy have enriched the language which is specific to this discipline, yet have also caused it to become more complex and often altered. In order to preserve its unity and its unifying role, this fundamental discipline, which has accumulated experience and knowledge over time, must maintain its linguistic unity and coherency of expression. Already aware of this issue, in 1962, the French Stratigraphy Committee published a booklet entitled

"Principles of stratigraphic classification and nomenclature", written by J. SIGAL and H. TINTANT. Since this date, no other works on nomenclature have been published in France. The international subcommission for stratigraphic nomenclature has produced reference works (the two successive editions of the "International Stratigraphic Guide", edited by H. HEDBERG in 1976 and A. SALVADOR in 1994) which we deemed to be incomplete or poorly adapted to stratigraphers' current needs.

The aim of this book is therefore to provide a reminder of stratigraphic terminology and to ensure its continuity while acknowledging recent developments in our knowledge. It proposes, defines and comments on the various terms in use within the different processes which come under the umbrella of stratigraphy. Its main objective is to clarify and standardize stratigraphic vocabulary, by introducing, advocating, or indeed advising against the use of certain words and expressions. This volume is therefore not a manual (such as POMEROL et al, 1987). It has been structured around the presentation of a vocabulary which is set out, explained and enhanced by examples and illustrations.

To propose a vocabulary means making certain choices. Two notions, sometimes antinomical, have served as a guide: precision and usage. The vocabulary used in stratigraphy must be rigorous, that is exact, precise, devoid of all ambiguity. However a word is only useful if it is accepted in common usage. The scientific community has often adopted – and commonly used – a term without respecting etymological rigor. Where this is the case, we have favoured the current use of the term, provided that the meaning was clear, rather than attempting to coin a new term whose acceptance would not be guaranteed. It can also happen that a word commonly used in stratigraphy adopts a different meaning according to the author, or that its definition may have altered with the progression of scientific thought, or even that its content has been changed by certain specialists as a result of this progression. Our aim therefore, while taking into account the opinion of experts, was to standardize the usage of terms by favouring etymological rigor and without striving after an unlikely consensus of opinion.

This volume is designed for geologists of all disciplines, for students beginning research, for teachers who wish to clarify documents whose language needs to be simplified for teaching purposes, for those who edit geological documents and whose duty it is to standardize linguistic usage.

The editors would be grateful to receive comments and suggestions from readers, which would allow us to improve, complete, simplify or modify the suggestions presented in this initial attempt to define stratigraphic terminology.

4. – STRATIGRAPHIC METHODS AND PROCESSES

4.1. FROM GEOMETRIC STRATIGRAPHY TO CHRONOMETRIC STRATIGRAPHY

Stratigraphic thought is organized into three basic stages which coincide essentially with the historical development of research over the last two centuries:

– **the first stage** consists of analyzing the geometric relations of geological bodies, according to the three dimensions. Time is not taken into account in terms of estimation or measurement and its passage is therefore disregarded. It is only considered from the point of view of a "time lapse", that which is necessary for example for a certain thickness of sediment to be deposited or for a given lithologic body to be created. The relations established between geological sequences on two different points – or **stratigraphic correlations** in the broad sense of the term – do not imply the identification of an era at this stage. This first phase leads to the elaboration of a **geometric stratigraphy**.

Geometric stratigraphy allows, if necessary, a given number of landmarks to be identified within the geological sequences, due to events which have an effect either on a regional scale (volcanic eruptions) or on a planetary scale (impact of meteorites, abrupt climatic changes, sudden opening of connective channels between oceanic basins which were previously separated...). If, in the subsequent stages, chronological or chronometric stratigraphy succeed in proving the synchronism of such landmarks, the latter will then lead to extremely accurate correlations between these events and the units which record them. This process is classed as **event stratigraphy**.

– **the second stage** allow the passage of time to be estimated (but not measured), that is the relative duration of the units and phenomena studied. It leads to a zoning of the time scale which allows the **relative dating** (without numbered values) of terrains or events to be established as well as simultaneous correlations between geological units. At the end of this stage, a relative **chronological stratigraphy** is achieved.

– **the third stage** is dedicated to the measurement of time, regardless of the nature or the thickness of the geological object and its geometric relations. The latter only appears as a time support, and possesses the crucial function of allowing time to be measured directly. From this, **absolute dating** is achieved, (so called in the sense that this stage may lead to measured units. This does not however mean that the measurement is definitive or perfect) on a time scale graduated in thousands, millions or billions of years. This final stage is a product of numerical, **chronometrical stratigraphy**.

4.2. STRATIGRAPHIC METHODS

In order to progress in stratigraphic processes, various methods which correspond to an equally numerous quantity of analytical and interpretative approaches of geological entities have been developed (Table 1.1). Of particular note are:

– **lithostratigraphy**, which concerns the lithologic characteristics of rocks;

– **seismic stratigraphy** which consists of reconstructing the geometry of deposits using the reflection of waves on surfaces of lithologic discontinuity;

– **genetic stratigraphies** based on the variations in the sedimentary environment (**sequence stratigraphy**) or of the Earth's orbit (**cyclostratigraphy**), gleaned from variations in the lithology in time and/or space;

– **chemostratigraphy**, which draws on the geochemical characteristics of sediments;

– **magnetostratigraphy**, using the residual magnetization of minerals in the rocks;

– **biostratigraphy**, which analyses the paleontological (fossiliferous content) and more precisely the succession of species over time (biochronology), in order to achieve the chronological objective of stratigraphy;

– **isotope geo(radio)chronology**, based on the properties of the natural radioactive decay of unstable isotopes trapped in certain mineral phases.

1. Stratigraphy: Foundations and perspectives

TABLE 1.1

Fundamentals and applications of stratigraphic methods (adapted from SIGAL, 1980 and REY, 1983).

SCIENCES OR DISCIPLINES	METHODS	NATURAL PHENOMENA → STRATIGRAPHIC SIGNALS					
		DISCONTINUOUS REVERSIBLE/REPETITIVE	CONTINUOUS REVERSIBLE	«INSTANTANEOUS» REPETITIVE	DISCONTINUOUS IRREVERSIBLE	CONTINUOUS VARIABLE SPEED	IRREVERSIBLE CONSTANT SPEED
Physics	ISOTOPIC GEOCHRONOLOGY						Natural radioactivity → geochronometers
Evolutive paleoecology	BIO-STRATIGRAPHY				Biological crises → wildlife regeneration	Biological evolution → fossil species	
Physics	MAGNETO-STRATIGRAPHY			Inversions of magnetic polarity			
Chemistry	CHEMIO-STRATIGRAPHY	Variations in the sedimentary environment → Geochemical trends and events			Meteorites → iridium		
Sedimentology Mineralogy Petrology Paleoecology Geophysics Astronomy	GENETIC STRATIGRAPHIES SEQUENCE STRATIGRAPHY	Variations in the sedimentary environment → Sedimentary sequences and discontinuities					
	CYCLOSTRATIGAPHY		Variations in orbital parameters → Cycles				
	SEISMO STRATIGRAPHY	Sedimentation → lithological discontinuities					
	LITHO-STRATIGRAPHY	Sedimentation → Lithostratigraphic units and sedimentary discontinuities			Volcanic eruptions → Tephra Meteorites → tectites magnetites extra-terrestrials		
PROCESSES		Geometric stratigraphy		Event stratigraphy	Chronological stratigraphy	Chronometric stratigraphy	
APPLICATIONS		Definition of descriptive units, local to regional correlations	Estimation of durations. Regional to planetary correlations	Regional to planetary correlations	Regional to planetary correlations	Dating. Relative. Regional to planetary correlations	Dating. Absolute. Planetary correlations

Some of these methods allow the descriptive units of stratigraphic classification to be defined (lithostratigraphic, sequential, magnetostratigraphic and biostratigraphic units) in which time is not a direct consideration, or is not given an exact figure.

For the reader's general interest, several other methods are mentioned which are less commonly used or whose use is more restricted, yet which may contribute to estimating time in certain, very specific cases: thermoluminescence, sclerostratigraphy (based on vegetable or animal growth stria), amino-stratigraphy (based on the racemization of amino acids). They will not be dealt with in this book, which does not claim to be exhaustive and which solely examines problems of terminology for those methods which are most commonly used.

4.3. STRATIGRAPHIC MARKERS

Natural phenomena (processes or events) leave a certain number of clues – or stratigraphic markers – whose temporal implications are extremely variable. Five basic types can be distinguished (REY, 1983):

– markers which are due to discontinuous, reversible or repetitive phenomena, such as the deposition processes or variations in the sedimentary environment. These markers are identified using lithostratigraphy, sequence stratigraphy and chemostratigraphy;

– markers which are due to continuous and reversible phenomena; such as the variations in the parameters of the Earth's orbit which create lithologic cycles. The latter are then analysed using cyclostratigraphy;

– markers which are due to "instantaneous" (with a duration of less than 10 000 years) and repetitive phenomena. These events are diverse in nature: volcanic eruptions and meteorite showers cause petrographic, mineralogical and chemical markers which are picked up by lithostratigraphy and chemostratigraphy; the inversion of magnetic polarity is picked up using magnetostratigraphy;

– markers which are due to discontinuous and irreversible phenomena. This category mainly includes biological crises, linked to major variations in the environment for reasons intrinsic or extrinsic to the Earth. These changes create significant gaps in the flora and fauna, which are identified using biostratigraphy.

– markers which are due to continuous and irreversible phenomena. There are two different types: the first of variable speed, are created by biological evolution and analysed using biostratigraphy. The second, of constant speed, are a result of the radioactive decay of certain natural isotopes and come under the domain of isotope geochronology.

These various types of signal, very different in nature, are also very different in terms of their duration, (and also therefore, in terms of accuracy), frequency and extremely variable geographical amplitude.

4.4. APPLICATIONS OF STRATIGRAPHIC METHODS

Each specific stratigraphic marker corresponds to a different method for the perception of time in geology:

– lithostratigraphy, the genetic stratigraphies (sequence stratigraphy, cyclostratigraphy), chemostratigraphy and magnetostratigraphy mainly allow a process of geometrical stratigraphy to be carried out, i.e. the definition of the relationships between geological entities and, by using the markers of brief events (event stratigraphy), allow regional or global correlations to be hypothesized. Nonetheless, the synchronism of these correlations and the identity of the era cannot be officially confirmed solely using these methods, since they are based on reversible or repetitive phenomena. They need therefore to be accompanied by chronostratigraphic **calibrations**;

– moreover, due to its continuous nature, cyclostratigraphy allows us to assess duration periods (astrochronology);

– biostratigraphy and isotope geochronology aid in the dating process: relative dating from chronological stratigraphy (biostratigraphy) or absolute dating from chronometric stratigraphy (isotope geochronology). Insofar as these methods enable us to attribute the same era to geological units located on different sites, they lead to the establishment of synchronous correlations.

Thus a complete stratigraphic process based on the use of a variety of methods corresponds to the transition from mainly descriptive geometrical stratigraphy, which defines local or regional units, to chronological stratigraphy, initially regional in nature, then planetary, and finally to chronometric absolute stratigraphy which gives the time continuum by calibrating the stratigraphic scales.

5. – STRATIGRAPHIC TRENDS

Three main trends stand out in the recent evolution of stratigraphy: growing tendencies towards a multidisciplinary system, universality and greater accuracy.

5.1. A MULTIDISCIPLINARY SYSTEM

The presentation of stratigraphic methods and processes, explained in the way we have done above, is very simplistic. In fact, one cannot attempt effective stratigraphic research using one method only: precise genetic stratigraphy is inconceivable without precise dating information (mainly biostratigraphic) which will avoid mistakes in correlation; in the same vein, it is difficult to draw up a solid biostratigraphy without the deposition geometry.

The various stratigraphic approaches are not cut and dried: each method may well – and should – intervene at each stage of the process. For example, an accurate biostratigraphy is a prerequisite for every magnetostratigraphic correlation, but the use of magnetostratigraphic data can considerably refine biostratigraphic data, above all in the conformation or invalidation of the synchronous extinction of various species. Thus the need for a more analytical view of stratigraphy, **integrated stratigraphy** became progressively more pressing. Integrated stratigraphy corresponds to the collective use of all available tools, taking into account the interdependence of the various geodynamic processes in order to gain a more coherent, uniform and universal perception of the stratigraphic recording. By way of reciprocation, stratigraphy becomes a fundamental discipline which is drawn upon by specialist areas as diverse as geological cartography, structural geology and applied geology.

5.2. UNIVERSALITY

Having originated Western Europe, stratigraphy has been considerably enriched over the past few decades by knowledge acquired on the various continents and from the ocean depths. It is thanks to this extension of stratigraphic data over the entire planet and to the consideration of orbital parameters that the time-scale has become progressively more manifest in geological groups which are more and more comprehensive, allowing it to move closer towards the time continuum… which it may perhaps reach one day.

The basic chronostratigraphic unit, the stage, cannot therefore be summarized in a simple profile concerning a specific place, its **stratotype**, but it embraces the entirety of the geological content (and recorded events) for a certain time interval, over the whole of the Earth's surface. From this perspective, the undertaking to define bottom edge stratotypes has proved to be a crucial advance in chronostratigraphy over the past few decades, facilitating correlations on a planetary scale and allowing all the deposits accumulated in the time interval of the stage to be taken into consideration.

5.3. ACCURACY

Perfecting stratigraphic tools has increased the resolution of this discipline, by defining stratigraphic units which are of increasingly better quality, or by distinguishing geological objects separated by progressively shorter time intervals. It is thus that over the past few years, the use of the expression **"High resolution stratigraphy"** has become more widespread. It consists of a process which aims to establish stratigraphic units which are of as high a quality as possible according to the tool used and/or the geological period taken into consideration: horizon in biostratigraphy, basic genetic unit in facies stratigraphy, parasequence in sequential stratigraphy, sub-magnetozone in magnetostratigraphy, "Gilbert" in cyclostratigraphy. For isotope geochronology, it could be an age distinction of around a million years for the Precambrian, of 0.1 Ma for the Cretaceous and of 0.01 Ma for the Miocene.

It is easy to see just to what extent this expression – which is currently enjoying increasing success – is ambiguous (since it adopts a different chronological meaning according to the methods and according to the age of the terrain), pretentious (because it implies a high degree of accuracy which, in reality, is rarely achieved and very rarely proved) and useless (the indication of the hierarchical level of the stratigraphic units which have been identified or uncertainty margins in the numerical dating, allowing the degree of accuracy achieved to be discerned immediately). *We therefore advise against the use of the expression "High resolution stratigraphy".*

6. – ORGANIZATION OF THE BOOK

This work presents, in separate chapters, the terminology which is specific to each stratigraphic method. The order of the chapters is modeled on the progression of the stratigraphic process, from the descriptive to the interpretative, from the methods of geometrical stratigraphy (lithostratigraphy and genetic stratigraphies grouped together due to the continuity and the complementary nature of the approaches; chemostratigraphy; magnetostratigraphy) to chronological stratigraphy (understood by biostratigraphy) and chronometric stratigraphy (essentially the domain of isotope geochronology).

The various approaches are presented in the same way: definition of the method; basic vocabulary; practical implications. Issues relating to rocks or geological groups which require specific processes or vocabulary (such as the stratigraphy of the Quaternary and surficial formations, of basements and of the Precambrian, of volcanic terrains) are grouped together in a separate chapter.

The two last chapters are dedicated to chronostratigraphy which combines the various methods and finalizes the stratigraphic process by providing the referential chronological framework upon which all geological objects can be placed and all the successive epochs in the Earth's history can be inscribed.

A glossary of the main terms (together with their French equivalents) is presented at the end of the book, as well as the standard methods for establishing and designating stratigraphic units.

Translated by C. BREWERTON

CHAPTER 2

LITHOSTRATIGRAPHY
from lithologic units to genetic stratigraphy

L. Courel (Coord.), J. Rey, P. Cotillon, J. Dumay, P. Mauriaud,
P. Rabiller, J.F. Raynaud & G. Rusciadelli

CONTENTS

1. – DEFINITION
   (L. COUREL & J. REY) 7
2. – LITHOSTRATIGRAPHIC UNITS 8
   2.1. Terminology 8
      2.1.1. Surface units
         (L. COUREL) 8
      2.1.2. Subsurface units
         (J.F. RAYNAUD, P. MAURIAUD & P. RABILLER) 11
   2.2. Practice 11
      2.2.1. Chronostratigraphic value
         of lithostratigraphic units
         (L. COUREL & J. REY) 11
      2.2.2. Specific issues in the definition
         of subsurface units
         (P. MAURIAUD & J. DUMAY) 12
3. – FROM SEQUENCES OF OBJECTS
   TO GENETIC SEQUENCES
   (J. REY) 17
   3.1. Terminology 17
      3.1.1. Sequences of objects 17
      3.1.2. Facies sequences 19
      3.1.3. Genetic sequences 19
   3.2. Practice 20
      3.2.1. Chronostratigraphic significance
         of sequences of objects 20
      3.2.2. Relations between lithostratigraphic
         units, sequences of objects
         and genetic sequences 20
4. – SEQUENCE STRATIGRAPHY
   (G. RUSCIADELLI) 20
   4.1. Definition 20
   4.2. Terminology and concepts 21
      4.2.1. Accomodation space 21
      4.2.2. The sedimentary signature
         of accomodation changes 22
      4.2.3. The sequence stratigraphic signature
         of the accomodation 24
   4.3. Practice 29
      4.3.1. The time-stratigraphic framework
         of sequences 29
      4.3.2. The hierarchy of relative sea
         level cycles 31
5. – LITHOLOGIC CYCLES
   (P. COTILLON) 32
   5.1. Definition 32
   5.2. Practice 33
      5.2.1. Cycles characterization 33
      5.2.2. Cycles, lithostratigraphic tools 34
6. – CONVENTIONS IN LITHOSTRATIGRAPHY 37
   6.1. Naming, definition, publication
      of lithostratigraphic units
      (L. COUREL) 37
   6.2. Procedure for codification of surface stratigraphic units
      (J. REY) 38
      6.2.1. The case of lithostratigraphic units 38
      6.2.2. The case of facies sequences 38
      6.2.3. The case of genetic sequences 38
   6.3. Procedure for codification of subsurface stratigraphic units
      (J.-F. RAYNAUD) 38
      6.3.1. Well logs and reference section
         of a lithostratigraphic unit 38
      6.3.2. Reference material, nature,
         preservation and availability 39
      6.3.3. Selection 39
      6.3.4. Definition 39

1. – DEFINITION

Lithostratigraphy is the branch of stratigraphy that organizes rock successions into units on the base of their lithologic properties. It involves the formal definition of such units (**lithostratigraphic units**) which must be recognizable in the field by observable physical features.

Wherever there are rocks, it is possible to develop a lithostratigraphic classification. By establishing the geometric relationships between rock units at different scales – e.g. from the single outcrop to a more or less extended area – lithostratigraphy provides an objective basis for the reconstruction of the geological history of any given area. The reconstruction of the spatial relationships of rock units expressed in maps and cross sections, in fact, provides a basic stratigraphic framework within which a more interpretative approach (**sequential analysis**) is used to infer **facies sequences** and discontinuities (Table 2.1). During this phase all the available tool including petrography, mineralogy, geochemistry, paleontology, sedimentology, and electric logs, are used to define the facies (lithofacies and biofacies) in outcrops and in the subsurface.

At this stage, the identification of cyclicities in the vertical succession of identified facies and lithologies is very helpful. On one hand it provides insights into the nature of the factors controlling the evolution of facies sequences. On the other hand, the identification of cycles may provide an element of correlation and a tool to assess the duration of a given interval provided that their periodicity is known.

Regional data should then be extended into a frame referring to global-scale events, for example of eustatic, geodynamic, climatic or astronomical origin, whose influence is reflected in the composition, depositional geometry and discontinuity within successions. Local/regional sequences are calibrated to supraregional/global stratigraphy with the aid of **genetic stratigraphy** (sequence stratigraphy, cyclostratigraphy) and **event stratigraphy**. These methods do not always give exhaustive information about the age of local units. However, they allow the identification of surfaces or depositional units, more or less synchronous on a regional scale, which can be dated by using other stratigraphic techniques (isotope

TABLE 2.1

The procedure and contents of lithostratigraphy

APPROACHES	UNITS	BOUNDARIES	HIERARCHY
BASIC ANALYSES	LITHOSTRATIGRAPHIC UNITS (homogeneous lithologic sets)	UNCONFORMITIES OR LITHOLOGIC CHANGES AT DIFFERENT SCALE	BED MEMBER FORMATION GROUP
SEQUENTIAL ANALYSES	FACIES SEQUENCES (Units characterized by the gradual transition of diagnostic features along a stratigraphic succession) Lithologic seq. Stratonimic seq. Mineralogic seq. Geochemical seq. Ecosequence Electric seq. Diagenetic seq. — Stratigraphic sequences at different scales	SEQUENCE DISCONTINUITIES (Sedimentary discontinuities corresponding to an abrupt change of a diagnostic feature along the sedimentary record, or assignificant chronostratigraphic change) Boundaries of facies sequences: change of the depositional environment or mechanism	SEQUENCE ELEMENTARY SEQUENCE (MESOSEQUENCE) MAJOR SEQUENCE or MEGASEQUENCE
STRATIGRAPHIC SYNTHESES AND INTERPRETATIONS (Sequence stratigraphy and facies stratigraphy)	GENETIC SEQUENCES Sedimentary units that reflect the evolution of the factors controlling the depositional process.	ANGULAR UNCONFORMITIES, UNCONFORMITIES, THIN SEDIMENTARY THICKNESS corresponding to an inversion point in the stratigraphic trend – abrupt sea level drop – transgressive surface – maximum flooding surface	5th AND 4th ORDER SEQUENCES (Parasequence, genetic unit) 3th ORDER SEQUENCE (Depositional sequence, Group of genetic sequences) 2nd ORDER SEQUENCE (Supercycle) 1st ORDER SEQUENCE (Megacycle)

geochronology, biostratigraphy, etc.) Once the synchronism of their boundaries is established with the aid of dating techniques, lithostratigraphic units provide a valuable time-correlation tool on the field.

Regional stratigraphic syntheses, which form the basis for global syntheses, are founded on lithostratigraphic analysis by integrating lithologic, biologic, genetic and sequential data. These data can be related to global forcing mechanisms of which they represent a local expression. Thus, lithostratigraphic units become the reference elements for the reconstruction of the geologic history.

2. – LITHOSTRATIGRAPHIC UNITS

2.1. TERMINOLOGY

2.1.1. *Surface units*

a – Definition

A **lithostratigraphic unit** is a body of rocks distinguished on the basis of its lithologic properties. Lithostratigraphic units may consist of sedimentary, or igneous, or metamorphic rocks. They are defined and recognized by observable physical features. Their geographic extent, therefore, is controlled entirely by the continuity and extent of their diagnostic lithologic features.

b – Lithologic units and lithostratigraphic units

Descriptive lithostratigraphic studies take into account different sets of units at different scales, according to their vertical and lateral extent. Terms like **laminae, horizon, bed, bank, series, sequence** are widely used to express lithologic units in a first descriptive step. When this lithologic units have an adequate lateral and vertical extent they are ranked as **lithostratigraphic units**. Their extent must be accurately described allowing their identification and correlation on a regional scale. Units representing significant geologic events of sedimentary, diagenetic or structural origin are worthy of being considered independently of their extent and thickness. The same principle applies to units showing a high potential for long distance correlation. A thin volcaniclastic bed (Fig. 2.1), for instance, can be very useful for regional correlation (BOURROZ et al., 1983) in spite of its limited thickness. On the other hand a barrier reef though showing a limited lateral extent and variable age is worthy of being mentioned in a regional synthesis for its great paleontologic and lithologic importance (reservoir potential).

Both lithostratigraphic and lithologic units, should be referred to by using a name in capital letters; e.g. "Grès de Fontainebleau".

c – Kinds of units

The various kinds of lithostratigraphic units can be ranked according to their thickness. Four formal units are considered: **bed, member, formation, group,** as suggested by the International stratigraphic Guide (HEDBERG, 1976; SALVADOR, 1994).

2. Lithostratigraphy

FIGURE 2.1
A Stephanian cinerite from
the Blanzy-Montceau-les-Mines basin.

It is advisable to use formal units in stratigraphical studies and official mapping, being aware that their definitions have to be published in easily accessible texts.

Formation

A formation is the primary unit of lithostratigraphic classification. It corresponds to a body of rocks of intermediate rank in the hierarchy of lithostratigraphic units. The thickness of a formation can vary from one to thousand of meters.

The formation is a reference unit in the reconstruction of the regional geological history and for comparison with other regions. Its name should have geographic derivation, and is sometimes combined with a brief descriptive term referring to its gross lithologic features. The original name of a formation must be maintained (anteriority rule) even when new studies give further information (nature or position of a boundary, new dating…).

The definition of a formation must be based on the description of different sections and/or wells and cartographic data, choosing references that is likely to be preserved as long as possible. The definition of a formation should include data about:

– Facies; petrographic and mineralogic nature of constituents; information on sedimentary structures and textures; biological content; geophysical and geochemical characterization;

– Geometry: vertical and lateral extent, boundaries, internal discontinuities and geometric relations with adjacent formations;

– Physical characteristics: permeability, hardness;

– Age (defined by biostratigraphy or isotope geochronology); it has to be remembered that a formation is defined by its content and its boundaries; the identification of its age, which can vary laterally, should not interfere with the definition of a formation; the age has to be considered as a supplementary information. The difference in age of the Calcaire de Comblanchien formation along a SW-NE transect in the Paris Basin, is an example (Fig. 2.2).

The other lithostratigraphic units can be defined as multiples or submultiples of the formation; the definition of these units, from the smallest to the largest is given in the following paragraphs.

Bed

The bed is the smallest formal unit in the hierarchy of lithostratigraphic units (SALVADOR, 1994). A bed is defined as a thin unit lithologically distinguishable from other layers above and below. It can often be used for correlation within the same depositional area (key beds, marker beds). Example: Couche blanche (Oligocene of the Paris Basin) or Couche

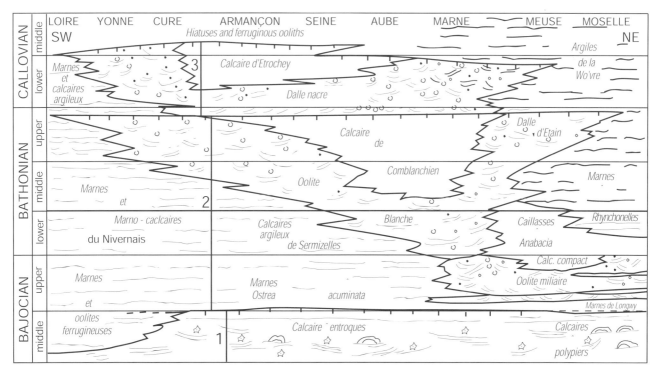

FIGURE 2.2
The stratigraphic relation between the various Dogger formations of the Paris Basin.
(*from* CAVELIER *et al.*, 1980, data by B. LAURIN, B. PURSER, J. THIERRY & H. TINTANT).

Louise (Westphalian of the Lorraine region). A bed can range in thickness from some centimeters to some meters.

The so-called **laminae** which can be organized to form laminations are the smallest, i.e. millimetre-scale, lithostratigraphic units.

The term **stratum** is "a layer characterized by lithologic properties allowing to distinguish it from adjacent layers" (HEDBERG, 1976), even though the boundaries may not be evident. Some sedimentologists use the term stratum in stratinomic descriptions instead of bed although the former does not have a formal significance.

Another informal lithologic unit is the bank, which is distinguished from the stratum and the bed by its geometry. However, as any other lithostratigraphic unit, the bank is defined on the base of its lithologic properties, e.g. a sandstone bank in a clayey succession, or the interruption of a homogeneous succession determined by a clay lamina (clay joint) or an interruption in the sedimentation processes (**diastem**).

A lithostratigraphic horizon (or lithohorizon) is typically a surface of lithostratigraphic change, commonly the boundary of a lithostratigraphic unit. However, it can be defined also as a lithologically distinctive thin marker bed within a lithostratigraphic unit, e.g. the uppermost Cenomanian Bonarelli Level within the Scaglia Bianca of the Umbria-Marche succession. The term horizon is frequently used in pedology, though with a different meaning, e.g. leached horizon.

A **flow** is a discrete extrusive volcanic body which can be distinguished from adjacent rock bodies on the base of its texture, composition, or other objective criteria. As suggested by the International Stratigraphic Guide the formal definition and naming of flows as lithostratigraphic units has to be limited to those that are distinctive and widespread.

Member

The member is "a lithostratigraphic formal unit of a rank immediately lower to the formation" and "no fixed rule is necessary to determine its lateral extension or thickness" (HEDBERG, 1976). Thus, the member has the function of allowing stratigraphers to subdivide the formation into units which are not beds (Fig. 2.3).

Different names not representing formal lithostratigraphic units can be used in case it is necessary to establish further subdivisions within a formation. From thicker to smaller lithologic units, the following can be used: series, **bundle**, layer, **assise**, unit...

Group

"The group is the formal lithostratigraphic unit next in rank above a formation" (HEDBERG, 1976). A group consists of a succession of formations sharing lithostratigraphic characteristics. When defining groups, much attention must be paid to the definition of the formations of which they are composed. A group can be subdivided into **subgroup**, and several groups are associated into a **supergroup**.

d – Subdivision in lithostratigraphic units

Such subdivision allows lithostratigraphic units to be arranged according to their size. As any other lithostratigraphic approach, however, it is also aimed at integrating regional data into a larger framework. Thus it has to be integrated with the analysis of facies, unconformities, the identification of the sequences and their hierarchy. Finally, it is necessary

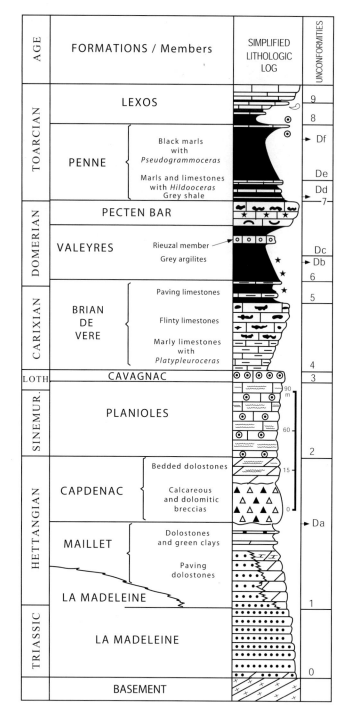

FIGURE 2.3
The lithostratigraphic units of the Lias from Quercy (Aquitaine Basin); formation names in capital letters, member names in lower case letters (modified after CUBAYNES et al., 1989).

to compare lithostratigraphic results with biostratigraphic and chronostratigraphic results.

e – Sedimentary unconformity

Lithostratigraphic units are often separated by sedimentary unconformities that represent an interruption in the sedimentation, no matter what their time duration. They can be characterized by interruptions of paleontological or lithologic nature, which may indicate diagenetic changes. Among sedimentary discontinuities, those related to autocyclic factors correspond to local facies changes caused by a modification in the equilibrium of the depositional environment, without the

2. Lithostratigraphy

intervention of external events, e.g. the migration of a river channel. Allocyclic changes, by contrast, reflect a modification of the sedimentary setting forced by factors external to the environment and have a different relevance. Sequential unconformities are treated in § 3.

2.1.2. Subsurface units

The definition of subsurface lithostratigraphic units follows the same principles used in the definition of surface units. These are transferred to the specific field of subsurface study and to the use of tools created for hydrocarbons exploration. In the practice, however, only the group, formation and member are used. In order to define these units, stratigraphers should follow the recommendations of the International Subcommission on Stratigraphic Classification of I.C.S, using the nomenclature of equivalent surface units. The following aspects are particularly relevent:

– the nature of type series. It is referred to an interval drilled by a particular company. The company (or a joint venture) that achieves the geological information, maintains the property rights for a certain period of time;

– the limited quantities of lithologic material (cores, lateral cores, cuttings...);

– the continuous record of physical data (well logs);

– the dating techniques (basically micropaleontologic and palynologic);

– the access to reference data.

Well log criteria (Fig. 2.4) are crucial in the definition of formation boundaries and, for practical reasons, are the primary tool in hydrocarbon exploration routine. The criteria of lithologic identification, preferably established in reference wells by means of cores, are sometimes difficult to use in nearby cutting wells.

2.2. PRACTICE

2.2.1. Chronostratigraphic value of stratigraphic units

As a result of complex sedimentary mechanisms controlled by a large number of factors (particularly relative fluctuations of the sea level), sediment accumulation occurs following vertical (aggradation) or lateral (progradation, retrogradation) accretion. For this reason, lithostratigraphic-unit boundaries do not have a chronostratigraphic value. Therefore, reliable elements allowing the identification of time lines should not be expected to be derived from descriptive lithostratigraphic analyses alone.

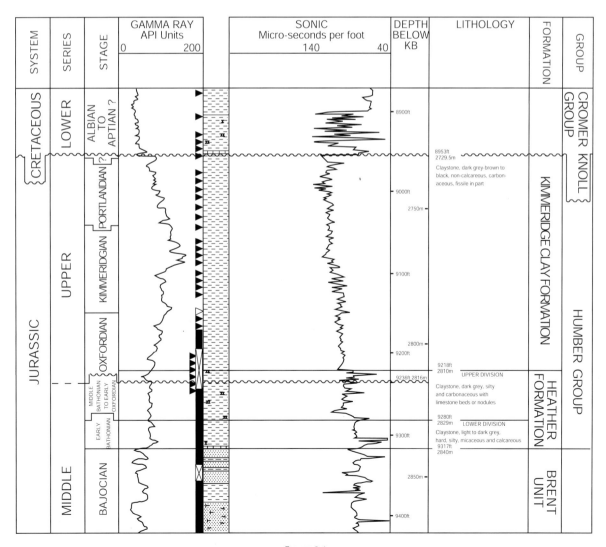

FIGURE 2.4
Jurassic UK Well 211/21 – A (Cormorant Field).

Biostratigraphic and isotopic geochronologic data should be obtained to attain a chronostratigraphic framework for lithostratigraphic data according to the following steps:

– simple dating of a single point in a 3D unit;

– use of several points to date unit boundaries at a regional scale;

– in this case, the identification of the synchronous or diachronous character of a boundary is a key element in reconstructing sedimentary dynamics and the stratigraphic history of the region within a wider picture;

– in particular cases, some units have a regional or global meaning such as volcaniclastic or euxinic horizon... In this case, the event recognized can be elevated to the rank of a marker for the entire area of its lateral extent.

2.2.2. Specific issues in the definition of subsurface units

a – Use of well logs

Information on the lithology of formations along well logs is obtained from samples collected during cutting or during specific operations (core sampling). This information are later extrapolated by using well log measurements to realize a site-to-site correlation.

Well logs acquired during or after drilling operations represent 90% of the whole information obtained and require 5-10% of the drilling budget. Geologists interpret the combined measurements of different specific instruments and adjust their interpretation on the basis of sample observation and analysis. For this goal, information on petrography, lithology, sedimentary structures, diagenesis, and on the nature and behaviour of the fluids including pressure and temperature, must be obtained. Traditionally, well logs are subdivided into three categories: classic well logs, high-resolution logs and geochemical logs (Fig. 2.5).

The so-called classic well logs have a stratigraphic resolution in the order of 0.5 meter. They allow a precise definition of lithology and reservoir quality. Lithologic interpretation is based on comparing the data obtained from a series of tools such as "*gamma ray*" (natural radioactivity measure), "density" (evaluation of rock density), "neutron porosity" (evaluation of rock porosity), "sonic" (measurement of the time taken for a sound to pass through a rock). Specific analyses for the characterization of reservoirs, such as the recognition of hydrocarbons, the evaluation of porosity, and oil and water saturation are carried out on two resistivity logs by means of different investigation rays. The comparison between the resistivity value in the area near the drilling site, saturated with bore-hole fluids, and a second value of resistivity closer to the actual resistivity of the formation, allows to evaluate the porosity of the rocks and the nature and saturation of the fluids present in the formation.

High-resolution logs have a resolution power in the order of one centimeter. They are obtained by using specific tools such as dipmeters and representation instruments that measure conductivity contrasts (SCHLUMBERGER FMI®), or seismic contrasts (Western Atlas CBIL®). The well logs acquired with these instruments give information on textural parameters, such as grain-cement relations, rock constituent elements (grain size and shape, type of cement and relations with porosity) and on structural parameters related to the geometry of constituents (lamination, homogeneity and heterogeneity...). In this way, structural and sedimentary dips are determined; fissures, cobbles, vugs and facies organization are identified. Other instruments give a comparable resolution through direct observation of cores.

Geochemical logs use various principles. They may consider the spectrometry of natural gamma rays, or they may record the percentages of the chemical elements present. The combined use of various tools allows also a measurement of the amount of organic matter in the sediments.

The interpretation of well-logs should be integrated with electric logs. The integration of any information obtained from the different kind of well logs with other information such as the drilling advancement velocity and other technical data of the well log, cutting description, calcimetry, core descriptions is crucial for a correct interpretation of subsurface units.

Well-log analysis has two main goals related to quantitative or economic aspects (e.g. dealing with saturation and porosity analysis) on one hand, and qualitative or geological aspects dealing with the lithologic and stratigraphic characterization of the drilled formations on the other hand. Reservoir engineers are responsible for the quantitative approach, structural and sedimentary geologists for the qualitative aspects. Only the qualitative approach will be dealt with in this chapter.

The analytical procedure (Fig. 2.6) consists in the subdivision of well log into electro-facies and electro-sequences. Electro-intervals are defined as depth intervals presenting homogeneous well-log characteristics that can be used for correlation. Electrofacies are defined as individual sets of log responses that are characteristic for a particular lithology (SERRA, 1984), allowing to differentiate them from

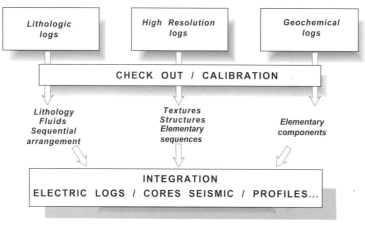

FIGURE 2.5
The procedure of well-log interpretation.

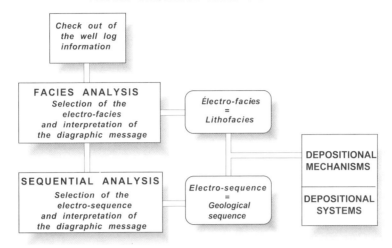

FIGURE 2.6
Block-diagramme of the facies and sequential analysis.

adjacent intervals. Usually it is necessary to calibrate logging data with core information from key intervals. Electro-sequences are depth intervals characterized by gradual changes of a measured parameter and bounded by two major shifts of the same parameter. Such a subdivision reflects a lithostratigraphic organization and integrates geological information derived from the analysis of cores and cuttings, later interpreted in a sequential organization (Fig. 2.7). Once the vertical interpretation has been completed, correlations with nearby well logs are defined, following the classic principles of stratigraphic correlation.

The use of high resolution physical data (reflection seismic and well logs) allows to, at least partly, overcome the main limitation in the subsurface stratigraphic analysis that is the limited access to discontinuous and small-sized samples. The importance of this tool and the economic aspects related to their use has resulted in a significant techonological progress since the 1970, with the development of several instruments that allow to solve specific problems.

The criteria used in the analysis of well logs are well defined and clearly expressed in a number of methodological publications. Particularly relevant is the work of SERRA (1979, 1984, 1985).

b – Seismic stratigraphy

Principles

Seismic is an echographic technique that provides information on subsurface geometry. Seismic stratigraphy follows the laws of acoustics and is based on acoustic properties of rocks. The application of this methodology follows three steps: data acquisition, processing and interpretation.

Data acquisition (Fig. 2.8)

From a point on the surface or near the sea level a vibration is sent; the wave produced propagate in all directions, is reflected and refracted while running through strata, and comes back to the surface after a certain time interval. This time depends on the length of the run-through way and the velocity of wave propagation which, in turn, is a function of the nature of the ground. This wave is recorded by receivers which translate the vibration into electric impulses.

Two types of waves are used in seismic profiling, P waves (primary waves) are mainly used but also S waves (secondary waves). Depending on the position of the recording machine relative in relation to the source, both reflected (P) waves (it is the commonest case) and refracted (S) waves are recorded. In the former case, the signals are called reflection waves and are used in particular situations, such as for the study of superficial levels, but also in the investigations of deeper rocks.

Data Processing

During this phase seismic data obtained in the field are translated into an interpretable seismic section.

For this goal, raw data, recorded on magnetic tape, are converted into:

– a seismic line (2D) which represents the series of traces corresponding to the acquisition profile. The seismic line (Fig. 2.9) shows the time image of the subsurface reflectors;

– a migrated seismic display (3D: x, y, time) which allows the reconstruction of the geologic setting (x, y, depth) and its actual geometry by placing seismic events in their true position.

Interpretation

Seismic horizons correspond to reflectors of significant extent, which are mainly represented by bedding surfaces or discontinuities. Reflectors may correspond to diachronous surfaces and are produced where vertical changes in acoustic impedance – commonly corresponding to a vertical change in rock type – occur. Bedding surfaces represent periods of non-deposition or a rapid change in the type of sediment deposited. A seismic reflection is caused by the lithologic contrast between the two types of deposits and/or the variation of physical characteristics along a stratal surface during a non-depositional event.

Discontinuities are stratigraphic surfaces created by erosion or non-deposition. **Unconformities** are characterized by erosive truncations and/or hiatus. An unconformity can correspond to a largely variable time interval, and the age of strata above and below it, can vary laterally according to the extent of erosion or non-deposition.

Although an unconformity represents a diachronous event, all the strata below it are older than those above it. Thus, unconformities will be referred to time lines which can be dated in the point where the unconformity changes into a conformity surface.

Unconformities generally create easily identifiable reflectors, determined by lithologic contrasts. Furthermore, angular differences can be marked by toplap truncation below

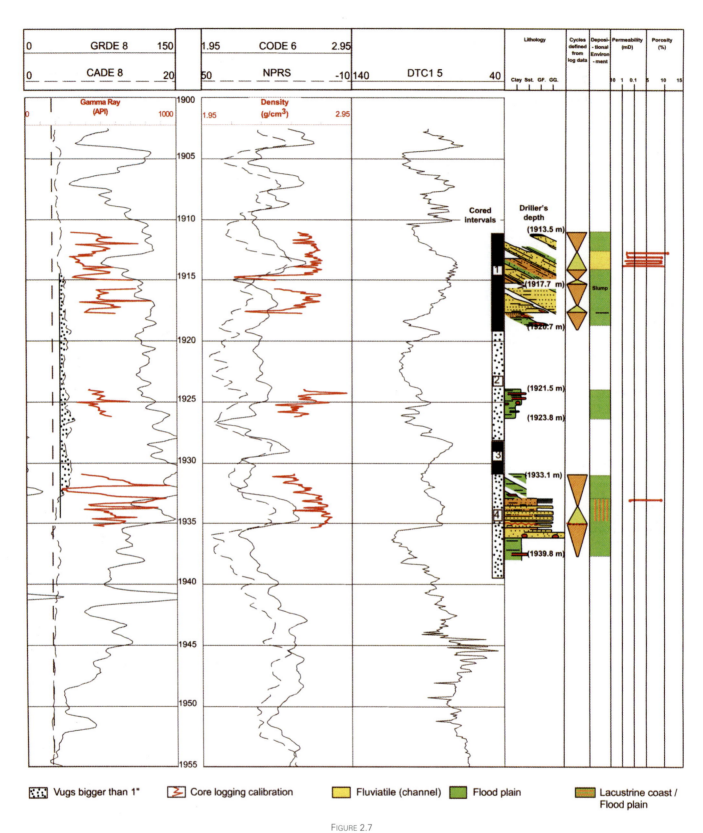

FIGURE 2.7
Example of the integration of various data: well logs, core description and sequential interpretation of facies.

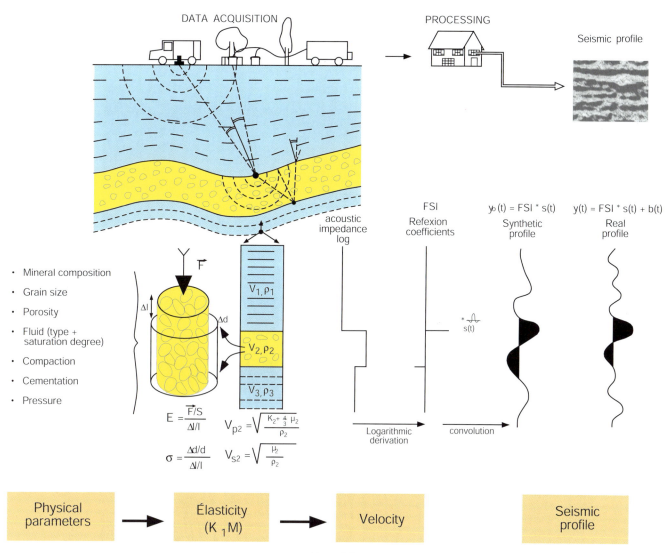

FIGURE 2.8
Acquisition and treatment of seismic data.

the unconformity and by onlap truncation and/or by downlap truncations above it.

Diagenetic surfaces, dykes, fluid contacts, low angle faults, permafrost and other physical surfaces, are diachronous surfaces which cut bedding surfaces.

Unit boundaries are often diachronous and discontinuous. As the boundaries of these units are not marked by physical surfaces, they do not always show a lateral continuity that can be recorded by seismic reflections. On the other hand, it is possible to observe changes in the features of reflections associated with stratal surfaces. The characters defining a reflection can cut reflection continuity, just as facies boundaries cut time lines. An example of stratigraphic and structural interpretation of seismic lines is reported in Fig. 2.9. Interpretation of seismic lines consists in the selection, based on local geological information, of sets of seismic horizons on different wells next to each other.

The calibration of well log record into the seismic line is obtained by comparing seismic profiles of vertical (PSV) or oblique (PSO) wells and synthetic profiles (synthetic seismic traces starting from acoustic and density logs).

Structural interpretation of seismic lines

The regional interpretation of a complete seismic survey allows to trace isochrons of the main mapped horizons in order to give a 3D representation (X, Y, time) of the geological setting. Information on sediment velocities allows to convert the isochron maps into isobath maps (X, Y, depth). The latter yield valuable paleogeographic information.

Isochrons, isobaths, and isopachs are derived from the structural interpretation of seismic lines.

Stratigraphic interpretation of seismic lines

Stratigraphic interpretation of seismic lines, whose basic principles have been defined in the AAPG Memoir 26 (PAYTON, 1977) "Seismic Stratigraphy – Applications to Hydrocarbon Exploration", contributes to the elaboration of a sedimentary model.

The objective is to grasp information on the nature of sedimentary deposits and the mechanisms for their formation.

Such information can be obtained from the analysis at the scale of the seismic reflection and/or of the seismic line:

1. seismic reflection

The interpretation is not only based on mapping reflection continuity, but also on evaluating the changes of amplitude which reproduce the influence of lithologic parameters (clay, porosity, fluid saturation) into seismic traces. Amplitude variations (and also the detailed study of the interval velocity)

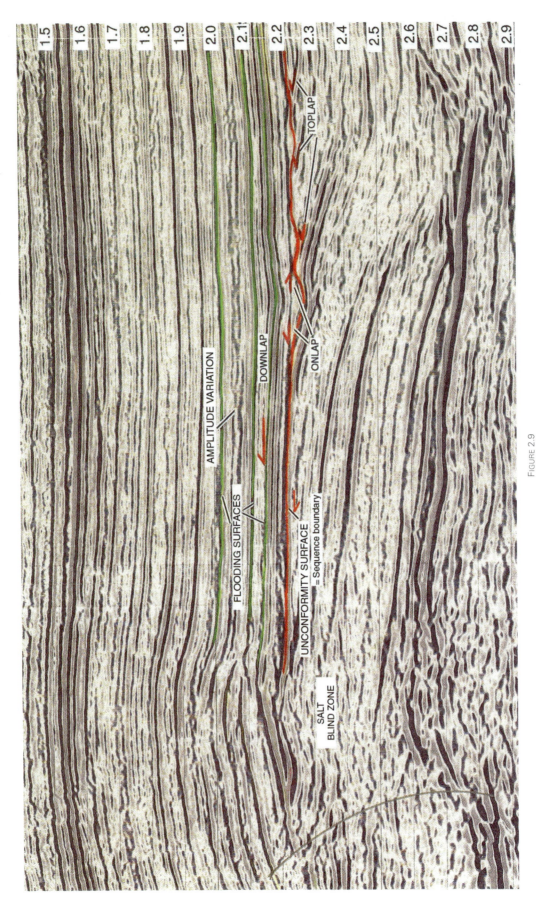

FIGURE 2.9

Example of structural and stratigraphic interpretation of a seismic profile.

are used to predict and quantify several characteristics of the rocks (lithology, fluid content, porosity, etc.).

2. seismic line

A first key step in the stratigraphic interpretation is the analysis of the seismic sequence. Seismic sequence analysis requires the identification, location and interpretation of consistent seismic horizons. The horizons are identified in the light of their geometric relations with overlying and underlying reflectors (onlap, downlap, toplap) and are interpreted in terms of stratigraphic unconformities (Fig. 2.9) as sequence boundaries, transgressive surfaces, and maximum flooding surfaces. The wells located along seismic lines allow a reinforcement of the interpretation and dating of the main horizons. The integration of information derived from biostratigraphic and sedimentologic analyses allows an optimal correlation between wells which is a prerequisite in the elaboration of a sequence model (identification and construction of depositional sequences) and in the establishment of a local chronostratigraphic framework.

The second key step in the stratigraphic interpretation of seismic lines is the analysis of seismic facies. This analysis is methodologically organized in three phases: definition of seismic facies units, their mapping and their interpretation. A seismic facies unit corresponds to a group of seismic reflections whose internal configuration, external shape and internal parameters (amplitude, frequency, etc.) differ from adjacent units. These features are related to the nature of deposits, including stratification and lithology. The information on seismic facies, together with well data, allows the analysis of sedimentary bodies. Stratigraphic interpretation of seismic lines, together with biostratigraphic, sedimentologic and structural studies, contributes to the reconstruction of the sedimentary history of a basin. The organization of sedimentary sets bounded by unconformities and/or their lateral equivalent, on the base of the obtained results, is carried out following the principles of sequence stratigraphy.

3. – FROM SEQUENCES OF OBJECTS TO GENETIC SEQUENCES

3.1. TERMINOLOGY

3.1.1. *Sequences of objects*

a – Definition

A "**sequence of objects**" is defined as a sedimentary unit characterized by a gradual evolution of the constituent elements (or objects) in a depositional succession of a-b-c, a-b-c-d... type. We use the expression "sequence of objects" instead of stratigraphic sequence to avoid confusion with "sequence stratigraphy".

The definition of a sequence of objects can be based on various elements and does not imply any genetic or causal interpretation. In this respect, a sequence of objects is an objective sequence. It is identified by the procedure of **sequential stratigraphy** (LOMBARD, 1956, 1972; DELFAUD, 1972).

The terms allostratigraphic unit (meaning sequence) and allostratigraphy (meaning sequence analysis) proposed by the North American Commission of stratigraphic nomenclature (1983) are ambiguous and must be revised. The prefix "allo", in fact, has two possible meanings. It can be used either when the stacking of deposits results from factors external to the sedimentary system or when referring to a stratigraphic technique which is different from that applied in a particular study.

b – Types

In the definition of this type of sequence an adjective illustrating the precise meaning of the elements which contribute to its definition must be used:

– lithologic sequence (sedimentologic features: composition or lithology (PURSER, 1972), texture, sedimentary structures...);

– stratinomic sequence (stratal stacking patterns and their thickness);

– mineralogic sequence (mineralogic components);

– geochemical sequence (chemical characteristics);

– ecosequence (paleontologic characteristics);

– diagenetic sequence (succession of evolutionary phases from a non consolidated sediment to a rock);

– electric-sequence (physical features observed in a well, etc.).

Examples: the three lithologic and mineralogic sequences described in the Vraconian-Turonian of Tunisia (Fig. 2.10); the ecosequences identified in the Albian of Portugal (Fig. 2.11).

The sequence of objects is a concrete datum. It can be modeled in a theoretical vertical (***idealized vertical profiles***) or horizontal (***idealized spatial profiles***) reconstruction starting from all the incomplete information collected about a region and for a defined age. Example: virtual ecosequence of the Albian setting in Portugal (Fig. 2.11).

c – Boundaries

The sequences of objects are bounded by **disconformities**. They correspond to abrupt changes in the sedimentary record (lithologic, paleontological, geochemical...) and to a significant chronostratigraphic gap (sedimentary hiatus due to erosion or non deposition). The boundaries of sequences of objects do not necessarily coincide with the sequence boundaries which can be based on other criteria. Among sedimentary unconformities, some called autocyclic, such as the migration of a river channel, correspond to a local facies change in the same environment, without the intervention of external factors. Others, allocyclic, testify changes in the surrounding environment due to factors external to the sedimentary environment. Being controlled by factors that can force the organization of sequences over wider areas (e.g. sea-level fluctuations), the latter have a different order of importance.

d – Hierarchy

The hierarchy of sequences of objects is based on their relative thickness, on their stacking order and on the information available, but not on their duration or hierarchy of lithostratigraphic units (member, formation) with which they are associated. According to their size, sequences are generally classified as follows:

– sequence unit, of millimeter to centimeter scale;

– elementary sequences, of decimeter to plurimeter scale;

– mesosequence, of decameter scale;

– major sequence or megasequence from decameter to hectometer scale.

This hierarchy has only a relative regional value.

The two fundamental sequence units are the **elementary sequence** (linked to the sedimentary dynamics at the scale of the depositional environment and possibly common to

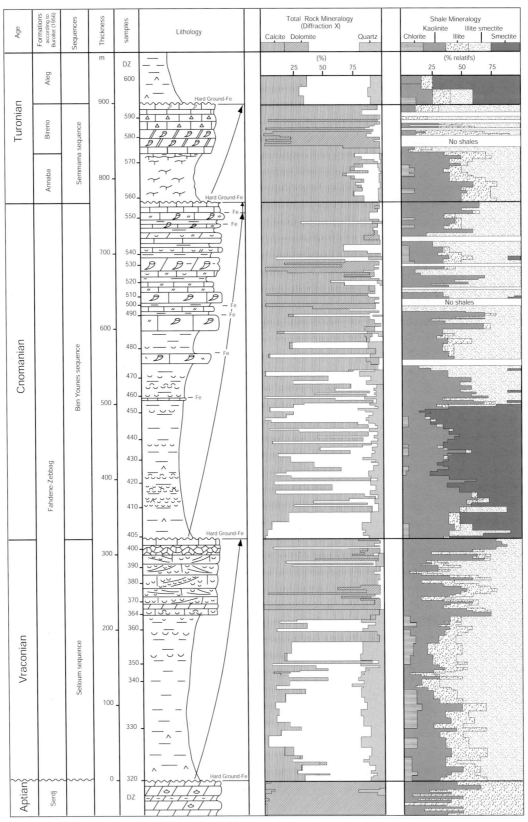

FIGURE 2.10

Example of a lithological (left column) and mineralogical (right column) sequences in the Vraconian-Turonian of Tunisia. Three major sequences can be recognized. The second major sequence can, in turn, be subdivided into sequences of lower rank. (after BISMUTH et al., 1981).

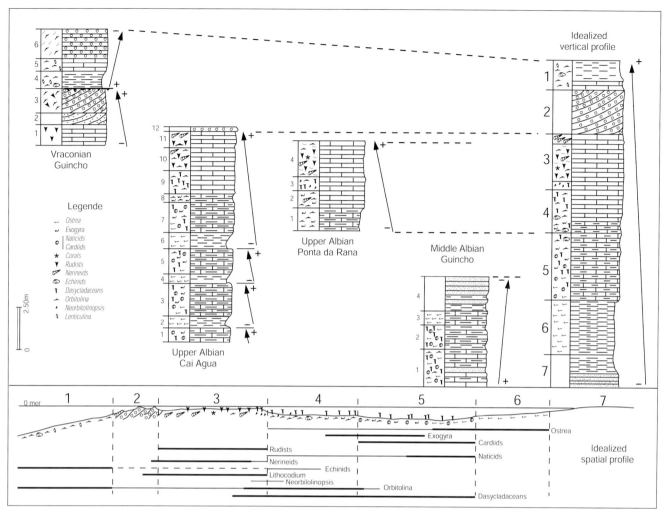

FIGURE 2.11
Example of ecosequences in the Albian of Estremadura (REY, 1979). Four local sections showing progressive variations in the paleontological associations allow to reconstruct a virtual composite sequence representing the theoretical vertical evolution of the fossil assemblages. Transposed on a horizontal axis, the virtual sequence describes a series of lateral sequence describing a series of 7 successive ecozones on a marine platform.

different basins) and the **major sequence** (associated to the sedimentary dynamics at the basin scale). The major sequence is always composed of elementary sequences. Example: mesosequences and elementary sequences in the Barremian in the SE of France (Fig. 2.12).

The **mesosequence** is a more subjective unit, of descriptive interest. Its use has to be strictly limited to exceptional cases where an intermediate order between the elementary sequence and the major sequence is present.

Sequence unconformities of different rank do not differ in their nature, but only in their magnitude: differences in constitutive elements present above and below the unconformity, geographic extension, and time interval recorded by the unconformity.

3.1.2. Facies sequences

The whole of lithologic, mineralogic, stratinomic, geochemical, paleontological and diagenetic features of rocks and of their evolution through time is included in the concept of **facies sequence**. This represents a succession of facies which allows the definition of the depositional environment and associated sub-environments and sedimentary mechanisms at different scales. The 3D reconstruction of these sets of deposits allows the development of a **facies model**.

At the basinal scale, or at the scale of the main domains and sedimentation types, facies sequences are genetically related to **depositional systems**. The identification of facies sequences and of depositional systems at different scales is, therefore, based on the analysis of the relations between sedimentary units as well as on the understanding of the factors controlling their evolution. The resulting sequences of facies and systems have a genetic meaning.

3.1.3. Genetic sequences

The **genetic sequence** is the sedimentary unit identified by considering the evolution of one or more variables which control sedimentation (tectonic, subsidence, absolute sea level, climate,...). This unit has a **genetic and causal meaning**.

Genetic stratigraphy is a branch of stratigraphy dedicated to analysis, definition and description of the processes controlling sedimentary processes and the geometric and chronological relationships between the resulting deposits. Thus, the expression "genetic stratigraphy" has a very general meaning, versus more restrictive meaning that should be avoided. It takes advantage from different approaches, such as **sequence stratigraphy** or **cyclostratigraphy**.

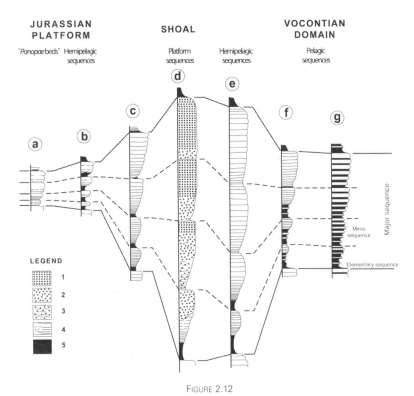

FIGURE 2.12

Subdivision into elementary sequences and mesosequences. Correlation of the Barremian of South-East France (H. ARNAUD, 1979).
1- infralittoral biosparites; 2- circalittoral biosparites; 3- biomicrites; 4- sponge spicule and foraminifera biomicrites;
radiolarian and ammonite bearing iron micrites (pelagic and hemipelagic limestones marly limestones); 5- marls.

3.2. PRACTICE

3.2.1. *Chronostratigraphic significance of sequences of objects*

Sequence analysis does not necessarily consider the chronologic aspects. The study of sequences of objects alone does not allow the identification of time markers. However, it is still possible to make the following considerations:

– the rocks above a sequence unconformity identified at more than one point are more recent than the ones below;

– some types of unconformities reveal a time gap (erosive surfaces) or a reduction in the time record of a successions;

– starting from stratinomic and seismic data, it is possible to recognize the growth direction of sedimentary bodies (aggradation, progradation, retrogradation) and to establish the theoretical relationships between sedimentary boundaries and time lines.

3.2.3. *Relations between lithostratigraphic units, sequences of objects and genetic sequences*

The boundaries between lithostratigraphic units (formations, members...) or facies sequences on one hand, and genetic sequences on the other hand, are not of the same nature. The former (formations, members, physical sequences) present a hierarchy based on thickness, while the others (genetic sequences) present a hierarchy based on the length of the processes controlling their development. There is a large distance between the meanings of these two kinds of lithostratigraphic units. A formation can be made up of more than one major facies sequence or it can correspond to only a fraction of a major sequence. Moreover it can be formed by one or more genetic sequences or by one or more systems tracts. A major facies sequence may include one or more genetic sequences, or represent only a fraction of a genetic sequence. The boundaries of stratigraphic units do not necessarily coincide with the boundaries of facies sequences, or with the boundaries of genetic sequences. Example: the formations, the systems tracts and depositional sequences of Aalenian-Bajocian in Normandy (Fig. 2.13).

4. – SEQUENCE STRATIGRAPHY

4.1. DEFINITION

Sequence stratigraphy is a relatively new method of sedimentary analysis that has considerably changed the way to approach stratigraphy and sedimentology (POSAMENTIER & VAIL, 1988, POSAMENTIER *et al.*, 1988, VAIL *et al.*, 1991, VAN WAGONER *et al.*, 1988). Sequence stratigraphy "evolved" from **seismic stratigraphy**, a method developed in the 1970's with the introduction of modern acquisition techniques of seismic data and based on the concept of "megasequence" of Sloss. Seismic stratigraphy was initially based on the observation of the spatial patterns of seismic reflectors in siliciclastic passive margins, and it was mainly addressed to the exploration of natural resources. The application to outcrops and the introduction of new concepts, such as the accommodation, improved the original principles and opened the method to academic researchers.

Sequence stratigraphy offers the opportunity to integrate a large spectrum of disciplines, such as seismic and log-stratigraphy, biostratigraphy, sedimentology, geochemistry and structural geology. It is a procedure in which the basin-fill is subdivided into a series of time and spatial related units, which offer a dynamic representation of lateral facies

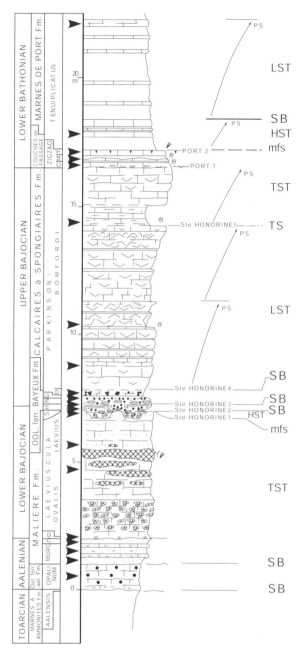

FIGURE 2.13
Lithostratigraphic units and depositional sequences in the Dogger of Normandy (after RIOULT et al., 1991; abbreviations as in figure 17. In this example the facies sequences can be interpreted as reflecting part of a system tract (the LST between Sainte Honorine 4 and Sainte Honorine 5), or as a complete system tract (the TST between Sainte Honorine 1 amd Sainte-Honorine 4). The boundaries between formations do not always coincide with the boundaries of system tracts (e.g. the upper boundary of the Calcaire à spongiaires Formation or with the boundaries of the 3rd order depositional sequences (e.g. the lower boundary of the Marnes de Port Formation).

relationships by tracing the evolution of depositional systems in response to transgressions and regressions forced by cycles of relative sea level change. The possibility to constrain the spatial pattern of deposition and erosion during a cycle of relative sea-level change gives a predictive value to the method. Sequence stratigraphy as a predictive tool for stratal patterns and lithofacies distribution in time and space has represented and still represents a fascinating call for the oil industry and academics. Since its introduction, sequence stratigraphy has been applied to a very large spectrum of depositional contexts, tectonic settings, and for various industrial and scientific purposes and, despite the criticism and still open debate on many questions, it has deeply improved our knowledge on sedimentary dynamics and their related processes.

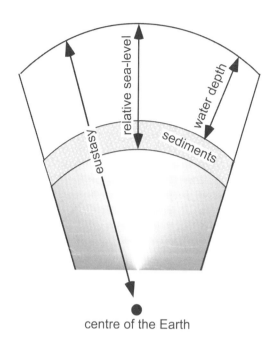

FIGURE 2.14
Definitions of relative sea level, eustasy and water depth.

4.2. TERMINOLOGY AND CONCEPTS

4.2.1. Accommodation space

A primary concept in sequence stratigraphy is that the distribution of sediments within a depositional system depends on the amount of space available for potential sediment accumulation. The space available for sediment accumulation is defined as the volume between the sea surface and a reference datum (i.e. the top of basement, or whatever other stratigraphic horizon), and it is called **accommodation space** or relative sea-level. Relative sea-level differs from water depth that represents the bathymetry, i.e. the distance between the sea surface and the sea-sediment interface, and eustasy that represents the distance between the sea surface and the center of the Earth (Fig. 2.14).

The accommodation is the algebraic sum of the tectonic subsidence and eustasy (Fig. 2.15); therefore, the amount of space created or destroyed is controlled by the balance between tectonics (subsidence and uplift) and eustasy, by varying the water depth and/or the topographic height or the ocean water volume. If this balance is negative or null, the relative sea level decreases and space available for sediments is destroyed or not created respectively; if the balance is positive the relative sea level increases and the space available for sediments being created.

The changes in the accommodation can be rapid or slow, depending on the type of driving mechanism and from the rates at which these mechanisms created or destroyed the available space. Tectonics can produce highly variable rates of accommodation changes depending on the type of process (Fig. 2.16). Tectonic subsidence

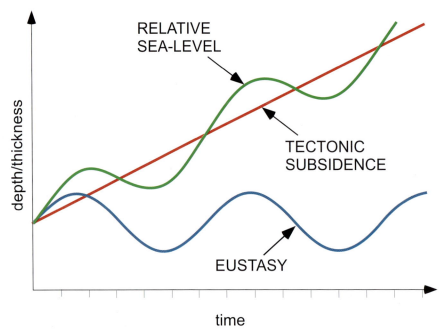

FIGURE 2.15
This diagram illustrates the graphic expression of eustasy, tectonics and relative sea level.
Eustasy varies throughout time according to a sinusoidal curve, whereas tectonic is considered constant during time.
The relative sea level represent the sum of these two parameters.

rates may vary, for example, from few meters to several hundred meters/Ma in foreland basins, or it may have rates up to 2 km/Ma in active intra-mountain basins; during rifting processes, subsidence rates may span between hundred and thousand ms/Ma. Also tectonically induced uplifts may record high rates due to processes such as the chain building (up to 15 km/My), or transpressional tectonics (between few hundred to few thousand ms/My) (Fig. 2.16). Short-term eustatic rates of rise, as derived from the third order fluctuations from the HAQ (1988) eustatic chart, may vary from a minimum of 27.5 m/Ma during the Dogger to a maximum of 687 m/Ma during the Paleocene (Fig. 2.16), whereas short-term eustatic rates of fall change from a minimum of 22 m/Ma during the Lias to a maximum of 316 m/Ma during the Upper Cretaceous. In contrast, long-term eustatic rates of rise vary from a minimum of 1.5 m Ma during the Miocene to a maximum of 10.3 m/Ma during the Lower Cretaceous, whereas long-term eustatic rates of fall vary from a minimum of 1.5 m/Ma during the Paleocene to 11 m/Ma during the Permian (Fig. 2.16).

Glacio-eustatic driven changes of the accommodation are generally characterized by higher rates than tectonics. The expansion and retreating of polar caps are able, in fact, to produce high amplitude (10 – 100 m) eustatic sea-level changes during relative short time intervals (1 – 100 ka). Thermal expansion and contraction of the ocean can produce a few meters of eustatic change over short time scales (0.1 – 10 ka), whereas water sequestrated on continents (rivers and lakes) may induce a few meters of eustatic change over time intervals of 0.1 – 100 ka. However, because glacio-eustatic changes have in general a higher frequency, the total amount of accommodation produced by the glacio-eustasy is generally smaller than tectonics. In conclusion, glacio-eustasy modulate tectonic driven changes of the accommodation, and the resultant effects can be enhanced or smoothed if their frequencies are in or out of phase.

Other mechanisms, such as compaction and large mass transport processes, can cause changes of the space available and they may produce stratigraphic architectures that mimic geometries and stratal stacking patterns of sequences.

4.2.2. *The sedimentary signature of accommodation changes*

Mechanisms driving changes of accommodation produce oscillations of the **base level** and induce erosion, transport or deposition of sediments from a source area to a basin throughout the depositional profile.

Sediments tend to fill the space available and move laterally or accumulate accordingly to the oscillations of the base level. The shifts of the base level are controlled by changes of the accommodation, and sediments are eroded if the base level falls below the depositional equilibrium profile (negative accommodation), whereas sediments accumulate if the base level rises above it (positive accommodation) (Fig. 2.17).

These simple relationships are complicated by the sediment supply, i.e. the amount of sediments produced and yielded to the system. Sediment supply is a function of several and more or less directly interrelated factors, such as climate, size and denudation characteristics of source areas (relief, lithology, vegetation cover, rivers and streams, etc.), size, geometry and water circulation characteristics of the basin, and obviously tectonics and eustasy.

For a given amount of relative sea level created, sediment supply mainly controls the infilling of the available space. Low rates of sediment supply allow the space available to increase. The base level rises above the depositional equilibrium profile and the water depth increases in the basin. Sediments are trapped landward, whereas starved conditions dominate basinward (Fig. 2.18a). With high rates of sediment supply, all the amount of space created is filled with sediments. The base level equals the depositional equilibrium profile, and the excess of sediments produced are by-passed toward areas with higher accommodation (Fig. 2.18b). Intermediate and more variable conditions occur with moderate sediment input. If the rates of accommodation space decrease during moderate or high rates of sediment input, the previously deposited sediments will be eroded (Fig. 2.18b right side). The sedimentary response resulting from the balance between accommodation and sediment supply can largely

2. Lithostratigraphy

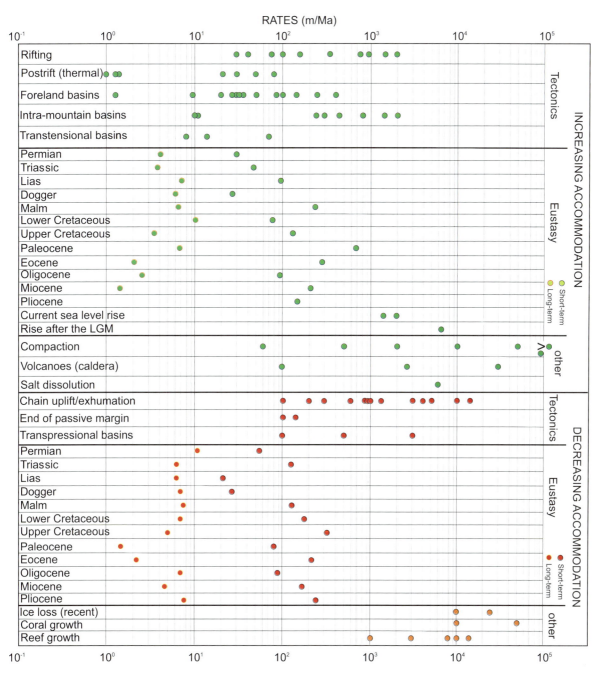

FIGURE 2.16

In this diagram, the accommodation rates due to different processes are displayed. Rates of different processes are reported in m/Ma and they are expressively plotted together to make the comparison of the amount of rates between different type of processes easier. Some of the reported processes display very high rates, such as the compaction that may reach over 400 km/Ma (subsidence induced by hydrocarbon extraction) or the growth of a single coral that may reach rates of 50km/Ma. Obviously, the amount of space created or filled by these processes over a period of several million of years is not realistic, because these are processes that act over very short time periods or they are limited by other factors. For example, because corals are not able to build above the seal level, the growth of a single coral, or a reef, is controlled by the rate of rise or fall of the relative sea level. Tectonic and other mechanism rates from several sources. Eustatic rates calculated on the base of the Eustatic chart of HAQ et al. (1988).

vary in time and space within a basin and from a basin to another, depending on the factors that control these two parameters.

An important issue is that the balance between sediment supply and accommodation controls the sediment distribution within a depositional system. Therefore, if the amount of sediment yielded and of space created varies with time, the sediment distribution within a sedimentary system will follows these changes. The effects will be recorded by the vertical superposition of strata ("stratal stacking pattern") reflecting different balances between sediment supply and accommodation rates, and expressed by peculiar geometric end members, termed progradational, retrogradational and aggradational (Fig. 2.19):

– **Progradation** occurs when, for successive time intervals, the amount of sediment produced (ds/dt) exceeds the amount of the space created (da/dt). As a consequence, facies belts and the coast line move basinward and, depending on the position along the depositional profile, the vertical stratal pattern is defined by coarsening, thickening and shallowing upward trends (Fig. 2.19a).

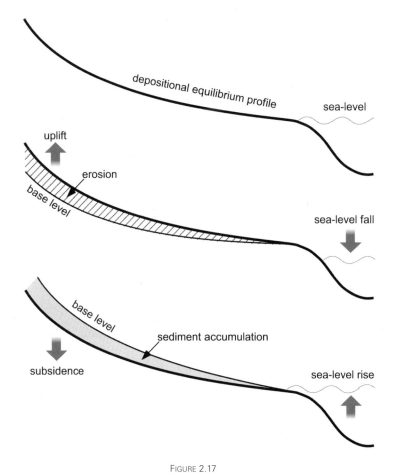

FIGURE 2.17
Definition of base level and its control on change of accommodation along an ideal depositional profile.

- **Backstepping** occurs when, for successive time intervals, the amount of the space created (da/dt) largely exceeds the amount of sediment produced (ds/dt). As a consequence, facies belts and the coast line move or backstep landward or in a higher position and, depending on the position along the depositional profile, the vertical stratal pattern is defined by fining, thinning and deepening upward trends (Fig. 2.19b).

- **Aggradation** occurs when for successive time intervals, the amount of the space created (da/dt) equals the amount of sediment produced (ds/dt). As a consequence, facies belts and the coast line accrete in a vertical position and no change in the upward trends of the stratal stacking pattern are observed (Fig. 2.19c).

These geometric end-members can be generally recognized in both carbonate and siliciclastic systems. However, as carbonates and siliciclastic react differently to environmental changes in terms of sediment production and distribution, similar processes may produce largely different sedimentary responses. In siliciclastic systems, sediment is produced by the weathering and erosion of an emerged area, whereas in carbonate platforms sediment is produced by the biological activity of living organisms ("carbonate factory"), whose growth potential depends on physical-chemical conditions of the water masses, such as light, temperature, salinity, circulations, etc. In carbonates, an emersion event produces chemical erosion (karst) and reduces the area extent of the potential sediment source. Therefore, during emersion, the sediment production is maximal in the siliciclastics and minimal in carbonates. Moreover, the angle of repose of siliciclastic sediments favours the development of low gradient topographic profiles, along which the progradation is promoted during relative sea level fall, throughout the basinward shifting of clinostratified deposits. In contrasts, the higher angle of repose of carbonate sediments and the ability of carbonate platforms to build upward with high reliefs promote steep and deep depositional profiles. When the relative sea level falls below the shelf break, the onward progradation is limited by the high topographic relief of the platform wall and by the decreasing potential of the light penetration with depth. Therefore, the only way to promote progradational patterns in carbonates systems is to have a gentle depositional profile and a healthy and highly productive carbonate factory able to export large amounts of carbonate grains downslope. This generally occurs when the platform top is flooded and the carbonate producing organisms expands over large areas of the platform top.

4.2.3. *The sequence stratigraphic signature of the accommodation*

The ideal evolution of a **depositional sequence** implies the return to conditions more or less similar to the initial ones. In terms of relative sea-level, this evolution is represented by a single sinusoidal cycle. Therefore, a cycle of relative sea level is defined as the complete evolution from a point to its successive equivalent along the relative sea level curve (Fig. 2.20). For practical scopes, this point should correspond to the most readily identifiable surface within a sequence and then it should be marked by significant changes in the sedimentary record in both continental and marine environments. There are four points that fulfill these

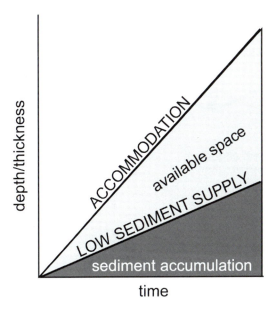

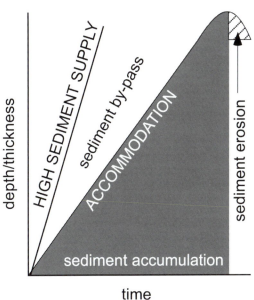

FIGURE 2.18
Relationships between accommodation, sediment supply and deposition.

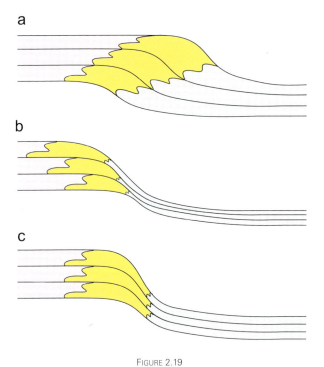

FIGURE 2.19
Geometric end members related to the interplay between accommodation and sediment supply.
a) Progradation; b) Retrogradation; c) Aggradation.

the surface forming at the time of increasing rates of sea level rise (points C-C_1 in Fig. 2.20) (Transgressive-Regressive Cycles), or the **maximum flooding surface**, corresponding to the time of maximum rate of rise of relative sea level (points D-D_1 in Fig. 2.20) (Genetic Sequences *sensu* GALLOWAY 1989, HOMEWOOD *et al.*, 1992).

Taking into account the major differences between siliciclastics and carbonates, the development of a sequence, **systems tracts**, key surfaces and related sedimentary processes during a complete evolution of a cycle of relative sea level will be illustrated hereinafter by considering the sedimentary response along an ideal depositional profile from proximal (continental) to distal (basin) areas. For simplicity, the sediment supply is considered constant during time. The time of initiation of relative sea level fall is here adopted as starting point (point A in the RSL curve of Figs. 2.20 and 2.21). A major turning point in the evolution of the space available for sediments occurs at this time. The amount of available space is minimum in continental and proximal areas and the deposition/erosion balance shifts toward erosion; a subaerial unconformity begins to form in proximal areas.

As the sea level continues to fall, the lowering of the base level produces the re-adjustment of the depositional profile through the exposure and erosion of previous deposits, forcing the downwards shift of the subaerial unconformity and the onwards migration of the facies belts. The retreating of the shoreline induces the deposition of shoreface deposits on previously deposited offshore sediments, with the formation of a shallowing and coarsening upward trend (Fig. 2.22).

The most part of the eroded sediments by-passes the shoreface and an increasing in sediment supply is recorded in the basin where space for sediment is still available. The entire period of sea level fall corresponds to the **falling stage system tract** (FSST). This system tract has been firstly defined as forced regressive wedge by HUNT & TUCKER (1993), and it can be considered the equivalent of the lowstand prograding wedge (early lowstand) of POSAMENTIER *et al.* (1992) formed during a forced regression. These two definitions vary, for

characteristics (A, B, C, D, in Fig. 2.20) and that can be considered as **sequence boundaries**. Their related surfaces are adopted by different sequence stratigraphic models to delineate basic units and internal subdivisions. According to the classic Exxon model, the sequence boundary has a different expression: on land, it is represented by the interval between the point A and B, corresponding to an **erosional surface** – or erosive **unconformity** – formed during the interval of relative sea level fall (points AB-A_1B_1 in Fig. 2.20); in the basin, where the stratigraphic record is more continuous, the sequence boundary corresponds to the correlative **conformity**, a surface within marine sediments representing the time of initiation of the relative sea level fall (points A-A_1 in Fig. 2.20). According to subsequent models, the sequence boundary in marine settings is always represented by a correlative conformity but it is assumed as the surface equivalent to the time at which the sea level fall reaches its minimum (points B-B_1 in Fig. 2.20). Other models consider the correlative **transgressive surface** as the reference point for the sequence development, i.e.

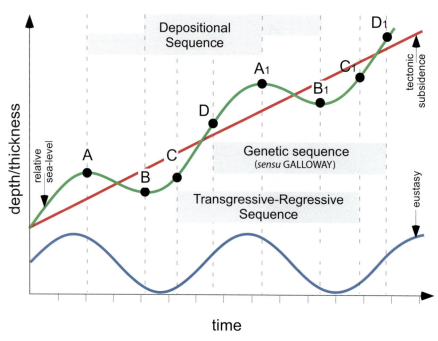

FIGURE 2.20
Relationships between eustasy, tectonics and relative sea level. Grey boxes refer to the four main different type of sequences adopted in the literature. Dotted lines represent sequence boundaries of different type of sequences. Depositional sequences are referred to the classical Exxon model (A-A1 sequence boundaries) and to the depositional sequence of HUNT and TUCKER (1992) (B-B1 sequence boundaries); genetic sequence according to the Galloway model (1989) (D-D1 sequence boundaries); transgressive-regressive sequence is referred to the Embry model (1993) (C-C1 sequence boundaries). The time of inversion points in the eustatic curve and the time of inversion points in the relative sea-level curve may vary as a function of the rate of sea level and the rates of the tectonic subsidence and sediment supply.

the essential, for the placement of the sequence boundary at the top and at the base respectively. The term falling stage system tract has been introduced by POSAMENTIER & ALLEN (1999) and by PLINT & NUMMENDAL (2000). Several examples have been documented in literature both in carbonate and siliciclastic systems from seismic and outcrop data. Among carbonate outcrop example are the Cretaceous Urgonian carbonate platform of the French Subalpine Chains (HUNT & TUCKER, 1993) and the upper Miocene reef complex of Mallorca (Spain) (POMAR, 1991; POMAR et al., 1996); various examples of forced regressions in siliciclastic systems are illustrated in POSAMENTIER et al. (1992) and HUNT & GAWTHORPE (2000). Several examples of falling stage systems tracts have been documented all over the world for the Quaternary. During this period, in fact, the expansion of glaciers up to low latitudes has produced large sea level falls during the Pleistocene glacial periods. This have favoured the large basinward displacement of the shoreline in most part of the continental shelves. In addition, the consequent dramatic fall of the base level has produced huge amount of sediment, favouring the preservation potential of these deposits.

At the end of the relative sea level fall (point B in the RSL curve of Figs. 2.20 and 2.22), the available space for sediments is at its minimum, the sea level reaches its lowest position and sediments prograde towards the basin. At this moment, the seaward position and the extent of the subaerial unconformity are maximal. In the basin, the marine equivalent of the maximal extent of subaerial erosion is the top of the falling stage system tract and represents a correlative conformity.

In carbonate settings with low gradient depositional profiles (carbonate ramps) and with production of loose carbonate particles (heterozoan association, i.e. "foramol facies"), a sedimentary response similar to siliciclastics may occur throughout periods of relative sea level fall. In contrast, carbonate platforms bounded by biogenic rims (photozoan association, i.e. "chlorozoan facies") or escarpments generally respond quite differently to periods of sea level fall. Once emerged, low topographic reliefs, early cementation and diagenesis protect carbonates from mechanical erosion. The main erosion process on platform top is due, in fact, to fluid circulation that produces chemical dissolution and karst structures. Moreover, the emersion of the flat platform top, i.e. of the area of the highest sediment production potential, inhibited the carbonate factory. At the same time, the steep slope prevents the seaward progradation of the margin and, during the entire phase of sea level fall, areas of sediment production are restricted to small-sized belts located along the seaward margin of the platform. As a consequence, a dramatic decrease of the volume of sediment produced and exported basinward is recorded during the entire period of relative sea level fall (Fig. 2.23). This setting will be maintained until the new flooding of the platform top.

The time of initiation of relative sea level rise (point B in the RSL curve of Figs. 2.20 and 2.24) represents a second major turning point in the evolution of the accommodation. The rise of sea level produces a slow but positive increase in the space available for sediments along the depositional profile. Rivers cease to incise and non-marine facies start to onlap the eroded substratum. However, far from the equilibrium state, sediments continue to supply the coastal zones and shoreface deposits develop at the new sea level position (Fig. 2.24). The period included between the points B and C corresponds to the **lowstand system tract** (LST).

At the point C (Figs. 2.20 and 2.25), the sea-level starts to rise with higher rates than sediment supply; the shoreline moves landward and sediments previously emerged and eroded are flooded (transgression). Currents and wave action erode and rework previously deposited sediments, producing a shallow ravinement surface. This surface, known as **transgressive surface**, represents a significant turning point in the stratigraphic organization of marine deposits,

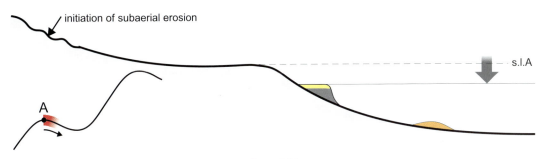

FIGURE 2.21

The relative sea level starts to fall. The base level fall below the depositional profile, reducing the available space for sediments. As a consequence, subaerial erosion begins landward while the shoreline and the related depositional systems migrate basinward. Abrupt facies changes occur along the shelf, whereas in the basin fan deltas overly offshore shales.

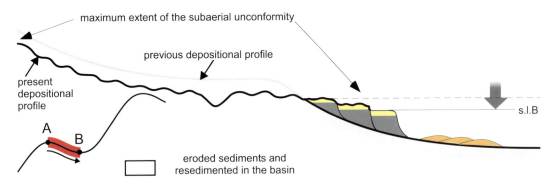

FIGURE 2.22

The relative sea level changes from increasing to decreasing rates of fall and reaches the maximum of fall at the point B in the relative sea level curve. During this interval, the base level continues to low and erosion truncates previous deposits up to the new equilibrium point along the depositional profile. At the point B, the base level reaches its lowest position and the accommodation is at its minimum. In marine settings stratigraphic successions display coarsening and thickening upward trends, whereas depositional patterns are defined by progradational geometries.
The sediments deposited during the interval of relative sea level fall correspond to the **falling stage system tract**.

and separates progradational from retrogradational stratal patterns. It develops throughout the entire interval of transgression and progressively shifts landward at the base of shoreface deposits following the shoreline ingression.

Throughout the points C and D of the relative sea level curve (Figs. 2.20 and 2.26), the relative sea level increases and sediments are forced to keep pace with and to fill the new space created. Marine and littoral environments shift landward and overly the previously eroded deposits, whereas fluvial sediments tend to infill the incised valleys on land. A generalized onlap characterizes the landward termination of marine strata, whereas starved conditions dominate in the basin. The resultant sedimentary response may vary, depending on the balance between the rates of sea level rise and the amount of sediments produced. If the rate of rise is greater than the rate of sediment supply, the backstepping of the facies belts (retrogradation) is recorded throughout the depositional profile (Fig. 2.26). In a fixed location, the vertical organization of successions will result in the progressive or abrupt superposition of facies, recording a deepening upward trend. If the rate of rise equals the rate of sediment supply, a generalized aggradation of the facies belts occurs and no particular changes are recorded in the long term signature of vertical stacked facies.

At the point D, the relative sea level reaches the maximum rate of rise; the shoreline and marine deposits achieve the most landward position recorded throughout the entire cycle of relative sea level. In the basin, which can be very far from the locus of deposition of coarse grained sediments, starved conditions dominate and favour the formation of a thin pelagic horizon ("condensed section") (Fig. 2.27), characterized by the concentration of marine fossils and authigenic minerals. The surface corresponding to the maximum landward shift of marine deposits and to the top of the condensed section in the basin represents the **maximum flooding surface**, and the interval corresponding to the entire period of transgression represents the **transgressive system tract**.

Some major differences in the sedimentary response can be recorded between siliciclastics and carbonates during the transgression. In carbonate platforms, the beginning of transgression is a crucial event for the successive

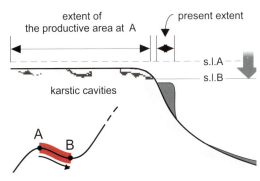

FIGURE 2.23

Carbonate platforms generally react differently to a phase of fall of sea level from siliciclastic, due to their peculiar organic origin and nature. The emersion of the platform top produces the reduction of the extent of the sediment production area, and erosion is generally of chemical nature. Progradation is often hampered by the steep rims of the platform.

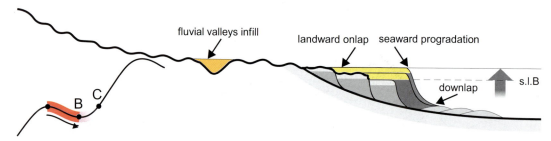

FIGURE 2.24

The relative sea level starts to increase with the consequent landward shifting of the shoreline. This produces the flooding of the previously emerged and eroded deposits, but the rates of rise are still too low and sediment supply still promote basinward progradational patterns.

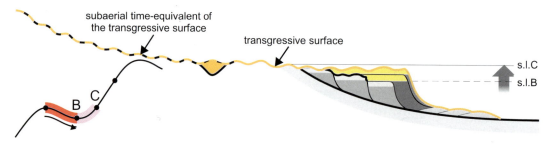

FIGURE 2.25

At the point C, the rates of relative sea level starts to rise with increasing rates and a major change in the sedimentary evolution occurs along the depositional profile. The sediments deposited during the initial rise of sea level (points B and C) correspond to the **lowstand system tract**. During the successive evolution, the increasing rates of rise of the sea level produce the rapid landward shifting of the shoreface. During the up dip movement, the shoreline erodes and reworks underlying sediments forming the **transgressive surface.**

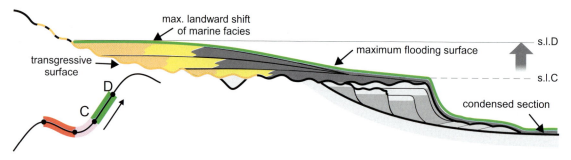

FIGURE 2.26

The relative sea level continues rise with increasing rates up to the D point. Sediments, which do not keep place with the rates of rise, tend to move landward, forming fining and thinning upward patterns in the stratigraphic successions. Depositional patterns are defined by a generalized retrogradation geometries. If a balance exists between the rates of rise of sea level and sediment supply, a stationary architecture of the stratal patterns can be recorded, and an aggradational pattern characterizes the depositional geometries. A the point D the maximum rate of sea level rise is recorded, and it marks the maximal landward position of the shoreline. At this time, source areas are located at the maximal distance from the basin, which is defined by starved conditions, and a condensed horizon may form. The interval corresponding to the period of increasing rates of rise corresponds to the **transgressive system tract**, and the point of the maximum rate of sea level rise corresponds to the **maximum flooding surface.**

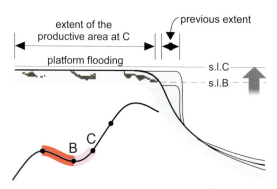

FIGURE 2.27

The time of increasing rates of sea level rise, corresponding to the transgressive surface, is crucial in carbonate platforms. Generally, benthic organisms re-colonize the flooded top of the platform; the development of a transgressive system tract occurs, and the carbonate production and the basinward exportation gradually resume and increase with time. But, if environmental conditions unfavorable for the reprise of the carbonate factory characterize the period of sea level rise, the drowning of the platform may occurs with the consequent deposition of pelagic sediments on the previously emerged platform top. This process may be produced also independent from the relative sea level, and associated to the input of large amounts of siliciclastic sediments.

2. Lithostratigraphy

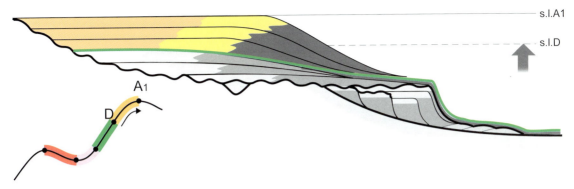

FIGURE 22.8

The rates of relative sea level rise start to decrease. Moving from the point D, new space continues to be created but, because the rate of relative sea level rise is decreasing, sediments move progressively basinward, where accommodation is still available. Stratigraphic successions record thickening and coarsening upward trends and depositional patterns are characterized by progradational geometries. Sediment accommodation will continues up to the point A1. The interval corresponding to the period of decreasing rate of sea level (points D-A1) corresponds to the **highstand system tract**. At the point A1 all the space available on the shelf will be filled and the successive fall of sea level will produce erosion of shelfal deposits and progradation of depositional systems toward the basin.

evolution, and the platform may die or grow, depending on environmental conditions. At the time of the transgressive surface, the sea level rises above the platform rim and the flat top of the platform is newly flooded (Fig. 2.27). Generally, the high growth potential of benthic communities allows the platform to keep pace with the rise of sea level, also during rapid and long-term periods of rise. In these cases, once the platform is flooded, the carbonate factory expands laterally and the area of carbonate production notably increases. This induces the shedding in the basin of an amount of sediment generally greater than in siliciclastics. During the entire interval of transgression, a healthy platform builds vertically keeping pace with sea level. In contrast, unfavourable environmental conditions may inhibit the development of the carbonate factory, with the consequent demise of the platform. The successive rise of sea level may induces the deposition of pelagic sediment on its top. This process is known as **platform drowning** and the related surface has been recently introduced as a particular type of sequence boundary for carbonate sequence stratigraphy.

The maximum flooding surface corresponds to a major turning point in the long term evolution of the accommodation, as it marks the time of inversion between increasing and decreasing rates of relative sea level rise. This produces the change from transgressive to regressive conditions, and the change from retrogradational/deepening to progradational/shallowing stratal stacking patterns. Once the rate of relative sea level rise starts to decrease, in fact, the volume of sediments exceeds the amount of accommodation created and sediments progressively tend to fill all the available space, expanding laterally. As a consequence, the shoreline starts to shift basinward, forcing depositional systems to build outwards. Shelfal deposition continues until the relative sea level rises creating new available space for sediments. Approaching the point A1 in the curve of the relative sea level, the rates of rise tend to decrease, but the space available for sediments continues to be created up to the point A1. This point corresponds to the time of maximum amount of space created during the entire cycle of the relative sea level. The interval between the maximum rate of relative sea level rise and the maximum of accommodation created corresponds to the **highstand system tract** (Fig. 2.28). Its upper boundary corresponds to the time of initiation of a new fall of relative sea level, and the evolution of a new cycle of relative sea level will start with the formation of a subaerial unconformity.

Despite the carbonate ramp systems may produce very similar sedimentary response than siliciclastics during sea level highstand, a major difference in terms of amount of sediment produced and exported basinward can be recorded in carbonate systems controlled by organic growth. In rimmed carbonate platforms, in fact, the carbonate factory reaches its maximum extent during this stage, because the decreasing of rates of relative sea level induces the benthic communities to expand laterally toward the lagoon and toward the basin, producing a bidirectional progradation. The resultant effect of this process is the attempt of depositional systems to build outward, where space for sediments is still available. The sediment produced in excess will be exported downslope, promoting the progradation of the platform margin and the increasing of the amount of resedimented carbonate particles basinward (Fig. 2.29).

4.3. PRACTICE

4.3.1. *The time-stratigraphic framework of sequences*

One important issue of sequence stratigraphy is the genetic meaning of the related stratigraphic units (sequences) and subunits (systems tracts). Differently from lithostratigraphy, which uses similar lithologic features for correlations, sequence stratigraphy correlates different lithologic features, genetically related and bounded by key surfaces that can be traced from continental to basinal environments. Example: Correlations in the Lusitanian Basin during the Lower Cretaceous (REY, 2006, Fig. 2.30).

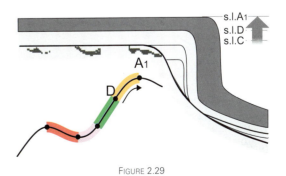

FIGURE 2.29

The decreasing rate of sea level rise induces the carbonate factory to expand laterally, promoting the basinward progradation and the increasing of sediment exportation.

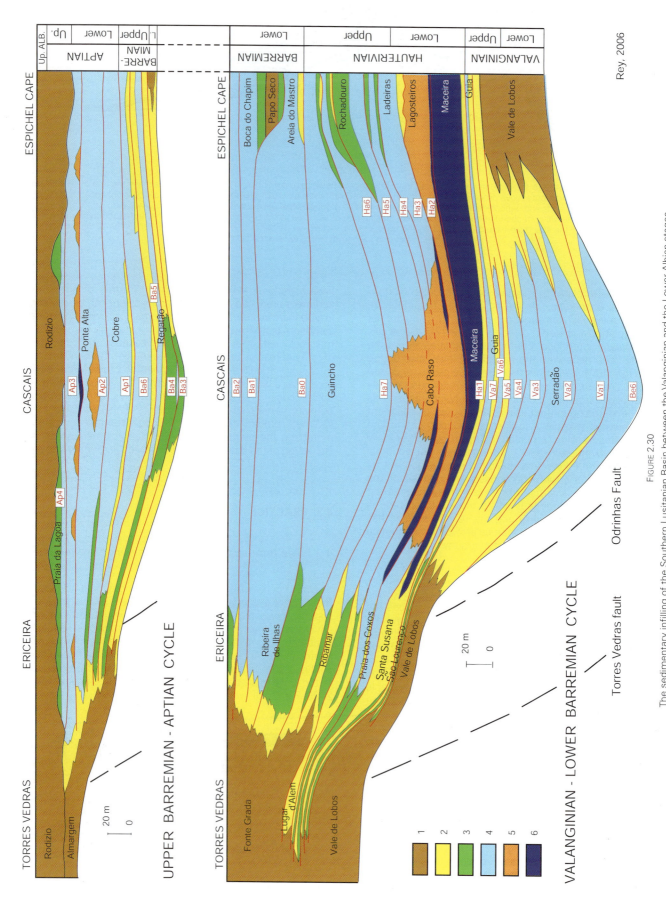

FIGURE 2.30

The sedimentary infilling of the Southern Lusitanian Basin between the Valanginian and the Lower Albian stages. The red lines indicate sequence boundaries (in REY, 2006).

1: fluvial deposits; 2: estuarine to nearshore deposits; 3: lagoon deposits; 4: inner platform deposits; 5: reefal buildups; 6: outer platform deposits.

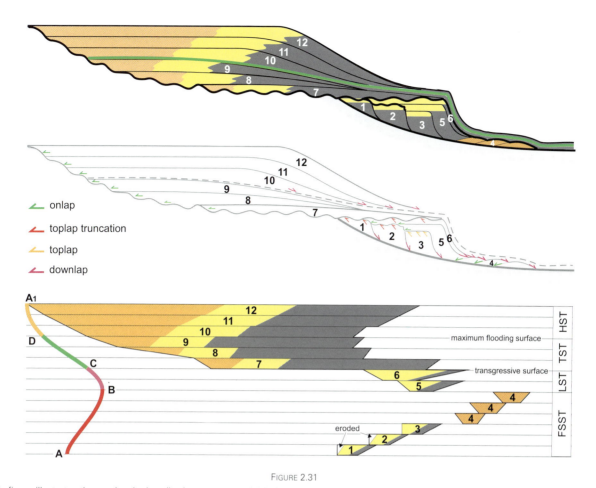

FIGURE 2.31

This figure illustrates the previously described sequence model (a) in terms of stratal terminations (b). In the diagram c, the spatial distribution of the depositional environment and stratal geometries have been used to define the sequence model in a chronostratigraphic framework (Wheeler diagram). This diagram allows to better appreciate the lateral relationships between adjacent depositional environments, the importance of hiatuses associated to the unconformities and the effects of the cycle of relative sea level during the sequence development.

The reconstruction of a stratigraphic framework, where facies and depositional systems are arranged according to principles of sequence stratigraphy, gives the possibility to relate the coeval evolution of adjacent depositional environments, and allows a deeper understanding of the relationships between sedimentary processes and mechanisms that control environmental changes.

Moreover, the relative chronostratigraphic meaning of sequence stratigraphic units and key surfaces allows to visualize the spatial relationships between adjacent depositional environments also in a time dimension. These relationships can be drawn in a chronostratigraphic diagram (WHEELER diagram) (Fig. 2.31) that allows to better appreciate the extent of time involved in key surfaces, generally, not or poorly represented by sedimentary record and, more in general, is useful to quantify variables such as sediment and subsidence rates and changes of relative sea level.

A chronostratigraphic diagram can be constructed from different types of data, such as outcrops, well logs and seismic lines. However, accordingly to the type of data source, different approaches have to be used to reconstruct the time relationships between depositional environments and key surfaces. The kilometric scale of the geological information depicted in seismic lines permits a geometric approach to define the relationships between different depositional events; moreover, the assumption that reflectors represent **isochronous** bedding surfaces allows to represent stratal (reflector) geometries as time lines. This makes the construction of chronostratigraphic diagram from seismic lines rather easily. A similar approach can be used in outcrops with kilometric scale exposures (seismic scale).

Differently, when only one dimensional data are available (stratigraphic logs, e.g. Fig. 2.32), the spatial and time relationships have to be reconstructed by lateral correlations of stratigraphic trends and key surfaces across different sections by means of synchronous horizons, such as a volcanoclastic level, or through bio-, magneto- or chemostratigraphy.

4.3.2. *The hierarchy of relative sea level cycles*

The change of the position along the relative sea level curve is not a steady-process, but it is modulated by the succeeding of small scale relative sea level fluctuations, induced by the continuous competition between sediment supply and accommodation. These short term fluctuations give rise to prograding units (coarsening and shallowing upward) known in sequence stratigraphy as **parasequences**. They are defined as the smallest units composed of relatively conformable successions of genetically related beds or bedsets bounded by marine flooding surfaces. Depending on the long-term pattern of the relative sea level, single parasequences stack forming progradational, retrogradational and aggradational parasequence sets. The stratigraphic signature of superimposed parasequence sets will define the internal architecture of the systems tracts within a sequence. The numbers in the diagrams of Fig. 2.31 may represent single parasequences or parasequence sets.

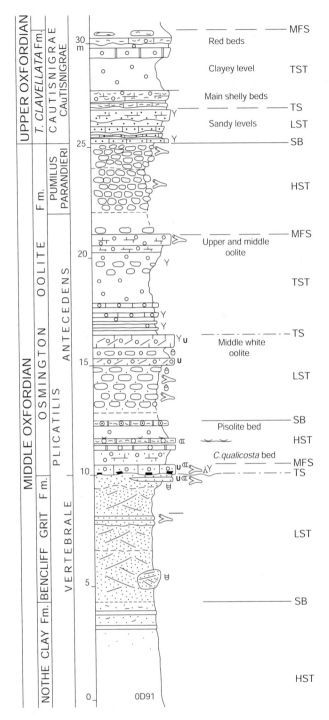

FIGURE 2.32

The 3rd order depositional sequences of the middle Oxfordian of Normandy (after RIOULT et al., 1991). LST: lowstand system tract; TST: transgressive system tract; HST: highstand system tract; SB: sequence boundary; TS: trangressive surface; MFS: maximum flooding surface. The thickness and lithology of the different system tracts varie from one sequence to another in response of the sedimenary dynamics and accomodation space.

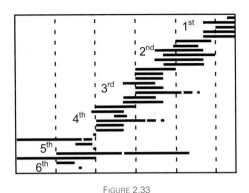

FIGURE 2.33

Duration of orders of stratigraphic sequences as defined by various authors (modified after SCHLAGER, 2004).

telescopic superposition of different scale cycles of relative sea level. The sedimentary record of a basin can be so subdivided into a relative hierarchical organization representing different orders of relative sea level cycles. In sequence stratigraphy about six orders of cycles are considered, starting from the sixth order that represents the highest cycle frequency to the first order that represents the lowest one.

Commonly, a different duration is attributed to the hierarchy of cycle orders so, for example, a time period of 50 to 200 Ma is considered for first order cycles, or a time period of 0.2 to 5 Ma for third order cycles. However, the attribution of the time length to a sequence or to a higher order cycle, simply on the base of the hierarchical order can be ambiguous and elusive. The duration of cycles, in fact, may largely varies and overlaps between different orders (Fig. 2.33, SCHLAGER 2004), because tectonic and eustasy are non periodic processes and they may produce changes in relative sea level of variable duration throughout time. Differently, cycles associated to astronomical mechanisms (**Milankovich cycles**) are linked to periodic processes that regularly repeat over time intervals of constant duration; therefore, their extrapolation to absolute ages can be made more confidently. Unfortunately, only relatively few stratigraphic sections and depositional environments fulfill conditions for high resolution subdivisions of the sedimentary record.

5. – LITHOLOGIC CYCLES

5.1. DEFINITION

Lithologic cycles are groups of at least two facies that are regularly alternated in the stratigraphic record. Depending on the number of recurring elements, lithologic cycles can be defined following a-b-a scheme, when only two lithologies alternate, or a-b-c-b-a-b-c when more the repetition of more two facies represent each single cycle. Lithological cyclicity are the sedimentological expression of the oscillation in environmental parameters controlling the facies (e.g. terrigenous supply, biogenic supply, etc.).The reconstruction of the depositional environment and the mechanisms controlling its oscillation from one condition to another, requires a thorough knowledge of the diagenetic processes that can largely alter the relations between facies.

Environmental changes controlling lithologic cycles can be related to internal factors which intervene in a repeated but non-periodic way or can induced by a forcing mechanism external to the system causing periodical oscillations. In the

Similarly to parasequences, sequences can stack forming larger scale units of one order higher magnitude, defined by progradational, retrogradational or aggradational sequence sets. These large scale units reflect relative sea level changes throughout a longer term cycle modulated by processes of a greater amplitude and wavelength. By considering the entire evolution of a sedimentary basin, during which processes act with different amplitude and frequency, the long term stratigraphic architecture of the basin-fill may be seen as the

FIGURE 2.34
Marl/limestone alternation in the Hauterivian pelagic facies of the Castillon (northward of Castellane, Alpes-de-Haute-Provence).

latter, provided that the periodicity of the forcing mechanism is known, lithologic cycles can be used for geochronological interpretations.

Although several parameters independent from lithology may show periodic fluctuations in the stratigraphic record, the word "cycle" without adjective is generally used to express its lithologic expression. The word "cycle", with this meaning, has been used in this text.

5.2. PRACTICE

5.2.1. *Cycles characterization*

Cycles can be described in terms of several parameters as follows:

– thickness. On the base of their thickness, inframillimeter-, millimeter-, centimeter-, meter- and decameter-scale cycles can be distinguished. Vertical changes of thickness define the regularity of the lithologic cyclicity, which is not necessarily related to the regularity of the forcing mechanism.

Millimeter- and sub millimeter-scale cycles are usually defined as **very high-frequency cycles**. Because of their little thickness (laminites), they are typically developed (or preserved) only in particular settings, such as protected depositional environments (lakes), or in the absence of bioturbational mixing (e.g. anoxic deposits). However, in the presence of high sedimentation rates, such as in evaporitic basins, centimeter- to decimeter- high-frequency cycles can occur (e.g. Permian basin of Zechstein, Germany).

Decimeter- to centimeter-scale cycles generally correspond to **high-frequency cycles**, well marked by lithologic alternations. These are typically preserved in depositional settings characterized by low sedimentation rates (some meters in some million years) and hypoxic-anoxic conditions unfavourable to bioturbation. These cycles are generally grouped in sets of several units (bundles) which define a higher order cyclicity;

– lateral extent. The lateral extent of lithologic cycles is usually proportional to their thickness, although several exceptions exist. Cycles resulting from fluctuations of primary factors (e.g. carbonate supply varying against a background constant fine clastic deposition) usually have a large lateral extent, particularly in pelagic or hemipelagic depositional settings. On the contrary, cycles controlled by secondary character (diagenetic cycles) are unlikely to be recognizable over wide areas. The identification of cyles characterized by large lateral extent enables high resolution regional or interbasinal correlation (see below);

– composition. Cycles can include 2, 3, or n elementary lithologies or lithologic end members. Cycles including only two terms (a-b, a-b), i.e. binary cycles or couplets, are the most frequent. In pelagic or hemipelagic environments, they usually produce decimeter- to meter- scale alternations of carbonate-dominated beds with clay-richer lithologies (Fig. 2.34). In this case a cyclical fluctuation in $CaCO_3$ content is the cause of the lithological alternation. However, alternations between carbonate and anhydride (Permian Formation of Castile, USA), between organic matter and eolic silt (Permian Formation of Bell Canyon, USA) and between sapropel and calcite (Upper Jurassic in New Mexico) also occur. An alternation is considered balanced when the two components have approximately identical thickness. When this is not the case, one or the other semicycle can thin to a millimetric thickness or disappear and create a hiatus.

A progressive shift between the two end-member of a cycle, sometime caused by bioturbation, allows identification of a third lithologic term, to form a tripartite cycle (a-c-b-c-a).

In cycles presenting more than three terms (up to ten for the cyclothems of the Pennsylvanian of North America (HECKEL, 1986)) cyclicity can be recognized only in rather large outcrops;

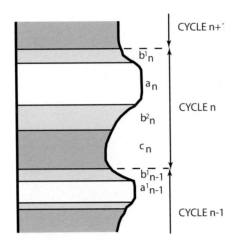

FIGURE 2.35
Standards of characterization for cycles composed by three lithologic end members:
a: limestones – b: marly-limestones – c: marls.

asymmetry index of the cycle n: $\dfrac{b1n}{b1n + b2n}$.

– asymmetry level. Cycle evolution may includes two alternating bed types known as rhytmic sequence (a-b) or the alternation of several sediment type forming a succession (e.g. A-B-C-n) which is repeated, known as cyclic sequence. The difference in the thickness of the various elements of a cycle defines the asymmetry level (Fig. 2.35) (COTILLON & RIO 1984). Maximum asymmetry characterizes a cycle where one of the elements is absent (incomplete cycle) as in the case of a deep truncation of part of the sequence. In this case the resulting sequence will be of type a-b-c-a-b-c;

– Type of control. **Autocycles** are primarily controlled by processes taking places within the depositional system, such as the migrations of a meandering river system, compaction or sediment load, from sediment supply, stacking rate, water chemistry, bioproduction, etc. These cycles are mainly found in continental, coastal and platform environments and their beds usually show only limited lateral extend.

Allocycles are mainly caused by variations external to the considered sedimentary system such as climate and eustatism. Allocycles are better preserved in low energy environments protected from surface dynamics, such as in open marine depositional settings. They recognition of cycle periodicity represent a fundamental tool in the reconstruction of environmental changes forced by global causes. For example, in pelagic limestone/marl or black shale alternations, astronomically controlled cyclicity characterizes not only lithofacies but also the microfaunal and nannofloral contents, the mineralogy (detrital clays), the geochemistry of the sediments, and the palynofacies (e.g. GALEOTTI et al., 2003). Changes in all these parameters provide evidences for a complex interaction of climate-related environmental changes on both the oceanic environment and continental areas (COTILLON et al., 1980).

5.2.2. Cycles, lithostratigraphic tools

The use of cycles in lithostratigraphy dates back to GILBERT (1895). As previously mentioned, a variety of sedimentary cycles can be recognized in the stratigraphic record depending on thickness, lateral extent and controlling mechanisms. Global climate related cycles, being controlled by factors acting over wide areas with regular periodicities, are the most useful in stratigraphy.

Metronomic variations of earth-sun and eart-moon orbital configuration result in a wide range of cycles of varying frequencies. Some examples are:

– the annual cycle related to the Earth's position around the sun;

– the lunar month Moon's position around the earth (lunar cycle = 27 days, 7 hours, 43 minutes) and semi-lunar;

– the solar magnetism (HALE cycles = 11 and 12 years);

– the so called **MILANKOVITCH cycles**, that induce **high resolution variations** of the climate, with a periodicity between 19 Ka and 2 Ma equivalent to the precession of the equinoxes (19 Ka and 23 Ka), the obliquity of the rotation axis of Earth (40 Ka and 53 Ka), the eccentricity of Earth's orbit (97 Ka, 123 Ka, 412 Ka, 2,35 Ka) (SCHWARZACHER, 1993) (Fig. 2.36).

One of the main goal of cyclostratigraphy is to establish the regularity of periodicity/ies of lithologic cycles and to relate them to known processes. The identification of lithostratigraphic cycles of known periodicity in a stratigraphic succession allows to use the recognized cyclicity as a chronostratigraphic tool.

Examples

Among **lithologic cycles**, varves represent rhythmic sequences of annual cycles occuring in lake deposits. These frequencies can be present also in evaporitic successions characterized by high rates of sedimentation (10 cm per year in the case of halite of the Permian of Zechstein), in present marine deposits (millimetric varves with diatoms in the California Gulf), and during the Lower Cretaceous of the Gulf of Mexico (10 to 30 μm thick cycles).

Tidal laminites show a silt-clay alternation in millimetric laminae which, under favourable conditions, can show the following succession characterizing each tide event: flood current, still stand phase, ebb current. Silt-clay couplets in semi-lunar cycles (Fig. 2.37) corresponding to relatively thin units (spring tides cycles) and to thicker units (neap tides cycles). Finally, the super cycles are bounded by thick couplets of equinox spring and autumn tide (TESSIER et al., 1989).

Among high frequency lithologic cycles, pelagic alternations can be used: calcareous bed/marl interbed or marl bed/shale interbed, with a decimetric to metric thickness related to MILANKOVITCH cycles (FISCHER, 1986).

Considering their characters, lithologic cycles can be used both for correlation and time measurement.

a – Cycles, correlation instruments

Very high-frequency cycles have a limited lateral continuity because of the fragility of the lamina compared to physical, chemical or biological diagenesis. Exceptionally, varves have been used for precise correlations over 105 km distances as in the Permian Fm, of the Bell Canyon (USA) (ANDERSON & DEAN, 1988).

High frequency cycle continuity on great distances presupposes a sedimentation in low energy, protected environments from the direct control of local phenomena: detrital supply, tectonics, reworking, erosion, benthic production, emersions, etc. This makes deeper and open marine environments more suitable for the preservation of high frequency cycles.

Bed to Bed correlations

Some examples: in the upper Cretaceous of the American interior basin, over distances of about 1 000 km and an area 388 000 Km^2 wide (HATTIN, 1971, 1985). Between central

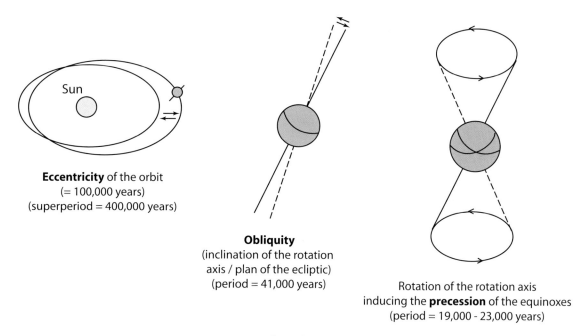

FIGURE 2.36
Schematic representation of the astronomical parameters controlling the insolation (after DE BOER, 1983).

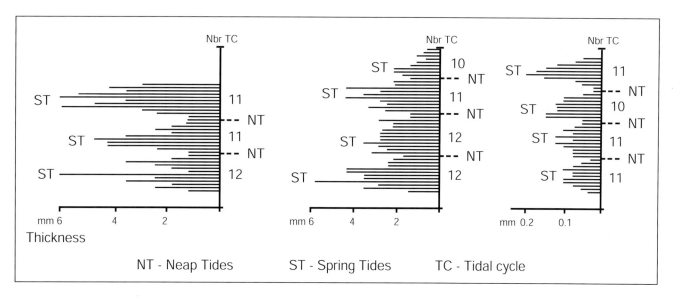

FIGURE 2.37
Graphic representation of the semi-lunar cycles defined by tidal rhytmites in the Bay of Mont Saint-Michel (after TESSIER et al., 1989).

Italy and Colorado, at the transition between Cenomanian and Turonian (DE BOER, 1983). Over the whole extension of the Vocontian basin (South-East of France) during Lower Cretaceous (about 12.000 Km²) (COTILLON et al., 1980).

Correlations based on vertical evolution and cycle thickness

This type of correlations is applicable where cycle periodicity does not change throughout and in the presence of comparable changes in the sedimentation rate. This conditions are satisfied in the presence of the so-called "productivity cycles" in pelagic limestone-marl alternations where changes in $CaCO_3$ content reflect fluctuations in the productivity of surface waters. Examples: Lower Cretaceous of Vocontian basin, Central Atlantic and Mexico Gulf.

*Correlations based on the amount of $CaCO_3$ of cycles (**carbonate cycles**)*

This approach used for the Cretaceous/Tertiary boundary in Spain (TEN KATE & SPRENGER, 1993, Fig. 2.38), is related to the previous one; it has been shown (COTILLON et al., 1980) that in pelagic limestone-marl alternations where the carbonate fraction is produced by planktonic organisms, a positive correlation exists between the thickness of strata and their carbonate content. A negative correlation exists instead in the so-called "dilution cycles" which are marl-limestone alternations produced by fluctuating clastic supply against a background constant carbonate productivity.

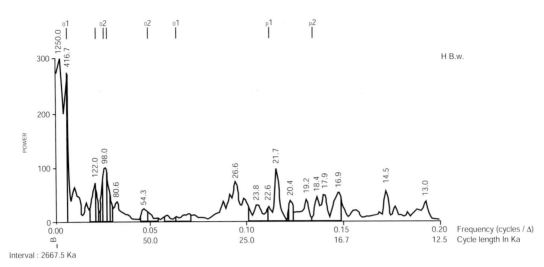

FIGURE 2.38

Spectral analysis of the CaCo3 content across the Cretaceous/Tertiary boundary at Zumaya (North Spain). The sampling resolution is 10 cm. X-axis: the number of cycle per sample or frequency (cycle/Δ) and correspondent time in thousands of years. Y-axis: the spectral power (amplitude). Some peaks have a period comparable to that of the Milankovitch cyles: 21,7 Ka (precession), 54,3 Ka (obliquity), 98, 122 (short eccentricity cycle) and 416 Ka (long eccentricity cycle)). (after TEN KATE et SPRENGER 1993).

Correlations based on the variation of cycle frequency throughout a succession

This type of correlation is based on the variations in sedimentation rates. It has been used to correlate cyclical series in a number of Hauterivian and Albian sites in the Atlantic and Pacific oceans (COTILLON, 1987). Correlations based on this parameter are possible only between successions deposited in similar depositional environments and at a comparable sedimentation rates. In this cases, the cyclicity is primarily controlled by fluctuations in just one parameter, that is changes in the sedimentation rate originating from climatically controlled cyclical fluctuations in the carbonate production.

Conclusions

Periodic, global or local phenomena control sedimentary processes and result in lithologic cycles whose characters and periodicity allow to infer the causes of their origination. These cycles can be used for high resolution lithostratigraphic correlations.

b – Astrochonology

Sedimentary archives record cyclical changes in sediment properties, fossil communities, chemical and isotopic characteristics resulting from past climate oscillations. The majority of **spectral analyses** (see Fig. 2.38) carried out on successions of Mesozoic limestone/marl alternations of hemipelagic and pelagic environments show that decimetric to metric bed/interbed couplets have a duration of ca. 21 Ka. On this base the duration of different chronostratigraphic and biostratigraphic units has been calculated.

Using this method, GILBERT attributed a duration of about 21 million years to the Upper Cretaceous of the American inner basin already in 1895. Since then an orbitally controlled lithologic cyclicity has been recognized in many Cretaceous succession of various areas including the lower Cretaceous of the Vocontian basin, the middle Cretaceous of Central Italy, the Cenomanian of England. The detailed study of some succession allows to recognize a hierarchy in the cycles, the grouping of precession-related limestone/marl couplets in bundles of 5 units being the expression of the short eccentricity (100,000 years) cycles; super bundles made up of 4 bundles are referred to the 400 ka-long eccentricity cycles (Fig. 2.39).

For younger intervals (i.e. Neogene), **astrochronology** utilizes the regular periodicities of past climate changes controlled by Milankovitch cyles and their expression in the sedimentary record to erect a chronostratigraphic time scale. As a first step, it is necessary to establish whether or not the cycles present in the sedimentary record under consideration are Milankovitch cycles. Cyclicities recognized in the stratigraphic record are compared to past variations in precession, obliquity and eccentricity of the Earth's orbit and rotation axis derived from computational models based on the mechanics of solar-planetary system and Earth-Moon systems. Sedimentary archives are therefore dated by matching patterns of paleoclimate variability with patterns of varying solar energy input computed from the astronomical model solutions. This procedure (**astronomical tuning**) has been combined with the additional correlation aids provided by magnetostratigraphy, chemostratigraphy and biostratigraphy to develop a detailed cycle scaling for the Neogene interval. The obtained scale (astronomical polarity time scale or APTS for short) is being extended back to about 23 Ma ago (see LOURENS *et al*. in GRADSTEIN *et al*., 2004).

The chronostratigraphic use of high frequency cycles must be accompanied by some precautions:

– Sedimentation range should be sufficiently high to allow the preservation of the bed-interbed couplets corresponding to 21,000 years cycles. Under lower sedimentation rate conditions, because of their reduced thickness, several cycles with a period of 21,000 years can be obliterated by bioturbation. Higher sedimentation rates, on the contrary, allow periodicity shorter than 21,000 years, to be recorded as centimeter to decimeter scale lithologic cycles;

– Biogenic and terrigenous fluxes which control the deposition of high frequency lithologic cycles must fluctuate in a regular way as it possible to see during highstand periods (COTILLON, 1992);

– The periodicity of orbital parameters varies though time depending on the continuous evolution of the Earth-Moon system. Example: by the Early Devonian, the fundamental periodicities of precession cycles were 17,000 and 20,000 years, therefore shorter than today (Fig. 2.40) (BERGER *et al*., 1992). As a consequence, an absolute duration should not be assigned to cyclostratigraphic unit. DE BOER & WONDERS

FIGURE 2.39
Marl/limestone alternations in the Albian Marne a Fucoidi of Piobbico (central Italy). The marl/limestone couplets reflect the precession cycle (21 kyr). The grouping of five couplets into bundles reflects the short eccentricity cycle (100 kyr) whereas the grouping of four bundles reflects the long eccentricity cycle (400 kyr).

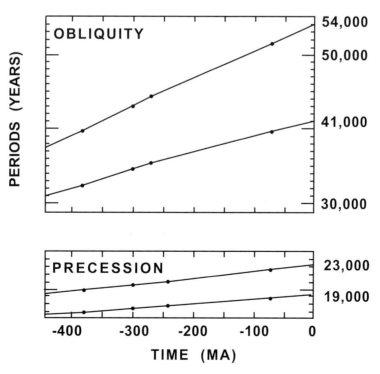

FIGURE 2.40
Evolution through time of the duration of the precession and obliquity cycles (after BERGER *et al.*, 1989).

(1984) proposed the name "**Gilbert**" to define a time unit of variable duration, corresponding is the periodicity of the precession cycle.

6. – CONVENTIONS IN LITHOSTRATIGRAPHY

6.1. NAMING, DEFINITION, PUBLICATION OF LITHOSTRATIGRAPHIC UNITS

As for any other stratigraphic unit, the proposal of a new formal lithostratigraphic unit requires a statement of intent to introduce the new unit and the reasons for the action. The definition of all the lithostratigraphic units necessitates to state several aspects including:

– Lithology: the distinguishing lithology(ies) of the new unit should be described, including information on colour, sedimentary structures, bed-thickness and cyclicity(ies).

– Lower and upper boundaries: well-defined lithologic criteria which should include biological content and data about well logs, seismic and chemistry, etc. The boundaries must be located in correspondence of clear lithologic changes. Lithologic criteria in the definition of boundaries provides the most objective and framework for successive geological mapping. Sequence stratigraphy criteria are preferred when dealing with successions representing a sequential organization. Sedimentologic/sequence stratigraphy criteria become the most important criteria when units are used as reference standard for a sedimentary domain.

– Age of boundaries: keeping in mind that boundaries of lithostratigraphic unit do not necessarily correspond to e.g. boundaries of biostratigraphic intervals, chronological elements should be given. The age of the unit should be indicated by reference to the standard schemes of periods/epoch/ages.

– Unit name: the name of the new unit has to combine information about the type of unit (Formation, Group,...), a geographic term and, if necessary, a lithologic term referring the type of dominant lithology (Grès Formation of Luxembourg). The geographical term used must refer to permanent features (localities, rivers...). "Upper, lower" should be avoided as their relative meaning could create confusion.

– **Lithostratigraphic stratotype:** a type and a reference sections should be indicated together with possible complementary information derived from subsurface data. The horizontal and vertical lithologic changes in the stratotype region must be described. The morphologic characters of the unit must be described. As to higher rank lithostratigraphic units of the Formation (Group,...), composite stratotypes can be grouped.

Information should be given also on the distribution (area of occurrence), the thickness and possible subdivisions.

6.2. PROCEDURE FOR CODIFICATION OF SURFACE STRATIGRAPHIC UNITS

6.2.1. *The case of lithostratigraphic units*

National Stratigraphy Committees formalize lithostratigraphic units of higher rank (Groups, Formations and Members). The Committees express their opinion on the conformity to the rules of nomenclature and principles of priority. Responsibility on the scientific validity of units rests with their authors. National Committee votes and list approved units which are yearly transmitted to IUGS. Details are made known to other workers possibly through publication in recognized scientific media.

6.2.2. *The case of facies sequences*

Facies sequences are informal units which are not submitted to any formal procedure for codification. However, at the scale of major sequences it is advisable to define these units with reference to a type locality (Ex: "sequence of FIGOLS, MUTTI et al., 1984"). Then, names already used to define formal lithostratigraphic units cannot be used in the definition of these informal unit.

6.2.3. *The case of genetic sequences*

Genetic sequences are informal units whose number in the literature is constantly and rapidly growing. At present, their subdivision lacks stability and it would be hasty to fix their number and rank. As an example, the French Stratigraphy Committee expressed the following recommendations as to 3rd order sequences:

– It is advisable not to define sequences using acronyms derived from a world chart, or by the numeric ages of their boundaries;

– Sequences can be defined in all sedimentary basins by acronyms illustrating: a) the basin; b) the stage followed by a number corresponding to chronological order, base to top (ex: B.P. Toa 1, B.P. Toa 2, B.P. Toa 3 for the first three sequences of the Toarcian of the Paris Basin). If a sequence spans two stages, only the stage age including the lower boundary of the sequences should be considered. It is clear that the number of sequences identified for the same time interval will vary from according to different authors and from basin to basin. It is desirable that a synthesis model will be published;

– Proper stratotypes cannot be proposed to genetic sequences as the facies distribution within each system tract has to be considered. The publication of reference successions, including reference sections in different environments for each genetic sequence, will be of great interest;

– It is advisable to propose correspondence tables between genetic and lithostratigraphic units for a better understanding of the equivalence between the sets defined by different stratigraphic approaches.

6.3. PROCEDURES FOR CODIFICATION OF SUBSURFACE STRATIGRAPHIC UNITS

6.3.1. *Well logs and reference section of a lithostratigraphic unit*

A reference well log must be defined. Its name is derived from its official acronym, the name of the stratigraphers(s) working on it, the geographic coordinates, beginning and end dates of well logging operations and by the altitude of the rotary table. For offshore wells, the water depth will be indicated as well. Additional reference wells can be indicated.

The description of the reference section has to include the following details:

– Lower and upper depths, the maximum thickness of the unit drilled within the same basin;

– Lithologic description, of the cores, with a description of physical characters (Ex: "The Formation [Kimmeridge Clay] is represented by non calcareous, grey-dark brown to black clay, coal rich deposits, characterized by a high radioactive response due to their content of organic material. Furthermore, it is typically characterized by low velocity, high resistivity, and low density" (Fig. 2.4);

– Boundaries: lithologic and well-logs criteria are systematically associated. Use of well logs becomes relevant in locating formation boundaries. Geologists preferably use shifts in radioactivity and acoustic logs;

– Lateral extent: it is defined by the perimeter containing the well logs farthest away from the reference well, where the lithostratigraphic unit has been drilled. The information framework is completed with seismic lines passing through the reference well;

– Age: an age interval encompassing the possible diachronism of the unit must be stated. At this goal, micropaleontologic and palynologic analyses, basic techniques for subsurface biostratigraphy, are carried out on cores samples.

6.3.2. *Reference material, nature, preservation and availability*

Though the standard section of subsurface units is, contrarily to outcrops, protected from natural or artificial degradation, the amount and availablity of reference material is obviously limited.

It is necessary that all the reference material of the lithostratigraphic unit – rock samples, different records, preparations, biostratigraphic information – be deposited at an institution with the proper curatorial facilities and assurance of perpetuity where the materials are available for study. The location of the depository for materials from the stratotype well should be given.

Rock samples

The reference collection must include a longitudinal section and a set of cuttings of the cores taken from the interval containing the lithostratigraphic unit. In principle, samples should not be used for destructive analysis.

Preparations

Thin sections, palynologic slides and micropaleontology residues are obtained from cores fragments or cuttings portions, and can be part of the reference material.

Availability of the material

After a period of time depending on different legislations, well logs and seismic lines are made available to official geologic institutions.

6.3.3. *Selection*

The choice of the reference well, the interval, and the general character of the standard section should be made on the base of the following criteria:

– high quality of the well logs, and if possible seismic lines. Clarity of criteria for the definition of unit boundary;

– continuous coring;

– Availability and preservation of documents and samples.

6.3.4. *Definition*

Name: institutions of coastal countries on the North Sea (Institute of Geological Sciences of Great Britain, Norway, Petroleum Direktorat of Norway, Rijks Geologisch Dienst of the Pays-Bas) issue lithostratigraphic nomenclatures to achieve a codification of the various units defined. The final aim is to obtain a practical nomenclature that can be shared by oil companies and with University institutions. Names derived from surface units, such as Kimmeridge Clay Formation, can be hardly applicable to subsurface unit terminology. Names are usually derived from geographic names of the nearby coast (Texel Chalk Formation), from marine birds as with the units of Texas, and famous navigators as for the upper Lias units. A number of procedures based on simple mnemonic techniques favours this practice; the Dunlin Group includes the formations of Amundsen, Burton, Cook and Drake.

Rank: the definition of the rank of a subsurface unit follows, similarly to the surface units, the general rules of stratigraphic nomenclature. In some cases hydrocarbon exploration geologists introduce criteria which follow industrial interests for reservoir definition. Accordingly, the latest edition of Lithostratigraphic Nomenclature of the North Sea (Part 1: Paleogene of the Central and Northern North Sea) defines sandstone units as members and clay units as formations. This is because these sandstones are generally included within clay units showing a larger lateral continuity. For this reason the latter deserve a higher rank in the classification.

The formalization of subsurface units follows the same procedure used for the surface units (see 6.2).

CHAPTER 3

CHEMOSTRATIGRAPHY

M. Renard, J.C. Corbin, V. Daux, L. Emmanuel, F. Baudin & F. Tamburini

CONTENTS

1. – DEFINITION 41
2. – TERMINOLOGY 41
3. – PRACTICE 44
    3.1. CaCO₃ fluctuations in pelagic carbonates 44
    3.2. Oxygen stable isotope variations 44
    3.3. Carbon stable isotope variations 45
    3.4. Sulfur stable isotopes 46
    3.5. Strontium stable isotopes 47
    3.6. Trace elements in carbonates 47
    3.7. Iridium anomalies 49
    3.8. REE, Rare Earth Elements 50
    3.9. Organic carbon and geochemical biomarkers 50
    3.10. New Frontier in Chemostratigraphy 52
        3.10.1 Oxygen isotopes 52
        3.10.2 Impact of gas hydrates for the carbon isotopes... 52
        3.10.3 Strontium isotopes and other isotopes
            (e.g., osmium) 52
4. – CONCLUSION 52

1. – DEFINITION

Chemostratigraphy is the application of sedimentary geochemistry to stratigraphy. In fact, chemostratigraphy is closely linked to **lithostratigraphy** since major lithologic variations relate to changes in the importance or balance between the different geochemical reservoirs. It is important to underline that geochemistry often reconsiders old stratigraphic terms, e.g., the Neocomian, which are no longer used because of the development of biostratigraphy. At the same time, mineralogical stratigraphy is clearly closely related to chemostratigraphy: the link between the mineralogical composition, elemental distribution and the isotopic composition of the sediment is just one example.

Chemostratigraphy is based on the theory that the chemical and physical composition of sea-water has changed through time and that these changes have been recorded by the chemical composition of sediments (by major and minor element distribution) and by the isotopic ratios of particular elements.

The widespread use of chemostratigraphy was initially impeded by three main problems. The first is a methodological one: in the early 1960s, geochemical studies of sediments were performed using petrologic bulk techniques. The mix of different mineralogical phases coming from various sources and of different ages led to difficult and frequent misinterpretation of results. Progress has been facilitated by undertaking analyses of purified phases (e.g., carbonates, clay, organic matter) and of determined biogenic components (e.g., planktonic foraminifera, benthic foraminifera, corals and ostracods).

The other two problems are more conceptual and concern the over-stressing of the diagenetic imprint on the original signal, and the application of the principle of uniformitarianism which requires that sea water characteristics have not changed in the past and that any chemical variation must be attributable to post-sedimentary processes. This last aspect was especially felt by European scientists.

Pelagic facies were the first to be studied, since they have many advantages over platform facies:

– (1) carbonate producers (i.e. planktonic foraminifera, cocco-liths) are present in smaller amounts in pelagic sediments, which therefore have a rather homogeneous mineralogy (i.e. low-Mg calcite); this aspect reduces the possibility of biological and mineralogical fractionation. The mineralogy of platform facies is complex, because stable (i.e. calcite and dolomite) and unstable (aragonite, Mg-calcite) carbonate minerals may coexist;

– (2) diagenesis of pelagic facies is controlled by sea water or by pore waters derived from sea water, while fresh water from the continent could impact early diagenesis of platform sediments.

Like other stratigraphic tools, sedimentary geochemistry has two principal approaches: one concerns the geochemical facies and provides information about paleoenvironment and paleogeography, the other concerns the temporal evolution and events. Geochemistry can thus be used as a tool to correlate stratigraphic series using either isotopic or chemical "events". These can be considered as time-synchronous either theoretically (because geochemical models predict an immediate response on geological timescales), or practically (because the events have been correlated by means of other chronostratigraphic tools).

Geochemical methods also facilitate relative dating once the geochemical records have been calibrated by geochronologic methods.

2. – TERMINOLOGY

The **residence time** in the ocean of a chemical element (time taken for an atom or molecule to complete one chemical cycle; that is, total oceanic inventory divided by external (i.e. riverine or eolian) input per year assuming cycle in long-term balance) is specific to each element (i.e. 1000 years for Mn, 2 Ma for Ca, 8 Ma for Sr, and 20 Ma for Mg). Thus the signature of a single event could be more or less dispersed through time, depending on the chemical element considered. However, because of the relatively short mixing time of oceanic water, sea water composition should be considered homogeneous over a certain geological instant, and geochemical variations should have a global meaning. Nevertheless, local factors, linked to the configuration of the basin and/or the distance from the continent, could impact the geochemical record, especially in platform settings.

Geochemical results can be expressed in either absolute values (e.g., for trace elements) or as ratios to a reference standard (e.g., stable isotopes, REE). Geochemical analyses may be performed on bulk samples, on isolated mineralogical fractions (carbonate, clays, siliceous fractions), on biological fractions (planktonic or benthonic foraminifera, ostracods), or on organic matter.

Trace elemental concentrations in rocks and sediments are generally expressed in **ppm** (parts per million, or mg/Kg), but for elements at low concentrations, such as iridium and platinoid minerals (Ni-Cu-Fe), the **ppb** notation (parts per billion or ng/Kg) is used. Isotopic ratios are expressed as parts per thousand and by the δ notation, as illustrated by the following formula:

$$\delta^{18}O = ([^{18}O/^{16}O] \text{ sample} - [^{18}O/^{16}O] \text{ standard}) / ([^{18}O/^{16}O] \text{ standard}) \times 1000.$$

Oxygen and carbon **isotopic ratios** in carbonates are expressed in relation to the ratio in the **PDB** standard (roster of the *Belemnitella americana* of the Cretaceous Pee Dee Formation, South Carolina, USA). A carbonate $\delta^{18}O$ value of 2 per thousand means that CO_2 derived from this carbonate is enriched in 2 parts per thousand compared with the CO_2 derived from the PDB Standard. Ratios measured in water are expressed in relation to **SMOW** (Standard Mean Oceanic Water), a theoretical seawater with properties close to the average of present-day oceanic sea waters. In the same way, some authors (e.g. DE PAOLO & INGAM, 1985; HESS et al., 1986) relate strontium isotopic ratios ($^{87}Sr/^{86}Sr$) of carbonate and phosphates (conodonts and fish debris) to sea water ($\delta^{87}Sr$), even though analyses are performed using the **NSB 987** standard (strontium carbonate). The **CDT** (Troleïte formation of the Canyon Diablo meteor) is used as a standard for the sulfur (FeS) isotopic ratio. Finally, REE concentrations in sedimentary rocks are related to continental crust value concentrations, generally the **NASC** (North American Shales Composite).

Since their inception, chemostratigraphic studies have tried to describe geochemical variations and, following the pioneer works by EMILIANI (1955), have employed variations in oxygen stable isotopes as a correlation tool. Several authors have tried to combine both isotopic (SHACKLETON & OPDYKE, 1973; WILLIAMS et al., 1988) and trace elemental methods (ODIN et al., 1982, 1994; RENARD, 1985).

To describe the variation in a geochemical proxy, three parameters are generally considered: trend, amplitude and rate of variation.

We can speak of increasing (positive) or decreasing (negative) trend. In terms of isotopic ratios, the term "heavy" indicates greater abundance of the heavier isotope, while the term "light" is used to indicate greater abundance of the lighter isotope or deficiency in the heavier isotope (heavy/light isotopic ratio).

To quantify the amplitude of variations, it is important to consider analytical errors (+/– 2 standard deviations) and dispersion (variability) of the data. The description of the amplitude of a signal is rather subjective, but a signal with strong amplitude is generally considered as being greater than 10 times the analytical error.

If the variation of the signal occurs over a period longer than 1 million of years, the change is considered slow. Variations occurring over a period between 100 ka and 1 Ma, and over a period less than 100 ka, are generally considered fast and very fast (high frequency variations) respectively.

A **long-term trend** is generally defined as evolution taking place over several million of years (Fig. 3.1). Short-term variations (shorter than 10 ka) are then superimposed on the general long-term evolution. The long-term evolution could also be characterized by breaks, defining periods when values are stable and/or fluctuate about an average value.

A **geochemical cycle** corresponds to repeated variations (which may be quasi-periodical) of an average signal over a certain period of time, uniform or variable. These cycles have facilitated the establishment of geochemical stratigraphic zonations (**Chemozones**). These zones have been defined by minima and maxima, or by the average point of the variation. Initially, geochemists used the term "isotopic age", but this was later abandoned for "isotopic stage", mainly because the length of **isotopic stages** (between 20 and 70 ka) is not comparable to the length of a classic stratigraphic stage (i.e. isotopic stages based on $\delta^{18}O$ fluctuations; EMILIANI, 1955; SHACKLETON & OPDYKE, 1973). At the present time, only the Quaternary isotopic zonation is really functional (Fig. 3.2).

Geochemical events (Fig. 3.1) represent unique and abrupt changes (either increases or decreases) during a generally long period during which they distinctly differ from

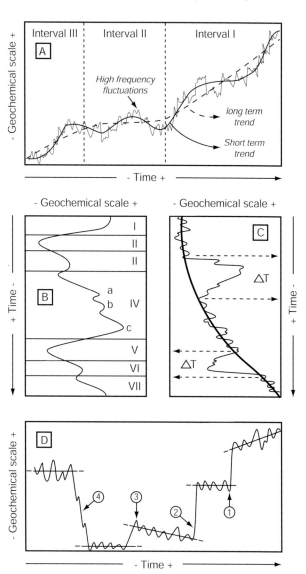

FIG. 3.1

Different kinds of geochemical variations used in chemostratigraphy

A. The evolution of Sr concentrations in pelagic carbonates during the last 150 million of years gives an example of the different kind of variations (long-, short-scale and high frequency).

B. Cyclic variations and definition of the chemozones associated with these cycles. Variations in the oxgen isotope ratio during the Quaternary and the associated isotope stages are a good example of this kind of variations (see Fig. 3.2).

C. Geochemical variations of Sr and $\delta^{13}C$ at the Paleocene/Eocene boundary. These are good examples of geochemical events (increase or decrease).

D. Geochemical shifts (increase or decrease, reversible or not). The evolution of the oxygen isotope ratio during the Tertiary gives several examples of this kind of variation (see Fig. 3.4).

3. Chemostratigraphy

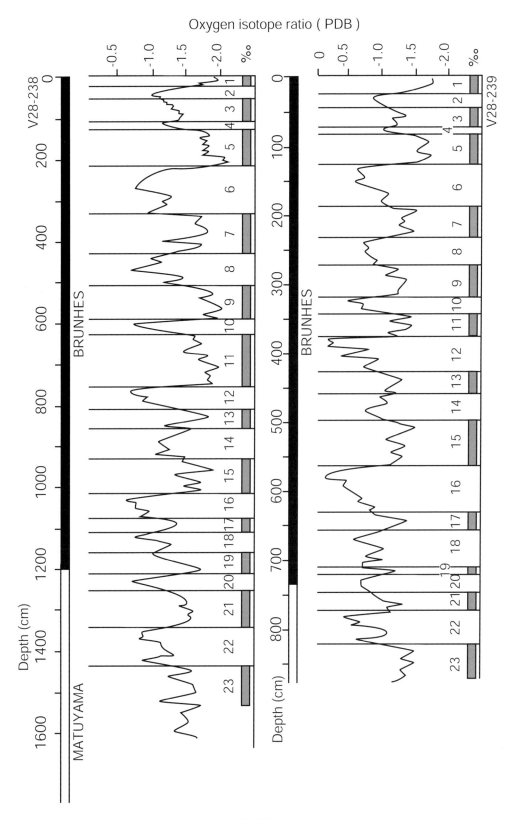

Fig. 3.2

Quaternary isotope stages. Comparison between two Pacific cores
(after Shackleton and Opdyke, 1973; modified by William et al., 1988).

the average signal (e.g. the $\delta^{13}C$ event of the Valanginian; Weissert & Lini, 1991; Hadji, 1991; Emmanuel & Renard, 1993). The length of these events, which characterize global modifications of the sedimentary environment, is variable (from thousands to millions of years). **Geochemical excursions** also characterize more or less rapid variations which are followed by the re-establishment of conditions relatively close to the initial ones. Some authors consider these two terminologies equivalent, while others use them for distinctly different durations of time.

Geochemical shifts represent changes (either increases or decreases), which are both really rapid and abrupt (about 200 ka) which might not be reversible.

Absolute values are also used. These permit definition of the source (i.e. for trace elements, strontium isotopes), glacial and/or inter-glacial conditions (i.e. for oxygen isotopes), or to estimate primary production and organic matter degradation (e.g., carbon isotopes). The terms **peak** or **anomaly** are generally used when extremely high values are observed (e.g., anomaly in iridium at the Cretaceous/Paleocene boundary). In this case, as for the description of events, it is important to observe the trend and morphology of the variation (e.g., symmetric, asymmetric, rate, etc.). It is also important to consider other processes and factors, which could impact the sedimentary record, such as the sedimentation rate or presence of a hiatus.

Indices are theoretical parameters. They include the sum, product or ratio of different geochemical signals (e.g., a salinity index, continental index, RENARD, 1975).

Last, when interpreting geochemical variations, one should be cautious about statistical methods (such as smoothing, moving average, etc.), since these are used to highlight general trends, thereby losing the stratigraphic information. For reference, WILLIAM et al. (1988) give a wide overview of different methods, which allow a statistical and mathematical comparison and deconvolution of geochemical and isotopic records.

3. – PRACTICE

3.1. CACO₃ FLUCTUATIONS IN PELAGIC CARBONATES

This is the first and, analytically, most simple chemostratigraphic approach. ARRHENIUS (1952) was the first to observe the apparent synchronicity of $CaCO_3$ fluctuations in core sediments from the East Pacific (for a review see BERGER & VINCENT, 1981; RENARD, 1985a). Each subsequent event has been numbered with the letter of the magnetozone to which it belongs, followed by a number indicating its rank (Fig. 3.3). $CaCO_3$ fluctuations in sediments represent one of the best correlation tools available for the marine Quaternary record, at least on a basin scale, even though the factors controlling the fluctuations are still debated. The relation to glacial and inter-glacial periods are complex and are generally out of phase between the Atlantic and Pacific oceans (see KENNETT, 1982). This method is still used for post-Oligocene sediments, but results are not reliable for older sediments because intervals characterized by minima in $CaCO_3$ are generally reduced by compaction and diagenesis to thin clayey layers between carbonates.

3.2. OXYGEN STABLE ISOTOPE VARIATIONS

The analysis and calculation of the oxygen isotopic ratio ($\delta^{18}O$) in carbonates is the basis of chemostratigraphy. The importance of this proxy is based on its potential use as a paleothermometer (BOWEN, 1991). During (bio)precipitation of carbonate in equilibrium with sea water, temperature-controlled isotopic fractionation takes place; the simplified equation for temperatures less than 16°C has been derived as:

$$T°C = 16.9 - 4 (\delta - \delta w)$$

where δ is the oxygen isotopic ratio in carbonate and δw is the ratio of CO_2 in equilibrium with the water from which the carbonate has precipitated. High values of the isotopic ratio indicate low temperatures, while low values indicate warmer temperatures. However, the interpretation of isotopic records is complicated by the unknown initial isotopic ratio of sea water. The ice volume effect due to ice cap formation is single most important factor to affect the isotopic composition of sea water. During ice formation, the

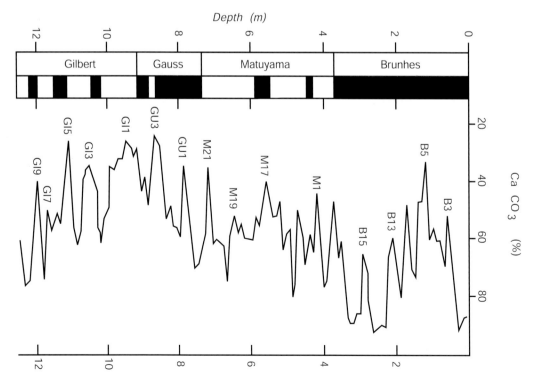

FIG. 3.3

Magnetostratigraphy and chemostratigraphy based on the CaCO₃ variations in an Equatorial Pacific core (after SAITO et al., 1975). Each decreasing shift is named after the corresponding magnetic epoch and by a number showing its rank.

lighter isotope is trapped in ice (–30‰ < δ.ice < –50‰), and sea water becomes correspondingly enriched in the heavier isotope (**frost effect**, EMILIANI, 1955). Consequently, sea water δ has been estimated at –1.28‰ for the middle Miocene, just before the formation of the Antarctic ice cap (SHACKLETON & KENNETT, 1975), at 1.3‰ during the last glaciation, while at present it is about –0.3 ‰.

EMILIANI proposed **isotopic stratigraphy** in 1955, while working on planktonic foraminifera sampled from oceanic sediments. This kind of setting lies far from the continent, and the associated effects of diagenesis and fluctuations linked to the coastal environment (e.g., changes in salinity). Observing a striking synchronicity of the isotopic fluctuations between cores from different basins, EMILIANI chose to divide the isotopic curves into "**isotopic stages**" (Fig. 3.2), each corresponding to a unique climatic fluctuation. These divisions were later revised by SHACKLETON & OPDYKE (1973, 22 stages) and then by WILLIAMS et al., (1988; 51 stages during the Lower Pleistocene).

EMILIANI first thought that the observed fluctuations were directly linked to changes in temperature; it has since been demonstrated that the principal process determining the isotopic fluctuations is the formation of ice caps. From this, several authors have proposed using $\delta^{18}O$ as an indicator of sea-level variation during glacial periods, with variations of 0.11‰ in $\delta^{18}O$ corresponding to a 10 m change in relative sea level.

Though not so precise as for the Quaternary period, isotopic analyses performed on foraminifera from throughout the Tertiary have proven useful both for stratigraphic correlations (RABUSSIER-LOINTIER, 1980; KEIGWIN 1980; VERGNAUD-GRAZZINI & OBERHAENSLI, 1986) and for paleoceanographic studies (e.g., stratification of sea water caused by changes in temperature as suggested by $\delta^{18}O$ in benthic and planktonic foraminifers, Fig. 3.4, SHACKLETON & KENNETT, 1975).

Several authors (MARGOLIS et al., 1975; ANDERSON & STEINMETZ, 1981) have since also observed that nannofossils and the less than 40μ fraction from marine sediments exhibit isotopic ratios similar to those of foraminifera. Finally, it has been shown (LETOLLE, 1979; RENARD, 1985; SHACKLETON, 1987a) that the isotopic ratios of bulk carbonates, though more variable, record both long and short-term changes in the oceanic realm (SHACKLETON et al., 1993; VOIGT & WIESE, 2000). From this, the oxygen isotopic ratio is now utilized extensively in the study of Mesozoic (Fig. 3.5) and upper Paleozoic sediments.

3.3. CARBON STABLE ISOTOPE VARIATIONS

The carbon isotopic ratio ($\delta^{13}C$) has recently become a widely used chemostratigraphic tool for studying pre-quaternary sediments, mainly because it is not really sensitive to late diagenesis. Moreover, its long-term record exhibits significant excursions, more often negative events, which correspond to stratigraphic limits. Several works have focused on those events which have been used as correlation points (KROOPNICK et al., 1977; WEISSERT et al., 1978; LETOLLE & RENARD, 1980; LETOLLE & POMEROL, 1980; SCHOLLE & ARTHUR, 1980; SHACKLETON & HALL, 1984; HILBRECHT & HOEFS, 1986; RENARD, 1986; SHACKLETON, 1987b; WEISSERT, 1989; WEISSERT & CHANNELL, 1989; WEISSERT & LINI, 1991; HADJI, 1991; MAGARITZ, 1991; PRATT et al., 1991; CORFIELD et al., 1992; GALE et al., 1993; EMMANUEL, 1993; ULICNY et al., 1993; JENKYNS et al., 1994; VOIGT & HILBRECHT, 1997; RÖHL et al., 2000; JENKYNS et al., 2002; WEISSERT & ERBA, 2004). Finally, $\delta^{13}C$ is also a potential paelodepth indicator (RENARD & LETOLLE, 1983).

The carbon isotopic ratio of ocean waters (between +1 and +2 ‰ in surface waters) is the product of the balance among three geochemical reservoirs, characterized by different volumes and $\delta^{13}C$ ratios. The atmosphere is a small reservoir (0.00054×10^{20} moles of C) with a slightly negative $\delta^{13}C$ ratio (–7‰). Carbonates represent the biggest reservoir (50.8×10^{20} moles of C), whose $\delta^{13}C$ is close to $\delta^{13}C$ of seawater (+2‰). Finally, organic matter (about 10.8×10^{20} moles of C) has a strongly negative $\delta^{13}C$ ratio, because of the fractionation during photosynthesis (about –25‰). Because of this highly negative ratio, organic matter controls the system and each variation in volume of this reservoir (primary

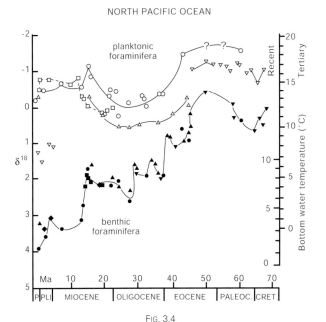

FIG. 3.4

Evolution of the oxygen isotope ratio in bentic and planktonic foraminifera during the Tertiary in low latitude North Pacific (after SHACKLETON and KENNETT, 1975).

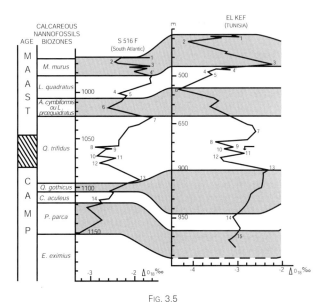

FIG. 3.5

Stratigraphic correlation based on the oxygen isotope ratio (measured on total carbonate) between an oceanic core (ODP Site 516F, South atlantic) and the El Kef section, in Tunisia. The oxygen isotope stratigraphy is compared to the nannofossil zonation (after CLAUSER, 1994).

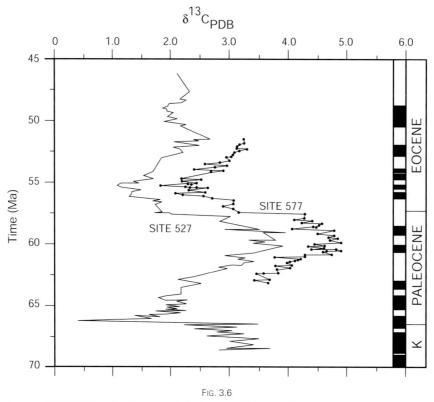

FIG. 3.6

Isotopic events (δ13C PDB) at the Cretaceous/Paleocene and Paleocene/Eocene boundaries (measured on total carbonate) in the Atlantic (Site 527) and Pacific oceans (Site 577), after Shackleton (1985). Outcrops of pelagic carbonates, such as the Gubbio section in Italy, record the same type of signal (RENARD, 1985).

production variation), or variation in flux within the ocean (i.e. organic matter oxidation) are recorded by $\delta^{13}C$ in carbonates. Thus, the carbon isotopic ratio in marine carbonate records: i) primary production fluctuations (for this reason we observe a correlation between $\delta^{13}C$ minima and biological crises, such as at the Permo-Triassic, Cretaceous-Tertiary, and Paleocene-Eocene boundaries; Fig. 3.6); and ii) changes in organic matter cycling within the sediments (i.e. preservation vs. regeneration), which relates $\delta^{13}C$ to anoxic events (WEISSERT et al., 1978; LETOLLE & POMEROL, 1980; DEAN et al., 1986; JENKYNS, 1988; BAUDIN, 1989; PRATT et al., 1991; ULICNY et al., 1993).

Hydrothermal activity, which produces CO_2 with a slightly negative $\delta^{13}C$ (between –5 and –7‰ for the East Pacific rise), could also influence seawater $\delta^{13}C$, but its contribution has been difficult to model.

Several dramatic events, which have been recorded by $\delta^{13}C$, permit stratigraphic correlations between pelagic series. Moreover, because of the interactions between the different reservoirs, CO_2 changes in the ocean have been recorded as impacts on the atmosphere (CERLING, 1992), and on continental carbonate (i.e. paleosol carbonate concretions), on shells of dinosaurs eggs, (IATZOURA et al., 1991; COJAN et al., 1994; 1995) and on mammal teeth (KOCH et al., 1995). For example, the excursion at the Paleocene/Eocene boundary has also been observed in the series of the Paris Basin (SINHA et al., 1995). These preliminary works illustrate the interesting possibilities of correlation between marine and continental sedimentary series. Finally, recent analyses on bone collagen (BOCHERENS et al., 1991) permit the reconstruction of alimentary chains and to understand paleoclimatic conditions (e.g., use of C3 or C4 plants).

3.4. SULFUR STABLE ISOTOPES

Despite a high degree of variability during a specific period of geological time, the ratio of $^{34}S/^{32}S$ in sulfates shows a clear long-term evolution (Fig. 3.7, CLAYPOOL et al., 1980; ODIN et al., 1982), with high values during the Cambrian (+30‰), a progressive decrease until the upper Permian (+10‰), and a steady increase through to recent values (+21‰). This evolution shows variations linked to the balance between evaporites (sulfates with high $\delta^{34}S$) and sulfurs (reservoir with low $\delta^{34}S$). The isotopic ratio is even lower once biochemical processes (e.g., bacterial activity) are taken into consideration. VEIZER et al. (1980) have shown a clear correlation between the temporal evolution of the $\delta^{34}S$ and $\delta^{13}C$ signals, which has been interpreted as the result of the interaction of the S and C cycle over time (ratios of oxidized S/reduced S and oxidized C/reduced C).

Three main dramatic events punctuate the general evolution (nominated Catastrophic Chemical Events by HOLSER): these occurred during the upper Precambrian (Yudomski event, about 635 Ma), the upper Devonian (Souris event, about 370 Ma), and the Triassic (Röt event, about 240 Ma). Two negative excursions are also observed during the upper Permian and the upper Paleogene (at about 55 Ma; PAYTAN et al., 1998). A detailed sulfur isotopes curve, generated using marine barite, covers now the Cenozoic, with a time resolution of about 1 million years (PAYTAN et al., 1998). Even if the high variability of the values (generally on the order of 15 ‰) strongly reduces the stratigraphic utility of this proxy, upper Primary formations have been correlated using the $\delta^{34}S$ ratio (PIERRE et al., 1984).

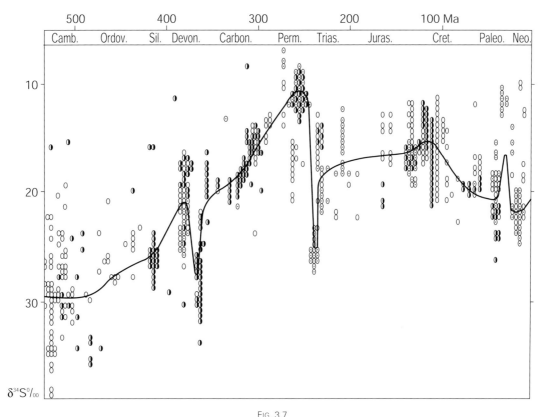

Fig. 3.7
Evolution of the sulfur isotope ratio (after Odin et al., 1982).

3.5. STRONTIUM STABLE ISOTOPES

Carbonate analyses provide insights into the evolution of the $^{87}Sr/^{86}Sr$ ratio of sea water. After the pioneering work by Peterman et al. (1970), several investigators have tried to establish a long-term curve. The strontium isotope curve shows a general decrease starting from the Precambrian ($^{87}Sr/^{86}Sr$ = 0.7090) to the Permian (0.7070; for a synthesis of these works, see Faure, 1982). This trend is interrupted by a decrease during the Devonian and the lower Carboniferous. The values increase abruptly during the Triassic (0.7082) and then decrease again to values characteristic of the upper Jurassic. Values increase, more or less steadily during the upper Cretaceous and mainly during the Tertiary, to attain present values (0.7092). This evolution is controlled by modification of the Sr flux and by the balance between continental erosion (which introduces strontium with a high isotopic ratio, about 0.7119) and hydrothermal activity (low strontium isotopic ratio, about 0.703). Several authors (Edmond, 1992; Richter et al., 1992; Godderis & Francois, 1995) attribute the increase in the $^{87}Sr/^{86}Sr$ ratio during the Cenozoic to uplift of the Himalayas. Technical progress has facilitated the generation of a large amount of data, mainly on foraminiferal samples from deep-sea cores (De Paolo & Ingram, 1985; Koepnick et al., 1985; Hess et al., 1986; Capo & De Paolo, 1990). The strontium isotopic record produced (Fig. 3.8) allows dating during the middle and upper Tertiary with a precision of one million of years. Nevertheless, correlations between platform and basin sedimentary series, for instance for Campanian and Maastrichtian stratotypes (Koepnick et al., 1985; Turpin et al., 1988; Clauser, 1994), are not in agreement with biostratigraphic data, suggesting that the strontium isotopic ratio may be affected by local perturbations.

3.6. TRACE ELEMENTS IN CARBONATES

The concentration of several dissolved elements in sea water is far from the saturation point. Thus, these elements do not directly precipitate, and they are generally removed from solution in combination with other mineral compounds (such as calcium carbonate) by a process called **coprecipitation**. Depending on the kind of mechanisms (co-precipitation and binding in the crystal, surface adsorption, or occlusion), the equations describing the processes are slightly different. But under natural conditions the processes are closely linked and it is possible to summarize them by the equation:

$$[X/Ca]_{mineral} = K_x [X/Ca]_{sea\ water}$$

where [X/Ca] is the molar ratio between the trace element and Ca and K_x is the apparent **mixing coefficient**.

Strontium, which can replace Ca in carbonate because of their similar properties, is widely used. K_{Sr} is a function of carbonate mineralogy, the organism vital effect, and temperature. This proxy is used in platform sediments to trace paleoenvironmental evolution, mainly for salinity studies, in terms of proximity of fresh water input. This approach has been tested in sediments of different ages and facies (for a review, see in Renard, 1985b,c).

The possible impact of diagenesis on the evolution and trend of strontium content in sediments has long been debated (see Morrow & Mayers, 1977; Veizer, 1977; Brand & Veizer, 1980; Baker et al., 1982; Renard, 1985a). This discussion has in part been resolved since strontium stable isotopic studies have shown that in upper Cretaceous sediments at least 80% of the original signal is still preserved (Richter & Liang, 1993).

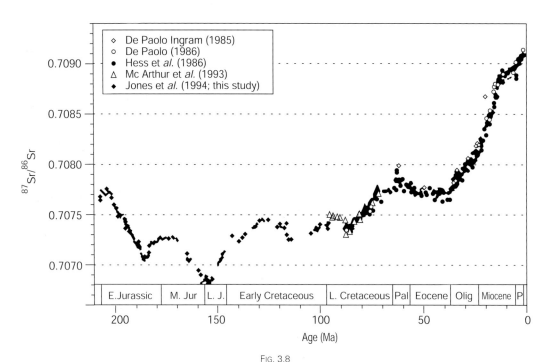

Fig. 3.8
Evolution of the strontium isotope ratio starting from the Jurassic (after Jones et al., 1994).

The proposed strontium and magnesium curves (Graham et al., 1982; Renard, 1985a; 1986) mainly express the Sr/Ca and Mg/Ca seawater ratio temporal evolution. Their long-term evolution has been linked to hydrothermal activity fluctuations. Short-term variations depend on transgression-regression cycles due to variations in pelagic carbonate production. In fact, Sr content in pelagic carbonate seems to be a reliable tracer of the evolution of carbonate platforms and of their crises, caused by sub-emersion, drowning or anoxic conditions (Renard et al., 1994). The temporal evolution of the strontium content is interrupted by several positive and/or negative excursions, which may have a stratigraphic meaning and might be used as a correlation tool (Fig. 3.9).

The manganese cycle is more complex (Bender et al., 1977), because this element can be either incorporated in low quantities in carbonates or can directly precipitate from seawater as Mn dioxide, which is then mixed with calcium carbonate deposits (Michard, 1969). In oxygenated environments, these two processes are active at the same time. In fact, these sediments are generally enriched in manganese, which is largely present as MnO_2. In anoxic environments, carbonates contain low amounts of manganese because only co-precipitation can occur.

Independent of this process, several authors have used manganese in platform facies as an indicator of continental input, but recent studies have shown that manganese could

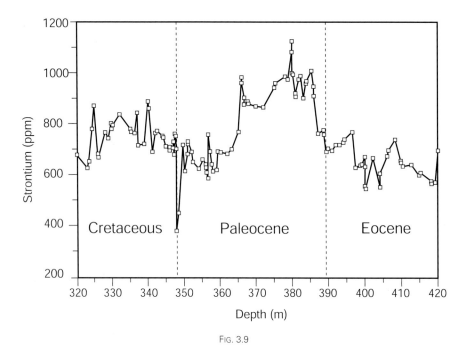

Fig. 3.9
Evolution of strontium concentrations in pelagic carbonate in the Gubbio section (Umbria, Italy), after Renard (1985c).

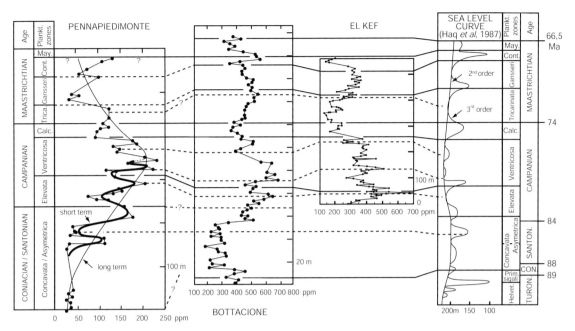

FIG. 3.10
Evolution of Mn concentrations in pelagic sediments from the Upper Cretaceous in Central Italy (Pennapiedimonte and Bottaccione), and in Tunisia (El Kef). Comparison with the eustatic curves (after ACCARIE et al., 1993).

be used in pelagic environments for sequential stratigraphy. In fact, following POMEROL's earliest works (1976, 1984) on Cenomanian sediments from the Paris basin, several authors have highlighted a relationship between high Mn contents and high sea-level stands (Fig. 3.10; ACCARIE et al., 1989; 1992). More recently, EMMANUEL (1993), EMMANUEL & RENARD (1993) and CORBIN (1994), CORBIN et al. (1995) have shown that in pelagic facies it is possible to characterize Tertiary sequences based on the Mn content of carbonates. Sediments deposited during low sea-level stands are characterized by low and stable Mn contents. During marine transgressions, Mn contents rise until they reach a maximum corresponding to the maximum transgression surface. The sediments deposited during high sea level stands are characterized by reduced Mn contents until the sequence boundary, which is characterized by lowest values (Fig. 3.11). Even if the use of this element proves really important for chemostratigraphy, we still lack the full understanding of the causal relationship between Mn content and sea level.

We still need to distinguish the changes in seawater contents of Mn (linked to sea floor spreading and hydrothermal activity; LYLE, 1976; KLINKHAMMER & BENDER, 1980) from the variations linked to carbonate dilution caused by changes in sedimentary production in response to the eustatic cycle.

3.7. IRIDIUM ANOMALIES

The concentration of **iridium** in the Earth's crust is considerably lower than the iridium concentration found in the mantle and in space. For this reason, Ir concentrations in sediments are interpreted to be the result of extraterrestrial material input. Employing this characteristic, ALVAREZ et al. (1980) showed the presence of an **Ir anomaly** (Ir = 9ppb, about 30-100 times higher than the surrounding sediments) in the clayey layers at the Cretaceous/Tertiary boundary in the Gubbio outcrop (Italy). This anomaly has been found in other outcrops where the sequence is complete (Fig. 3.12, BONTE et al., 1984; ROCCHIA et al., 1987) and initiated a long debate focusing on whether the origin of this anomaly was volcanic or cosmic.

The discovery of nickelifers spinels with an extraterrestrial signature in the layer (ROCCHIA et al., 1992), and the dating of a reopening phase at about 65 Ma in zircons from the ejecta of the Chicxulub crater (Yucatan peninsula, Central America) and of the clayey layer in Haiti and in the Raton Basin in the USA (KROGH et al., 1993), clearly suggested a cosmic origin.

This tracer has a wide stratigraphic interest because of the possibility of correlating ocean and continental domains, independently of the relation with a biological crash at the K/T boundary. However, its use in periods other than the Cretaceous is still equivocal. It is important to emphasize that with a constant cosmic input, lower sedimentation rates

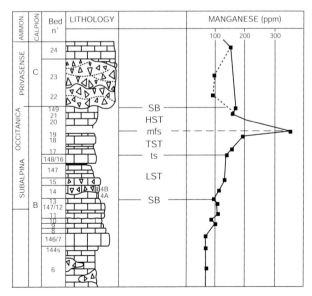

FIG. 3.11
Fluctuations of the Mn content in the pelagic carbonates from the Be3 sequence of the Berriasian at Berrias (France) SB = sequence boundary, LST = lowstand system tract, TS = transgressive surface, TST = transgressive system tract; MFS = Maximum flooding surface, et HST = highstand system tract (after EMMANUEL et al., 1993).

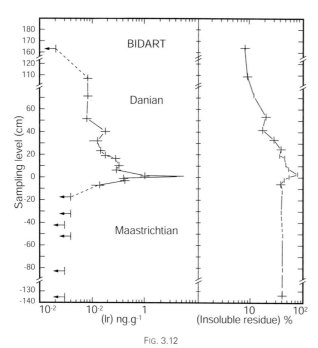

Fig. 3.12
Anomaly in the iridium concentration at the Maastrichtiam/Danian boundary in the Bidart outcrop (France). The clay layer at 0 m represents the boundary (after Rocchia et al., 1987).

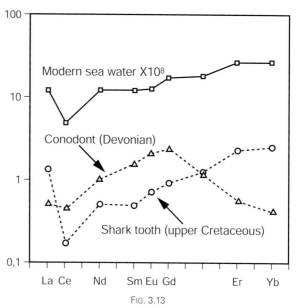

Fig. 3.13
Comparison of the REE serie in modern sea water, in apatite from a Devonian Conodont, and apatite from a Cretaceous shark tooth (after Albarède, 1990).

(forming a condensed layer) can cause an increase in the apparent concentration of cosmic matter. Rocchia et al., (1986) measured an Ir peak of about 3 ppb in a condensed layer of a Bajocian age in Venetia (Italy). Assuming present accumulation rates of cosmic material, this layer represents a period of about 750,000 years. Iridium is thus a potential tool to evaluate the time extension of condensed sections (condensed sedimentation); on the other hand, hiatuses (lack of sedimentation) do not show any enrichment, because in such cases iridium has been transported away.

Other lower-amplitude Iridium anomalies have been observed at different time intervals. Some of these are not truly well defined mainly because they have been observed only at one site or by a single research group (e.g., at the Cenomanian/Turonian, Permo/Triassic, Frasnian/Famenian, and Precambrian/Cambrian boundaries). On the other hand, the anomaly in the upper Pleistocene (Kyte & Brownlee, 1985) and the one just below the Eocene/Oligocene boundary (Alvarez et al., 1982; Montanari et al., 1993) appear to be consistently real. Montanari et al. (1993) discovered an iridium peak of about 159 ±19 ppt (parts per trillion) at 35.7 ± 0.4 Ma (NP 19/20 zone) in the Massignano outcrop (Italy) and another peak of about 156 ± 10 ppt in slightly younger sediments (by about 0.2 to 0.5 Ma) from ODP Site 689 (Weddell sea). The comparison between results found in the Pacific and Indian oceans and in the Caribbean region suggests a period of closely spaced impacts in the upper Eocene (of about 1 Ma, between 35.7 and 34.7 Ma).

3.8. REE, RARE EARTH ELEMENTS

Rare earth elements (**REE**; a series of 14 elements, from La to Lu) are still not extensively used in stratigraphy. They can precipitate as trivalent hydroxides and carbonates, but, with the exception of Ce, these phases are rarely close to saturation in seawater. They are generally extracted from sea water by adsorption processes, ion exchanges and co-precipitation (generally with Ca in carbonates and with Fe(III) in manganese nodules). The normalized distribution spectra of REE in relation to **NASC** (ordered by their atomic numbers) in modern sea water has a peculiar shape, with a deficit in Ce (mainly because of the reduced solubility of trivalent cerium), and relative enrichment of heavy rare earth elements (e.g., Er, Yb) compared with light rare earth elements (La, Nd, Sm).

The REE spectra of pelagic carbonates, phosphatic debris and authigenic clays precipitated from sea water share the same signature as sea water (Grandjean-Lecuyer et al., 1993), while detrital clays have a continental signature (the normalized spectra in relation to NASC are flat). However, diagenetic redistribution may modify the original signal (Milodowski & Zalasiewicz, 1991). Stratigraphic correlations have been attempted in Turonian carbonates (Wray, 1995).

REE analyses in biogenic phosphate from Cretaceous age shark teeth (Albarede, 1990) show spectra similar to modern sea water, while Devonian conodonts have a completely different signature, characterized by low light and heavy rare earth elements, and by the absence of the cerium deficit (Fig. 3.13). This suggests that the ancient ocean was more anoxic (Ce mainly in a trivalent form), and that primary production patterns were different (light REE are preferentially adsorbed on biogenic material). Rare earth elements are potentially a proxy for biologic evolution and a marker for anoxic episodes in the ocean.

3.9. ORGANIC CARBON AND GEOCHEMICAL BIOMARKERS

Organic matter is an ordinary component of sedimentary rocks, even though it is generally present in low amounts (lower than 1%). So many different components typically constitute the bulk organic matter that a complete and detailed analysis is practically impossible. Therefore, this organic matter is studied either by means of general analyses or by isolating certain fractions using specific chemical procedures (e.g. acid dissolution or use of organic solvents).

Total organic carbon content (**TOC %**) and determining the **type of organic matter** are typically the first parameters characterized. However, these two proxies are not truly

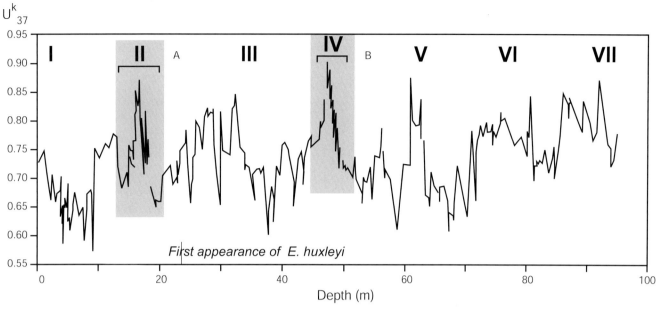

Fig. 3.14
Use of organic geochemical markers in a high-resolution stratigraphic study of Quaternary sediments from ODP Site 658 (Leg 112, Peru upwelling margin). The Uk37 index is calculated as the ratio between the concentration of two different alkenons, produced by coccolithophorids. I-IV: glacial terminations; shaded areas: extremely rapid variations of the Uk37 index.

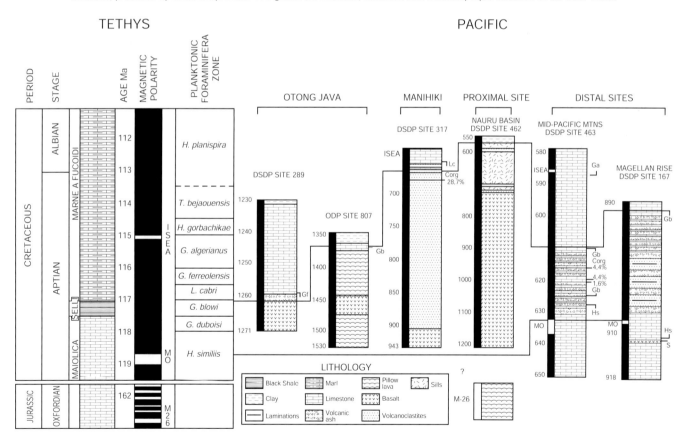

Fig. 3.15
Correlation of organic rich sediments of the lower Aptian between the Pacific Ocean and the Tethys domain (Umbria-Marchean basin, Italy). These deposits belong to the blowi foraminiferal zone. The Selli level (Italy) and pelagic sediments in the Pacific are characterized by high concentration of marine organic matter. Depth in meters below sea floor.

stratigraphic tools, since variations in content and kind of organic matter do not usually reflect global environmental changes, but are linked to local conditions, which determine organic matter production and preservation. For use as a stratigraphic tool, organic molecules need to satisfy some fundamental conditions:

– they need to be simple enough to be easily analysed,

– they need to be complex enough to derive interesting information and,

– they need to be refractory enough to be unaltered over geological timescales, because of the effect of temperature and other processes that tend to alter and modify organic matter composition.

Such molecules are called **geochemical fossils** or **geochemical biomarkers**. Because of the requirements listed above, these molecules cannot be common metabolites, which are generally present in large number but lack specificity, nor can they be too specific, because they would be too labile and unstable over long geological time.

Typically organic chemistry uses:

– the spectra of the distribution of carbon chain length in n-alkanes;

– the presence of certain kind of complexes (n-alkanes, hopanes, steranes, alkenones, etc.) whose origin is known;

– ratios of certain compounds (Fig. 3.14);

– carbon stable isotopic ratios of specific organic compounds.

More recently, the study of carbon isotopes in marine organic matter seems to open new possibilities for the use of organic matter as a stratigraphic tool (HAYES et al., 1999).

Evidently, the best stratigraphic approach is one which combines more than one proxy (the so-called 'multi-proxy' approach). In this way, organic geochemistry permits recognition of specific time markers at a local or even more global scale, once that their stratigraphic position has been established by other absolute stratigraphic tools (Fig. 3.15).

3.10. NEW FRONTIERS IN CHEMOSTRATIGRAPHY

3.10.1. Oxygen isotopes

During the last decades, several studies have widened the horizons of chemostratigraphy. Thanks to new analytical instruments and techniques (e.g., AMS ^{14}C dating technique, that requires milligram sized samples; or laser ablation ICP-MS, used for elemental and isotopic micro analyses in solid samples), which allow higher precision analysis, high-resolution studies of the marine and continental records have been published. Depending on the sedimentation rates, the achieved time resolution is on the order of less than 100 years for the Quaternary (i.e. WANG et al., 1999), and about 10^4 years for older time frames (i.e. Eocene, see ZACHOS et al., 1996). These progresses have allowed the study of high-frequency climatic changes, such as Heinrich and Dansgaard-Oeschger events in the North Atlantic (BOND & LOTTI, 1995; BOND et al., 1997), during the last 70 ka, or such as El-Niño-Southern Oscillation during the last 130 ka (HUNGHEN et al., 1999; TUDHOPE et al., 2001).

The use of chemostratigraphy is now not only restricted to marine and lacustrine sediments, but in the last 15 years important progresses have also been made in studying ice cores. These studies have provided long climatic history records (BARNOLA et al., 1987; GROOTES et al., 2001; PETIT et al., 1999; STUIVER & GROOTES, 2000) that have been key to discover high-frequency climatic changes.

The comparison of isotope records from ice cores from both Greenland and Antarctica ice sheets, and marine sediments (e.g., WANG & OBA, 1998) has started the understanding of the relative timing between Northern and Southern Hemisphere climate dynamics (e.g., leads and lags; BLUNIER and BROOK, 1998, 2001).

The most widely used chemostratigraphic tool, at least for the last 3 Ma, remains $\delta^{18}O$ records. Almost 100 stages have been identified for the last 2.8 Ma, and boundaries for about 50 stages have been defined (GRADSTEIN et al., 2004), also thanks to new U-Th dating (e.g., ROBINSON et al, 2002). The recovery of sediment cores and the sampling of new outcropping sections have also provided new continuous stratigraphy sequences. Thanks to these studies, oxygen isotope stratigraphy is proving to be reliable also for geological periods older than 3 Ma (HODELL et al., 1994; TIEDEMANN et al., 1994; MILLER et al., 1991; VOIGT & WIESE, 2000; WANG et al., 2003; ZACHOS et al., 1996, 2001). An orbitally-tuned stratigraphy is now available for the last 23 Ma (GRADSTEIN et al., 2004).

3.10.2. Impact of gas hydrates for the carbon isotopes

Release of clathrates (methane hydrates), which have a negative $\delta^{13}C$ signature, could influence the carbon isotopic signature of seawater. Recent studies have seen in the sudden dissociation of clathrates the mechanism for abrupt perturbations of climate in the past, recorded by negative excursion of the $\delta^{13}C$ (NORRIS & RÖHL, 1999; BERNER, 2002; BICE & MAROTZKE, 2002; TRIPATI & ELDERFIELD, 2004).

3.10.3. Strontium isotopes and other isotopes (e.g., osmium)

Recently, MCARTHUR et al. (2001) have published an improved version of the LOWESS fit for the marine 87Sr/86Sr record. This includes also the Triassic and Paleozoic record (0-509 Ma).

Because of the recovery of continuous sedimentary sections from different regions of the world oceans, revised and more accurate stratigraphic reference curves of marine 87Sr/86Sr are now available (EBNETH et al., 2001; FARRELL et al, 1995; HODELL & WOODRUFF, 1994; JONES et al., 1994, 2003; KORTE et al., 2003; MARTIN et al., 1999; OSLICK et al., 1994; REILLY et al., 2002).

Isotopes of other trace elements are now tested as paleoenvironment indicators and, most important, as stratigraphic tools.

The osmium isotope signal, recorded in bulk and metalli-ferous sediments, provide a stratigraphic tool comparable in use to strontium isotopes. The osmium isotope composition of seawater ($^{187}Os/^{186}Os$) is in fact controlled by the balance between weathering of continental rocks and the input from cosmic dust and/or hydrothermal activity. In the last decade, many studies have been published, and now long-term (PEGRAM et al., 1992; RAVIZZA, 1993; PEUCKER-EHRENBRINK et al., 1995) and higher resolution records are available (OXBURGH, 1998; RAVIZZA & PEUCKER-EHRENBRINK, 2003a, 2003b).

4. – CONCLUSION

After a difficult beginning, chemostratigraphy now has several tools, whose utility is determined by technical progress (e.g. the use of automatic spectrometers) and generally lower analytical costs. This flaw is mainly based on the necessity of a huge amount of data to obtain sufficient resolution for stratigraphic studies. Nevertheless, during the last years geochemistry has become a principal part of stratigraphy, along with more classical approaches like paleontology. The increasing number of geochemical parameters made available by technical progress, has also allowed the development of a multi-proxy approach, which is indeed helpful in building a more precise stratigraphic framework of the geological past (JENKYINS et al., 2002; WEISSERT & ERBA, 2004).

Chapter 4
MAGNETOSTRATIGRAPHY

B. Galbrun (Coord.), N.K. Belkaaloul & L. Lanci

CONTENTS

1. – DEFINITION 53
   1.1. The earth's magnetic field 53
   1.2. The time-averaged geomagnetic field: the axial dipole hypothesis 54
   1.3. Reversals of the earth's magnetic field 54
2. – TERMINOLOGY 55
   2.1. Magnetostratigraphic polarity units 55
   2.2. Nomenclature 55
3. – PRACTICE 56
   3.1. Field sampling 56
   3.2. Sampling in sediments cores 58
   3.3. Measurement of remanent magnetization 58
   3.4. Samples demagnetization 58
      3.4.1. Alternating field (AF) demagnetization 59
      3.4.2. Thermal demagnetization 59
   3.5. Rock magnetism 59
      3.5.1 Identification of ferrimagnetic minerals in a rock 59
      3.5.2 Curie Temperature analysis 59
      3.5.3. Acquisition of thermal demagnetization of IRM ... 59
   3.6. Analysis of magnetization components 59
      3.6.1. Analysis of vector diagrams 60
      3.6.2. Analysis of remagnetization circles 60
   3.7. Corrections of the measured direction of remanent magnetization 61
      3.7.1. Correction for orientation of sample (geographic correction) 61
      3.7.2. Correction for tilt of bedding (tectonic correction) 61
   3.8. Field tests of magnetization stability 61
      3.8.1. Reversal test 62
      3.8.2. Fold test 62
      3.8.3. Conglomerate test 62
   3.9. Statistical analysis of directions and poles 62
      3.9.1. The Fisher distribution and its use in paleomagnetism 62
      3.9.2. Calculation of virtual geomagnetic pole (VGP) position and its confidence limits 64
      3.9.3. VGP latitude 64
4. – CONCLUSIONS 64

1. – DEFINITION

A large part of sedimentary and volcanic rocks contains a fraction of magnetic minerals. These minerals of different nature, dimensions and shape, give their natural magnetic properties to the rocks. In a stratified series, the analysis of the evolution of the natural magnetic characteristics (magnetic susceptibility, intensity and direction of the **Natural Remanent Magnetization** (NRM), etc.) constitutes the **magnetostratigraphy** sensu lato. The purpose of magnetostratigraphy is to organize the rock strata according to their magnetic properties that were acquired at the time of deposition. Since the polarity of the Earth's magnetic field has reversed repeatedly in the geological past, the most useful magnetic property in stratigraphic work is the change in the direction of the NRM of the rocks caused by reversals in the polarity of the Earth's magnetic field. The **magnetic polarity transition zones** have a duration of only a few thousand years and are synchronous over the entire globe, therefore the geomagnetic reversals record in marine or land-based sediments can provide isochron units applicable to worldwide correlation.

The study of the magnetization acquired by the rocks at the time of their formation and their magnetic properties are the objectives of paleomagnetism. Paleomagnetism and magnetostratigraphy are based on two fundamental assumptions:

– (1) the Earth's magnetic field direction was recorded in a rock at the time of its formation or at a subsequent known time;

– (2) the **Earth magnetic field**, averaged over periods of several thousand years, corresponds to that of a geocentric axial dipole. This last assumption means that the calculated paleomagnetic pole coincides with the paleogeographic axis of the Earth.

1.1. THE EARTH'S MAGNETIC FIELD

More than 99% of the magnetic field measured at the surface of the Earth originates inside the Earth. The strongest component of the field, which accounts for 95% of the magnetic field of internal origin, can be described as a **magnetic dipole** located at the center of the Earth and inclined at about 10.6° to the Earth's rotation axis.

The remaining 5% of the field of internal origin, obtained by subtracting the field of the inclined geocentric dipole from the total field, is collectively called the **non-dipole field**. It consists of magnetic fields that are due to even-numbered combinations of opposite poles. The coefficients of degree $n = 2$ in the Gauss expansion of the magnetic potential correspond to magnetic quadrapoles, the next higher terms (of degree $n = 3$) to magnetic octapoles, etc. The coefficients of degree 8 and higher are very small and the calculation of Gauss coefficients must usually be truncated. A global model of the field, which is updated every 5 years, is provided by the **International Geomagnetic Reference Field** (IGRF), which is based on coefficients up to $n = 10$. Measurements of the magnetic field from the MAGSAT and OERSTED Earth-orbiting satellites allowed more accurate field survey including precise measurements of lithospheric anomalies.

The dipolar component of the Earth's magnetic field is referred to as **inclined geocentric dipole**. The **geomagnetic poles**, sometimes referred to as the geomagnetic dipole poles, are the two points on the Earth's surface formed by the axis of the inclined geocentric dipole. Since the dipole is centered, the geomagnetic poles are situated exactly one opposite to each other. However, the observed magnetic field is not vertical at the geomagnetic poles. The North geomagnetic pole position can be calculated from the first three coefficients of the IGRF.

$$\theta = 90 - \cos^{-1}\left(\frac{-g_1^0}{\sqrt{(g_1^0)^2 + (g_1^1)^2 + (h_1^1)^2}}\right)$$

$$\lambda = \tan^{-1}\left(\frac{h_1^1}{g_1^1}\right)$$

The **magnetic poles**, sometimes called as the Magnetic Dip Poles, are the points on the Earth's surface at which the magnetic field is directed vertically downward (north) or upward (south). Alternatively, their inclination is ±90° and their horizontal intensity is zero. The magnetic poles are observable; it is possible to make observations in the vicinity of one of the magnetic poles, from which its location can be determined. This is periodically done for both poles, more regularly and frequently for the North magnetic pole. It is also possible to estimate the position of the magnetic poles from a magnetic reference field model such as the IGRF. The positions calculated using the IGRF coefficients will differ from the observed positions, but the differences are relatively small. The discrepancy between magnetic and geomagnetic poles is a consequence of the non-dipolar field. The positions of the Earth's magnetic and geomagnetic poles for 2001 are shown in Table 4.1.

TABLE 4.1.

Positions of Different Magnetic Poles for 2001

TYPE OF POLE	GEOMAGNETIC	MAGNETIC (IGRF)	MAGNETIC (OBSERVED)
Position (North)	79.6° N, 71.6 W	81.0° N, 110.0 W	81.3° N, 110.8 W
Position (South)	79.6° S, 108.4° E	64.6° S, 138.3° E	64.7° S, 138.0° E

In its present configuration the Earth magnetic field is directed downward (inclination = 90°) at the magnetic pole located close to the geographic North Pole and is directed upward (inclination = –90) at the magnetic pole close to the geographic South Pole.

This configuration of the magnetic field, where the magnetic north pole of the geocentric dipole is close to the geographic South Pole, is said of **normal polarity**. It alternated numerous times during the geologic past with an **inverse polarity**, where the magnetic north pole is about the geographic North Pole.

The magnetic B field of a dipole is symmetrical about the axis of the dipole. At any point at a distance r from the centre of a dipole with a moment m, the radial component Br and a tangential component θ can be obtained by differentiating the potential with respect to r and θ, respectively:

$$Br = \frac{\partial W}{\partial r} = \frac{\mu_o}{4\pi}\frac{2\ m\cos\theta}{r^3}$$

$$B_\theta = \frac{1}{r}\frac{\partial W}{\partial \theta} = \frac{\mu_o}{4\pi}\frac{2\ m\sin\theta}{r^3}$$

On the surface of a sphere whose centre coincides with that of the dipole, the magnetic field is inclined to the surface with an angle I, given by:

$$\tan I = \frac{Br}{B_\theta} = 2\cot\theta$$

On the Earth, the angle I is the inclination of the field, and θ is the angular distance from the magnetic axis. This last equation has an important application in paleomagnetism.

1.2. THE TIME-AVERAGED GEOMAGNETIC FIELD: THE AXIAL DIPOLE HYPOTHESIS

The fact that the Earth's magnetic field varies with time was well established several centuries ago. The slow changes of the field, which became evident only over decades or centuries of observation, are known as **secular variations**. The dipole field exhibits slow variations of intensity and direction. Analyses of the Gauss coefficients and extrapolated archeomagnetic results show a near-linear decay of the strength of the dipole moment at a rate of about 3.2% per century between about 1500 AD and 1900 AD. At the start of the 20th century, the decay speeded up and has averaged about 5.8% per century during the last 30 yr. The dipole axis also changes its position due to secular variations. Precise measurements, useful for spherical harmonic analyses, have been available only since the early 18th century and prior to this the pole position has been inferred from less reliable data. The most relevant aspect of pole motion is the westward drift of its longitude at a rate that would correspond to a period of about 2000–3000 yr, while the tilt remained between the 11° and 12°.

The records of the geomagnetic observatory are not long enough to enable to verify if the changes due to secular variations are cyclical, but the record of earlier magnetic field can be inferred from dated archeological materials, lava flows, and young sedimentary sequences. If quasi-cyclical, the directional changes of the field imply that the mean long-term strength of the non-axial component of the measured dipole and non-dipole field would average to zero within a few cycles. The only long-term component of the geomagnetic field that persists and is not averaged to zero is the axial component of the dipole.

This implies that the Earth's magnetic field averaged for a few thousand years is equivalent to a magnetic dipole whose axis corresponds to the Earth's rotation axis and therefore the mean paleomagnetic pole position will coincide with the axis of rotation of the Earth. That is, all terms but $g_{1,0}$ cancel out. This is a fundamental assumption of paleomagnetism known as the geocentric axial dipole **(GAD) hypothesis**. The validity of the GAD hypothesis is of extreme importance in paleomagnetism and magnetostratigraphy and has been carefully investigated (e.g., EVANS 1976; SCHNEIDER & KENT, 1990).

1.3. REVERSALS OF THE EARTH'S MAGNETIC FIELD

The occurrence of lavas magnetized in the opposite direction of the Earth's magnetic field was discovered in the early part of 1900. BERNARD BRUNHES (1867-1910) found ancient lava flows in France whose magnetization appeared to be reversed (BRUNHES, 1906). Other examples were then found, and the Japanese geophysicist MOTONORI MATUYAMA (1884-1958) examined the evidence and suggested that the magnetic signatures were evidence of actual reversals (MATUYAMA, 1929). Matuyama proposed that long periods existed in the history of Earth during which the polarity of the magnetic poles was the opposite of what it is now. The **Earth magnetic field** polarity reversal phenomenon was recognized as global by the French physicist PAUL-LOUIS MERCANTON, who found reverse polarity in Australian rocks that were of the same age as reversely magnetized European rocks. He suggested that magnetic polarity reversals could make an excellent stratigraphic marker for correlation purposes. Geomagnetic field reversals occur in the very short time of about 3500-5000 yr (e.g., MERRILL & MCFADDEN, 1999). This is much shorter than the length of a polarity chron or subchron and geologically, they can be considered like instantaneous events.

Although the polarity reversal of the Earth's magnetic field was first documented in radiometricly dated continental lava sequences, the same polarity pattern was later found in the interpretation of the record of the oceanic magnetic anomalies near ocean ridge systems. The record of the oceanic magnetic anomalies is since considered the best record of the history of geomagnetic reversals.

Older parts of oceanic anomalies reversal record have been positively correlated to the reversal polarity sequences found on sedimentary rocks. Magnetostratigraphy has been used to correlate the oceanic magnetic anomalies record in numerous sections. The ages of key sections of the record of geomagnetic polarity have been determined with different methods (radiometric, biostratigraphic, cyclostratigraphic) in many correlating sections. The ages of magnetic anomalies between these dated tie-points can be computed by interpolation or extrapolation. What results is a dated polarity sequence, called a **geomagnetic polarity time scale** (GPTS), which can be directly correlated to the paleontological record.

The GPTS is well established for the time interval from the present back to the Late Jurassic (Oxfordian), a period in which the oceanic record of polarity is well defined. The determination of a GPTS for earlier eras is complicated because no oceanic record exists and the sequence of magnetic reversals can be determined solely from the sedimentary record, which is often discontinuous. It is known that, during the Triassic, there were numerous reversals, the Permian and Late Carboniferous were dominated by constant reverse polarity, and earlier in the Paleozoic reversals were common. An unprecedented sedimentary record of the geomagnetic reversals during the Upper Triassic was establish in the lacustrine sequence of the Newark basin (OLSEN & KENT 1990; KENT, et al. 1995a, b; OLSEN & KENT 1999; KENT & OLSEN, 1999) which gave continuous record of about 20 Ma and could be precisely dated using astrochronologic techniques.

2. – TERMINOLOGY

2.1. MAGNETOSTRATIGRAPHIC POLARITY UNITS

Magnetostratigraphy organizes systematically rock strata into units that are identifiable based on the stratigraphic variations of their paleomagnetic polarity. Thus, a magnetostratigraphic polarity unit is defined as an objective unit ideally based on a directly determinable property of the rocks, which is their magnetic polarity. Strictly, the presence of the unit can be assured only where this property can be identified and in these respects, it is similar to a lithostratigraphic or a biostratigraphic unit. However, lithostratigraphic and biostratigraphic units are usually geographically restricted, whereas magnetostratigraphic polarity units are extended globally and from this point of view they are more similar to chronostratigraphic units even if they are defined primarily, not by time, but by the polarity of remanent magnetism of the rocks.

When comparing **polarity horizons** and **magnetozones** with other stratigraphic units and particularly with biostratigraphic units, it must be kept in mind that they have a very little individuality (i.e. each reversal is very much the same as the others). Therefore, although they are very useful isochronous horizons, which can be correlated among completely different environments, they usually need some supporting evidence, such as paleontological or radiometric data to be unequivocally identified.

Magnetic polarity units have been established in two different ways: i) through the use of magnetic profiles from ocean crust to identify and correlate linear magnetic anomalies, which are interpreted as reversals of the Earth's magnetic field and ii) through a combination of radiometric, biostratigraphic or astrochronologic age data with magnetic polarity determinations on outcropping or cored volcanic and sedimentary rocks.

Profiles of the magnetic anomalies preserved in the pattern of sea-floor spreading offer the best sequential record of reversals of the Earth's magnetic field for the past 150 Ma. This record has been dated by radiometric (e.g. BAKSI et al., 1992; SPELL & MCDOUGALL, 1992; BAKSI, 1993), paleontological and astrochronologic techniques (e.g., SHACKLETON et al., 1990; HILGEN, 1991) in correlating stratigraphic sections or cores. However, the nature of these magnetic anomalies does not allow designating any satisfactory type intervals or type boundaries. Ideally, a definition of a magnetostratigraphic polarity unit should be a stratotype in a continuous sequence of rock strata where the polarity pattern of the unit and its upper and lower limits are clearly defined by means of boundary stratotypes. A similar attempt was made, for instance, by ALVAREZ et al. (1977) in establishing an upper Cretaceous stratotype section of this kind in the Gubbio section, Italy, or by LANCI et al. (1996) at the Oligocene/Paleocene boundary at Massignano, Italy. Here the polarity units have been identified geologically and geographically and related to lithostratigraphic and biostratigraphic data.

2.2. NOMENCLATURE

In principle, the nomenclature system should be flexible enough to allow for polarity intervals discovered at a later

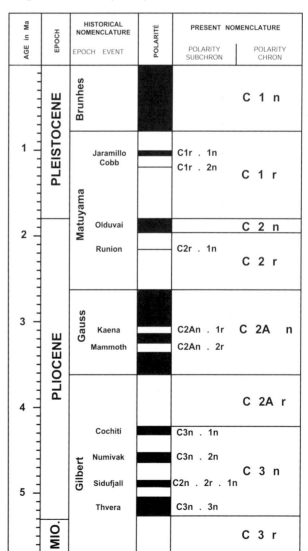

FIG. 4.1

Geomagnetic polarity scale of Plio-Pleistocene (normal or reversed polarity periods are plotted in black or white respectively).

date to be conveniently and unambiguously inserted into the existing system. The first four main polarity periods (called epochs) were named after eminent researchers in geomagnetism (Fig. 4.1). From younger to older they are named, BRUNHES (mostly normal), MATUYAMA (mostly reversed), GAUSS, (mostly normal) and GILBERT (mostly reversed). Shorter polarity intervals that occurred within the above periods (called events) were named after the locations where they were first observed. This system of nomenclature is obviously impractical for earlier chrons because of their large number; however, these names are retained because of historical precedent. They are superseded by a nomenclature based on the **polarity chrons** established by the International Stratigraphic Guide following the reports of the subcommission on a **Magnetic Polarity Time Scale** and the International Subcommission on Stratigraphic Classification (Fig. 4.2).

The terms recommended for describing subdivisions of time based on geomagnetic polarity are **polarity subchrons**, **polarity chrons**, and **polarity superchrons**. Polarity chrons describe the main subdivisions and replace the older term "epoch". The term subchron describes very short (~ 0.1 Ma) polarity intervals occurring within a chron and replaces the older term "event". Superchrons are used to describe very long periods of single polarity such as the Cretaceous Quite Zone.

Polarity chrons are numbered according to a scheme that was originally set up by HEIRTZLER et al. (1968), based on the positive marine magnetic anomalies from oceanic magnetic profiles. The numbers go from I at currently spreading mid-ocean ridges to 32 (PITMAN et al., 1968). However, many chrons have been added to the initial set of 32 numbered anomalies. Additional chrons representing smaller positive anomalies are labeled by adding letters and decimals to the original set of 32 numbers (LABRECQUE et al., 1977; NESS et al., 1980). HARLAND et al. (1982) identified unlabelled chrons by adding letters to the next youngest numbered chron, e.g. 5A and 5B follow 5, and 5AA and 5AB follow 5A. To avoid confusion between the chron numbering scheme based on the paleomagnetism in sedimentary sequences and ocean floor, LABRECQUE et al. (1983) suggested prefixing ocean floor spreading chrons by the letter C.

In the Geologic Time Scale (1982) of HARLAND et al. previously unlabelled reversed chrons were given the number of the preceding normal chron with the letter **r** appended. In CANDE & KENT (1992a, 1995) geomagnetic polarity timescale for the Late Cretaceous and Cenozoic, authors make some small modifications to that of HARLAND et al. (1990). When chrons are subdivided into shorter polarity intervals (subchrons), CANDE & KENT (1992a) identify them by appending, from youngest to oldest, a 0.1, 0.2, etc., to the primary chron name, and adding an **n** for a normal polarity interval or an **r** for a reversed interval. For example, the three normal polarity intervals composing anomaly 6C (chron C6Cn) are called subchrons C6Cn.1n, C6Cn.2n and C6Cn.3n, whereas HARLAND et al. (1990) refer to them as chrons C6C.1n, C6C.2n and C6C.3n (Fig. 4.3).

CANDE & KENT use the designation –1, –2, etc., following the primary chron or subchron designation to denote small-scale oceanic magnetic anomalies sometime referred as **tiny wiggles.** Thus, for example, the tiny wiggles between anomalies 12 and 13 (i.e. within chron C12r) are called from youngest to oldest C12r-l, C12r-2, etc. Since the origin of these small-scale magnetic anomalies is uncertain, CANDE & KENT refer to these geomagnetic features as cryptochrons. Not all the tiny wiggles in the oceanic record have been reported in the CANDE & KENT (1995) GPTS, but many of those reported have been proved to be a global feature of the geomagnetic filed. However, the origin of the tiny wiggles is still debated. CANDE & KENT (1992 b) suggested that the most probable origin of tiny wiggles is intensity fluctuations of the geomagnetic field rather than field reversals and several attempt to find such short reversals in the sedimentary record have failed (e.g., LANCI & LOWRIE 1997).

Polarity chrons prior to the Cretaceous quite zone (pre-Aptian in age) are described by the letter M going from M0 to M29 assigned to marine anomalies in order of increasing age (LARSON & PITMAN, 1972; LARSON & HILDE, 1975; CANDE et al., 1978). The prefix M stands far Mesozoic. Compared with the Cenozoic reversal sequence, where positive anomalies correspond to normal chrons, in the Mesozoic reversal sequence (**M-sequence**) the numbered oceanic anomalies correspond to reversed chrons, although M2 and M4 chrons have normal polarity.

3. – PRACTICE

This paragraph introduces shortly to the techniques of paleomagnetism used in magnetostratigraphic studies in sedimentary series. The detail of the techniques of analysis of the paleomagnetism and rock-magnetism can be found in many texts, e.g., TAUXE (1998), OPDYKE & CHANNELL (1996), BUTLER (1992), COLLINSON et al. (1967), COLLINSON (1983). The applications of the magnetostratigraphy have been reviewed by HAILWOOD (1989a), LOWRIE (1990).

3.1. FIELD SAMPLING

As a general scheme, paleomagnetic sampling is based on a hierarchical sampling designed to eliminate or minimize non-systematic errors and to average out the effects of secular

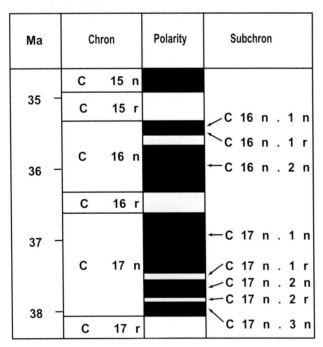

FIG. 4.2

General scheme for magnetochrons nomenclature taken from the Eocene geomagnetic polarity time scale. For the Cenozoic and upper Cretaceous sequence the chrons names are prefixed by the letter C followed by the number of the corresponding positive oceanic magnetic anomaly. The polarity is indicated with the letters **n**, for normal and **r** for reversed, which precede the normal chron with the same number. Subchrons are named starting from the oldest with number 1 and the letter **n** or **r** indicating their polarity.

4. Magnetostratigraphy

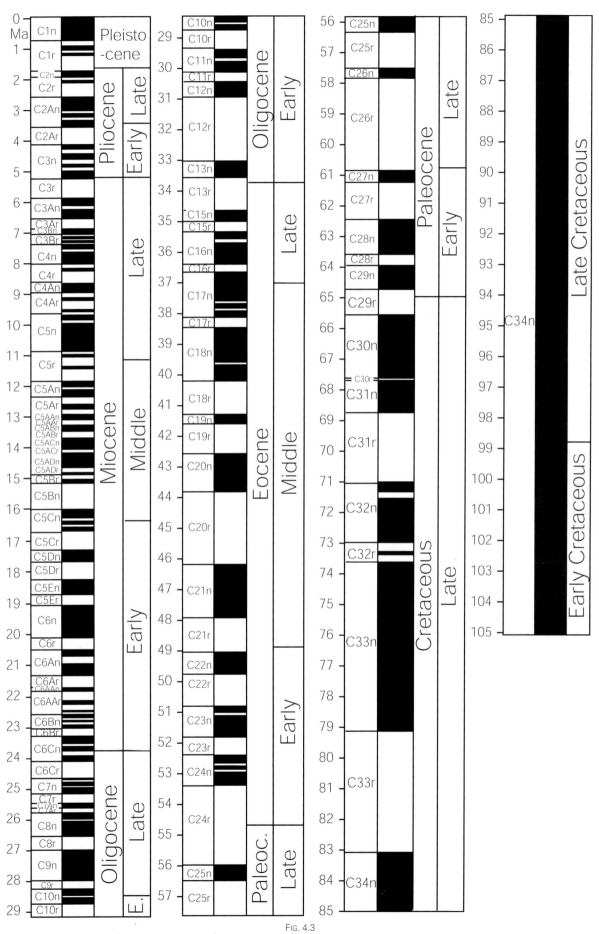

Fig. 4.3

Geomagnetic time scale for Cenozoic and Cretaceous from CANDE and KENT 1995. Only the chrons' numbers are indicated.

variation of the paleomagnetic field. At each hierarchical level, averaging and statistical analyses are carried out on the remanent magnetization vectors. The data from a few specimens from the same sample may be averaged to give a mean value for the sample itself, although this step is omitted in several studies. If the rock-type is suitable for paleomagnetic analysis, the directional scatter between samples at a site will be small or moderate and about 6-10 samples are enough to define a site-mean direction with acceptably small error. As in other statistical analyses, the error of the mean decreases as $\sqrt{N}$, therefore, to halve the error of the site-mean, the number of samples must be increased fourfold. The mean values of typically 10-20 sites from the same formation are averaged to get a mean value for the formation, from which the position of the paleomagnetic pole at the age of the formation may be calculated.

Samples for a magnetostratigraphic study are taken in stratigraphic sequence. A sampling interval based on the estimated sedimentation rate must be chosen so as to give the desired resolution (e.g. at 10 m/ Ma, a 1-meter sample interval corresponds to 100 ka). The correct sampling interval can vary from the centimeters to the meters and it is chosen accordingly to the project's goals and the quality of the outcrop. If possible, the sampling should be designed to allow some of the field tests of the magnetization stability over geological time, such as the fold or reversals tests described below. The two most common ways of field sampling are by taking oriented blocks and by drilling oriented cores.

An oriented block sample must be large enough for one or more standard paleomagnetic samples (a cylinder 2.5 cm in diameter and 2.2 cm long) to be drilled out of it later. A large 'T' indicating the azimuth and the plunge directions is marked on the oriented surface. Normally the azimuth and plunge line are recorded (but the orientation convention may change in different laboratory). Samples are drilled in the laboratory normal to the oriented surface and the azimuth A and dip D of the sample axis are calculated.

More often standard samples are drilled directly from the outcrop with a portable drill. The axis of the cylinder must be oriented with a device that measures the dip of the hole while the azimuth of the hole is measured with a magnetic compass. A sun compass may be used in outcrops with strongly magnetic volcanic rocks. The orientation marks transferred to the sample normally indicate the downhole direction of the cylinder axis and the upward direction of one cylinder face. The sample marking may change according to different laboratory conventions; anyhow, a precise sample orientation is always needed to relate the magnetization direction measured in the rock sample to the geographic reference coordinates at the site (**sample coordinates**).

3.2. SAMPLING IN SEDIMENTS CORES

The sampling techniques of sediment cores may vary, but usually they do not involve the drilling or the orientation of the samples. Sediments cores are split in two halves and the samples are usually taken by pushing a plastic box or cylinder of 7-10 cm^3 volume directly in the sediment with axis perpendicular to the sediment surface. If cores have been oriented during the drilling, the sample azimuth can be used, in most cases, anyhow, one can only count on the sample inclination to infer the magnetic polarity.

A recent technique is the u-channel sampling that consists of taking a single continuous sample usually $2 \times 2 \times 150$ cm along the whole core section. These long and thin samples can be measured at high resolution up to every cm length, using cryogenic magnetometers equipped with in line AF demagnetizer that are able to automatically perform a full demagnetizing cycle. With a few limitations that include the impossibility to use thermal demagnetization, u-channels represent a major improvement in the measurements speed of paleomagnetic samples.

3.3. MEASUREMENT OF REMANENT MAGNETIZATION

Two types of magnetometer are commonly used in modern paleomagnetic laboratories, the spinner magnetometer and cryogenic magnetometer. The latter are the fastest and most sensitive, capable to measure magnetic moment as low as to 1×10^{-8} emu, and are use in most of laboratory.

Measurament procedures

Measurement techniques in discrete samples are usually based on over-definition of the magnetic vector. This means that more measurements of the X, Y and Z components of magnetization are made changing the samples orientation with respect the magnetometer measurement axis. The redundancy allows averaging out the possible imperfect positioning of the sample in the magnetometer and a control of the quality of the measurement.

Spinner magnetometers and old fashion 2-axis cryogenic magnetometers measure 2 magnetization components in each plane of measurement. It is customary to repeat the measurements in three orthogonal planes (e.g., the $X-Y$, $Y-Z$ and $Z-X$ planes in sample coordinates). Modern 3-axis cryogenic magnetometers can measure all three orthogonal magnetization components simultaneously. However, 3 to 8 repetitions are usually made, with different alignments of the sample and magnetometer axes each time, such that each of the X, Y and Z components is measured along each instrument axis. U-channel samples cannot be repositioned in the magnetometer and therefore only a single measurement is taken.

The averaged values of the X, Y and Z components are then transformed to the polar coordinates traditionally used in paleomagnetism that are: the intensity of remanent magnetization (Mr), the **magnetic declination** (D) and **magnetic inclination** (I) of the paleomagnetic vector, where:

$$Mr = \sqrt{(X^2 + Y^2 + Z^2)} \quad \tan D = Y/X, \quad \sin I = Z/Mr$$

The standard deviation (or the variance) of the measured X, Y and Z component can be taken as an estimate of the quality of each single measurements, that can also be expressed as the standard angular deviation.

3.4. SAMPLES DEMAGNETIZATION

Rocks contain millions of tiny magnetic grains that give them their magnetic properties and that are able to record the magnetization acquired when the rock was formed. Part of these grains may not be stable enough to retain their original magnetization through geologic time and may change their direction of magnetization. In this way, rocks can acquire a viscous magnetization in the direction of the ambient field. The grains carrying the viscous magnetization have lower anisotropy energies (they are magnetically "softer") and their contribution to the sample magnetization can be erased (randomized) without affecting the more stable ("harder") grains carrying the ancient remanent magnetization. There are several laboratory techniques available for erasing viscous components that exploit its soft nature. The most used techniques are alternating field (AF) and thermal (Th) demagnetization that can isolate the component of the magnetization based on their coercivity or unblocking temperature respectively. Mixed treatments (usually AF followed by Th demagnetization) are possible but rarely used. The basis for alternating field demagnetization is that soft components also have low coercivities and that for thermal demagnetization is that they also have low blocking temperatures. To investigate these components, the NRM is stepwise demagnetized to progressively "erase" the most unstable fraction of magnetization, which is the most likely to have acquired a secondary direction after the rock formation.

3.4.1. Alternating field (AF) demagnetization

If a rock sample is placed in an alternating magnetic field, the magnetizations of all grains or domains with coercivity smaller than the peak alternating field (AF) are remagnetized (i.e. reoriented along the field direction). If the peak field is slowly reduced to zero in a field-free space the part of the rock magnetization with coercivity smaller than the peak AF, is randomized (i.e. the directions of individual grains are oriented randomly). Since the random directions sum to a null component, they are effectively removed and the part that remains has been "magnetically cleaned". The procedure is carried out in progressively increasing peak fields such that the NRM components of increasing coercivity can be isolated. This method is technically limited by the field strengths that can be produced in the demagnetizing coil. Commonly the peak field is around 0.1 T, some equipment can reach around 0.3 T, and therefore **AF demagnetization** is effective in demagnetizing rocks that contain magnetite. In most laboratories the sample is AF demagnetized successively along three orthogonal axes alternatively discrete samples can be rotated ("tumbled") as randomly as possibly within the demagnetizing coil. Many modern magnetometers have an AF demagnetizer mounted "on line" such that each sample can be measured and progressively demagnetized without intervention of the operator. This is particularly useful in u-channel samples where AF is the only possible demagnetization technique.

3.4.2. Thermal demagnetization

If a rock sample is heated to a given temperature (T), the magnetization directions of grains or domains that have lower unblocking temperatures than T are thermally randomized. The random directions are "locked" when the sample is cooled below the unblocking temperature in field-free space and consequently this fraction of the NRM remains demagnetized. In stepwise **thermal demagnetization**, the heating and cooling cycle is repeated with progressively higher maximum temperatures and allows isolating the component of NRM based on their unblocking temperature. This method is more flexible and powerful than AF demagnetization because the range of temperatures needed to destroy the NRM in all magnetic minerals found in natural rocks is below 700 °C, which is an easily achieved temperature. However, the thermal alteration of the rock sample can change the magnetic mineralogy and complicate the demagnetization results. This technique is also not usable for u-channel samples.

3.5. ROCK MAGNETISM

3.5.1. Identification of ferrimagnetic minerals in a rock

Natural rocks may contain several magnetic minerals, or different phases of the same mineral. In a paleomagnetism it is helpful to recognize what ferrimagnetic mineral is the carrier of the paleomagnetic signal, and there are several laboratory methods allow the identification of magnetic minerals on the basis of their coercivity and thermo-magnetic properties.

3.5.2. Curie Temperature analysis

The **Curie temperature** is the temperature where the ferromagnetic behaviour disappears and the material became paramagnetic. Each mineral has a distinctive Curie temperature therefore its measurement allows the identification of a magnetic mineral. Measurements are made using a Curie balance that measures the saturation magnetization of a mineral sample placed in the strong field gradient of an electromagnet and slowly heated to increasing temperature. The saturation magnetization usually shows weak temperature dependence and suddenly decreases to its paramagnetic value when the Curie temperature is reached. Modern Curie balances have a horizontal movement that is not susceptible to weight loss of the sample during heating and are sensitive enough to measure whole samples even in sedimentary rocks. When samples are too weak to be measured directly it is possible to extract ferrimagnetic minerals although extraction is sometimes difficult and it is often not certain that the extract is representative of the rock as a whole. Unfortunately in many cases the presence of strong paramagnetic components, complicated mineralogy, and thermal alteration of the samples may give results very difficult to interpret if not meaningless.

3.5.3. Acquisition of thermal demagnetization of IRM

Remanent magnetization resulting from short-term exposure to strong magnetic fields at constant temperature is referred to as isothermal remanent magnetization (IRM). In the laboratory, IRM is imparted by exposure to a magnetizing field generated by an electromagnet. IRM is the form of remanence acquired by ferromagnetic grains with coercive force smaller than the applied field; thus, a stepwise acquisition of IRM can be used to investigate the coercivity spectrum of the samples.

A combination of coercivity spectrum and unblocking temperature, which is usually close to the Curie temperature, may offer a convenient method to identify the magnetic minerals in a rock. Each minerals, in fact, can be characterized by a combination of these two parameters. These properties can be exploit by applying a stepwise increasing IRM to the sample and observing the magnetic field at which the magnetization reaches the saturation (i.e. does not increase anymore). These measurements are followed by a stepwise thermal demagnetization of the IRM acquired to observe the unblocking temperature at with the magnetization is destroyed by the thermal energy. The combination of these two parameters should serve to identify the magnetic mineral(s) carrying the magnetization in the sample. Examples of IRM acquisition and thermal demagnetization are shown in Fig. 4.4.

A more sophisticated technique (LOWRIE, 1990) involves the use of magnetizing field of different strengths induced at orthogonal directions immediately before the thermal demagnetization.

3.6. ANALYSIS OF MAGNETIZATION COMPONENTS

The stability of a remanent magnetization during stepwise demagnetization is analysed by graphical inspection of the remanent directions and the so-called **"characteristic**

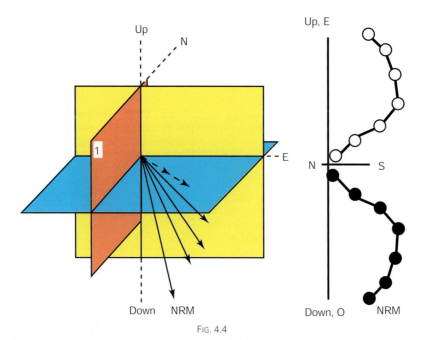

FIG. 4.4
Vector diagrams of progressive demagnetization of Natural Remanent Magnetization (NRM).
The magnetic vectors are projected in to the horizontal and the north-south or east-west vertical planes.
The choice of vertical projection usually depends on the magnitudes of N and E components.
Inclination (open symbols) and declination (filled symbols) can be seen on the same diagram, often referred
as Zijderveld plots (ZIJDERVELD, 1967), obtained by folding the vertical plane in the horizontal.

remanent magnetization" (ChRM) is found with the aid of statistical best-fitting methods.

Paleomagnetic directions are usually plotted using vector diagrams or stereographic projection. In order to avoid visually distorting the grouping of directions, the Lambert equal-area stereonet is usually preferred to the equal-angle stereograms.

3.6.1. Analysis of vector diagrams

The most used method of analysis of the structure and stability of a remanent magnetization is by constructing a vector diagram (ZIJDERVELD, 1967). The magnetic vectors are projected into north (N), east (E) and vertical (V) components. Plots are then made of the N component against the E component, and of one of the horizontal components (N or E) against the V component (Fig. 4.5). In other words, the vectors are projected in to the horizontal and the north-south or east-west vertical planes. The choice of vertical projection usually depends on the magnitudes of N and E components.

On vector diagrams (often called **Zijderveld plots**) the NRM components with different directions can be recognized as straight-line segments. When the different components have overlapping spectra of coercivity or blocking temperature the NRM demagnetization path may show curved lines. In the simplest case where only a single stable component of remanent magnetization is present, progressive demagnetization will produce a straight line to the origin of each part of the vector diagram. Unstable components are removed at lower temperature or demagnetizing field and show on a vector plot as lines that are not directed towards the origin. If a stable vector remains after demagnetization of the less stable fraction, it is usually shown by a straight line to the origin of the vector diagram.

Modern techniques of analysis of the direction components involve the use of principal component analysis (PCA) (KIRSCHVINK, 1980) to compute the best-fit line between several demagnetization steps aligned to the same straight-line segment. To calculate the PCA, the directions of a sequence of data points, usually 4 or more, which form a single component are converted to corresponding cartesian (x, y and z) coordinates, then their center of mass is calculated as follows:

$$\bar{x} = \frac{1}{N}(\sum x_i); \quad \bar{y} = \frac{1}{N}(\sum y_i); \quad \bar{z} = \frac{1}{N}(\sum z_i)$$

where N is the number of data points involved.

The origin of the data cluster is then transferred to the center of mass of these data by subtracting the mean of all the components from each individual component as follows:

$$x'_i = x_i - \bar{x}; \quad y'_i = y_i - \bar{y}; \quad z'_i = z_i - \bar{z}$$

Then we compute the covariance matrix C of the data in transformed coordinates, which is the matrix of sums of squares and products

$$C = \begin{bmatrix} \sum x'_i x'_i & \sum x'_i y'_i & \sum x'_i z'_i \\ \sum x'_i y'_i & \sum y'_i y'_i & \sum y'_i z'_i \\ \sum x'_i z'_i & \sum y'_i z'_i & \sum z'_i z'_i \end{bmatrix}$$

and we find the eigenvalues and eigenvectors of C. The direction of the eigenvector associated with the greatest scatter in the data (the principal eigenvalue) corresponds to a best-fit line through the data. This is usually taken to be the direction of the component in question.

3.6.2. Analysis of remagnetization circles

If more than one magnetization component is present, it is possible that the coercivity or blocking temperature spectra of the components will overlap completely. When the overlap is being demagnetized, the vector diagram may show a curved rather than a straight segment, and if the overlap is complete, no straight segment will be found. When two components have overlapping spectra of coercivity or blocking temperature, the technique of the analysis of the remagnetization circles may help to find the ChRM direction.

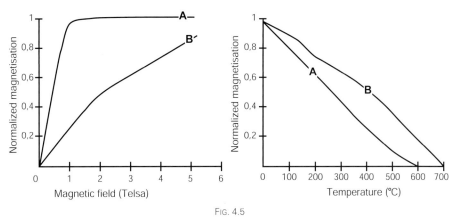

FIG. 4.5
Diagram of Isothermal Remanent Magnetization (IRM) acquisition (left) and thermal demagnetization (right). Curve A is for magnetite and curve B for hematite.

In this case, although changing in direction, the path of the demagnetized direction will fall in the plane defined by the two vectors. Using the PCA analysis, the eigenvector associated with the least eigenvalue can be taken as the pole to the best-fit plane wherein the component of interest must lie. Plotted on a stereogram the best-fit plane will define a great circle (the remagnetization circle of the sample) converging to the direction of the stable component as the more unstable (secondary) direction is partially removed. If the secondary direction is different in each sample from the same site, there will be a set of great circles that point at the primary direction. The method of analysis therefore consists of determining the intermediate directions during demagnetization of each sample and fitting their trajectory with a great circle the common intersection point of all remagnetization circles determines the common primary direction for the suite of samples.

3.7. CORRECTIONS OF THE MEASURED DIRECTION OF REMANENT MAGNETIZATION

In the laboratory the direction of remanent magnetization is determined relative to reference axes marked on the sample that were oriented relative to present day **geographic coordinates** (horizontal and north direction) at sampling time. The direction of the sample magnetization must obviously be reoriented to the original geographic coordinates of the site by applying suitable corrections. Moreover, often samples are taken in the field from sedimentary rocks whose bedding may now be tilted as a result of tectonic effects. If the bedding attitude has been recorded in the field, the sample direction can be restored to the original direction assuming that bedding was originally flat and only a simple horizontal bending as occurred.

3.7.1. Correction for orientation of sample (geographic correction)

If a ChRM direction measured relative to sample axes has declination D and inclination I, its direction cosines (λ, μ, ν) are given by:

$$\lambda = \cos D \cos I \qquad \mu = \sin D \cos I \qquad \nu = \sin I$$

If the sample comes from a core drilled with azimuth A and dip C, the direction cosines (λs, μs, νs) of the ChRM relative to geographic axes are calculated as follow:

$$\lambda s = \lambda \sin C \cos A \quad - \mu \sin A \quad + \nu \cos C \cos A$$
$$\mu s = \lambda \sin C \sin A \quad + \mu \cos A \quad + \nu \cos C \sin A$$
$$\nu s = -\lambda \cos C \quad\quad\quad\quad\quad\quad + \nu \sin C$$

The ChRM direction (Ds, Is) relative to the present geographic coordinates is given by:

$$\tan Ds = \mu s / \lambda s \qquad \sin Is = \nu s.$$

3.7.2. Correction for tilt of bedding (tectonic correction)

Tectonic correction is applied when the bedding is not horizontal as a result of tectonic tilting. This is made by applying a simple 'rigid-body rotation' about the horizontal strike line to restore the inclined bedding to a horizontal position. Given the bedding plunging at an angle P below the horizontal in a direction with azimuth B, the direction cosines (λt, μt, νt) of the rotated ChRM are calculated as follow. (Note: if bedding is overturned, P must be replaced by ($P + 180°$).

$$\lambda_1 = \lambda s \cos B + \mu s \sin B;$$
$$\mu_1 = -\lambda s \sin B + \mu s \cos B;$$
$$\nu_1 = \nu s$$

$$\lambda_2 = \lambda_1 \cos P + \nu_1 \sin P;$$
$$\mu_2 = \mu_1;$$
$$\nu_2 = -\lambda_1 \sin P + \nu_1 \cos P$$

$$\lambda t = \lambda_2 \cos B - \mu_2 \sin B;$$
$$\mu t = \lambda_2 \sin B + \mu_2 \cos B;$$
$$\nu t = \nu_2$$

The **tectonic coordinates** corrected ChRM direction is given by:

$$\tan Dt = \mu t / \lambda t \qquad \sin It = \nu t$$

3.8. FIELD TESTS OF MAGNETIZATION STABILITY

Paleomagnetic field test for the stability of the remanent magnetization were developed in the early days of paleomagnetism to prove that a ChRM measured in a certain rock was primary (i.e. was acquired at the time of rock formation). In fact, while modern laboratory technique may reveal precisely the components of NRM and the mineralogy of the ferromagnetic grains carrying a ChRM, they cannot prove that the ChRM is primary.

Field tests of paleomagnetic stability are probably the best way to demonstrate the stability of a remanent magnetization through the geological time and can offer crucial information about the timing of ChRM acquisition. Especially in studies of old rocks and orogenic zones, field tests of paleomagnetic stability can provide critical information.

3.8.1. Reversal test

Because of the GAD hypothesis, at all locations, the time-averaged geomagnetic field directions during a normal-polarity interval and during a reversed-polarity interval are antipodal.

If in a set of paleomagnetic sites or samples the ChRM have been properly isolated and the secular variations have been averaged (which is often the case in sedimentary sequences), the mean ChRM direction for the normal polarity samples should be antipodal to the mean ChRM direction for the reversed polarity samples (Fig. 4.6). In this case, the ChRM directions are said to "pass the reversals test" (McFADDEN & McELHINNY, 1990). A positive reversals test indicates that ChRM directions are free of secondary NRM components and that the geomagnetic secular variation has been adequately averaged. Furthermore, if the sets of normal and reversed polarity sites conform to stratigraphic layering, as expected in a magnetostratigraphy, the ChRM is probably a primary NRM. The paleomagnetic data set "fails the reversals test," if the average directions for the normal and reversed polarity sites differ by an angle that is significantly less than 180°. This indicates either presence of an unremoved secondary NRM component or inadequate sampling for geomagnetic secular variation.

This test is often applicable because, unlike the conglomerate or fold test, does not require special geologic settings, and is particularly useful in the magnetostratigraphy studies where several polarity reversals are usually found.

3.8.2. Fold test

The fold test is perhaps the most important field test made in a paleomagnetic study but it requires sites sampled on both limbs of a fold. In the fold test, relative timing of acquisition of the ChRM and folding can be evaluated. If a ChRM was acquired prior to folding, directions of ChRM from sites on opposing limbs of a fold are rotated away from its original direction when plotted in geographic coordinates but the original direction should be recovered when the tectonic correction is made and the bedding is restored to horizontal (Fig. 4.7). The ChRM directions are said to "pass the fold test" if clustering increases after application of the tectonic correction. On the converse they "fail the fold test" if the ChRM directions become more dispersed after application of the tectonic correction. Statistical methods (e.g., McELHINNY, 1964; McFADDEN & JONES, 1981; McFADDEN, 1990; TAUXE & WATSON, 1994) are used to verify the significance of the grouping of the paleomagnetic directions at various degree of "unfolding".

3.8.3. Conglomerate test

This test requires the existence of a conglomerate containing large pebbles of the investigated rock and it assumes that these cobbles have been randomly re-oriented by the sedimentary processes that formed the conglomerate. If ChRM in cobbles from a conglomerate has been stable since before deposition of the conglomerate, ChRM directions from numerous cobbles or boulders should be randomly distributed (positive conglomerate test). A non-random distribution indicates that ChRM was acquired

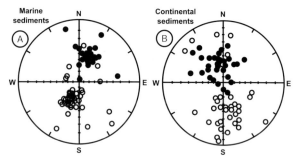

FIG. 4.6

Typical example of paleomagnetic directions obtained from a sedimentary sequence plotted on an equal area stereographic projection. Open symbols indicate projections on the upper hemisphere (negative inclinations) and filled symbols projections on lower hemisphere. A) Excellent grouping of normal and reversed directions from a sedimentary pelagic sequence (Toarcian ammonitico-rosso limestones from Greece, GALBRUN 1993); B) Paleomagnetic directions from the continental sequence of the Ager basin (Spain) showing a not so good grouping (GALBRUN et al., 1993).

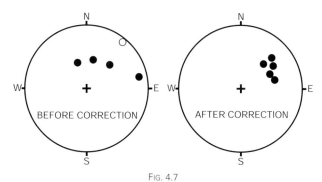

FIG. 4.7

Example of positive fold test. The mean paleomagnetic directions of 5 sites, one with a negative inclination (open symbol), are better grouped after the bedding is restored to horizontal. This is an evidence that the rock magnetization have been acquired prior to folding.

after deposition of the conglomerate (negative conglomerate test). A positive conglomerate test from an intraformational conglomerate provides very strong evidence that the ChRM is a primary NRM.

3.9. STATISTICAL ANALYSIS OF DIRECTIONS AND POLES

3.9.1. The Fisher distribution and its use in paleomagnetism

Paleomagnetic directions, obtained from PCA of the most stable component of the magnetization, are represented as unit vectors. A probability density function applicable to directions was developed by FISHER (1953), and it is known as Fisher distribution. Directions are distributed according to the probability density function:

$$P(\theta, \kappa) = \frac{\kappa}{4\pi \sinh \kappa} \exp(\kappa \cos \theta)$$

The "precision parameter" k describes the dispersion of the points. If $k = 0$, the directions are uniformly or randomly distributed and approaches ∞ as the points cluster tightly about their mean direction.

The best estimate of the mean direction of a population of directions (unit vectors) is the vector mean of the unit vectors. This can be calculated from the sum of the north, east and down components (direction cosines) of an individual direction (Di, Ii) given by:

4. Magnetostratigraphy

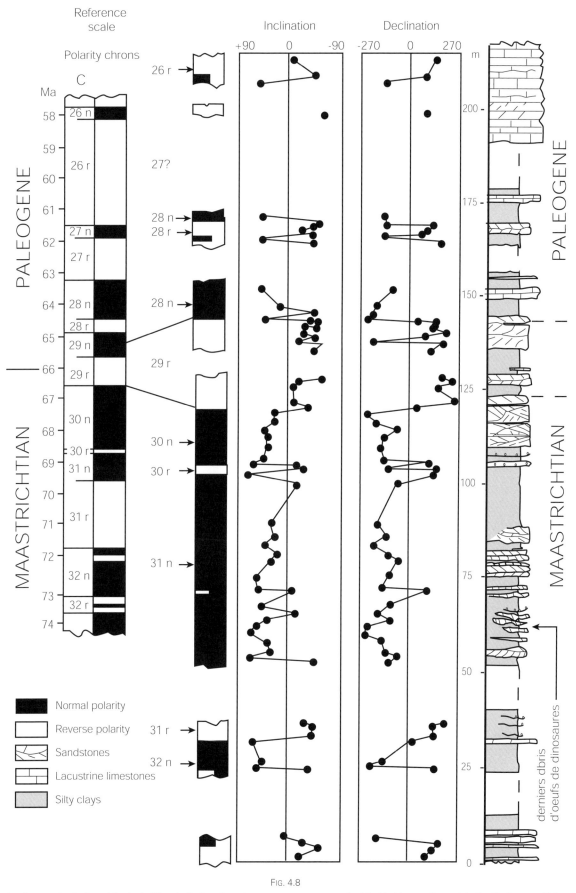

Fig. 4.8

Example of magnetostratigraphic study. The studied sedimentary sequence was poorly dated but known to contain the Cretaceous/Paleogene boundary. Concurrent changes of inclination and declination the ChRM directions define the polarity reversals and identify the magnetic polarity sequence. Inclination and declination can also be combined calculating the VGP latitude, as described in the text, which expresses the magnetic polarity with a single parameter. Magnetozones are defined by at least two consecutive samples with the same polarity. No interpolation is carried out between too distant specimens resulting in an interruption of the inclination and declination records. Correlations with the geomagnetic polarity time scale are made with the help of the few biostratigraphic data available. (Modified after GALBRUN et al., 1993).

North component, $xi = \cos li \cos Di$

East component, $yi = \cos li \sin Di$

Down component, $zi = \sin li$

The direction cosines of the N unit vectors are summed to give:

$$X = \Sigma xi; \quad Y = \Sigma yi; \quad Z = \Sigma zi$$

The length R of the vector sum of the N unit vectors is given by:

$$R^2 = X^2 + Y^2 + Z^2$$

The mean direction (D, I) is then given by:

$$\tan D = Y/X; \quad \sin I = Z/R.$$

Since only a small number of directions are sampled from what is assumed to be an infinite population, the calculated parameters are approximate estimates of the true. Fisher showed that the best estimate (k) of the precision parameter k (valid for $k > 3$) is given by:

$$k = (N-1)/(R-R)$$

where R is the vector sum of the N unit vectors, as above.

The k–parameter is equivalent to the inverse of the variance of the distribution. If k is large, the distribution approximates a two-dimensional Gaussian distribution.

In paleomagnetism the 95% level confidence is commonly used to define a circle with radius 95 around the estimated mean direction to measure the dispersion of the data. This define the area where there is a 95% probability to find the true mean of the distribution. An approximate formula for the radius of the 95% circle of confidence is:

$$\alpha_{95} = 140/(\sqrt{Nk})$$

A more exact formula (valid for $k > 3$) is:

$$\cos \alpha_{95} = 1 - \left[20^{\frac{1}{N-1}} - 1 \right] - \left(\frac{N-R}{R} \right)$$

3.9.2. *Calculation of virtual geomagnetic pole (VGP) position and its confidence limits*

Any pole position that is calculated from a single observation of the direction of the geomagnetic field is called a **virtual geomagnetic pole** (**VGP**). In paleomagnetism, a site-mean ChRM direction is a record of the past geomagnetic field direction at the sampling site location during the short time over which the ChRM was acquired. Thus, a pole position calculated from a single site-mean ChRM direction is a VGP. The position of the VGP can be calculated from the site mean declination (D), inclination (I) and site coordinates under the assumption of a dipolar geomagnetic field. The angular distance (p) to the magnetic pole can be calculated from the field inclination (I) using the dipole equation:

$$\tan p = 2/\tan I$$

The value of p determines the radius of a minor circle centered on the sampling site location that is the locus of all possible magnetic pole positions. The declination defines a direction along a great circle that passes through the sampling site and makes an angle D with the magnetic meridian. The place where this great circle intersects the minor circle with radius p is the VGP location.

The latitude (λ) and longitude (ϕ) of the VGP that gives the field direction (D, I) measured at a given site (latitude λs, longitude ϕs) can be calculated by:

$$\sin \lambda = \sin \lambda s \cos p + \cos \lambda s \sin p \cos D$$

$\phi = \phi s + \beta$, for $\cos p > \sin \lambda s \sin \lambda$

$\phi = \phi s + 180 - \beta$, for $\cos p < \sin \lambda s \sin \lambda$

where $\sin \beta = \sin p \sin D / \cos \lambda$

The confidence level for the VGP position can be calculated from the radius of the site confidence circle α_{95}. An oval of confidence around the VGP position, having latitude error $\delta\lambda$ and longitude error $\delta\phi$ given by:

$$\delta\lambda = \alpha_{95}\frac{(1+3\cos^2 p)}{2}; \quad \delta\phi = \alpha_{95}\frac{\sin p}{\cos I}$$

3.9.3. *VGP latitude*

The VGP latitude is the latitude of VGP position of a single sample or site with respect to the paleomagnetic pole for a given location and time. Because of the GAD hypothesis the VGP position is expected to be always close to the paleomagnetic pole, thus in a good paleomagnetic study the VGP latitude should always be close to +90°, for normal polarity or –90° for reverse polarity.

In magnetostratigraphy, the VGP latitude is a convenient way to express the magnetic polarity of a sample combining the information from the paleomagnetic declination and inclinations. In practice in a magnetostratigraphic study the paleomagnetic reference North pole is calculated from the mean ChRM directions of the whole set of samples and the VGP latitude is computed for all the individual samples. Normal polarity samples will have VGP latitude close to 90° and reversed samples close to –90°.

It should be stressed that in absence of other information the actual polarity of the sequence is somewhat arbitrary. In fact, continents can be transported from one hemisphere to the other by plate movements and the paleomagnetic sites can be tectonically rotated during their geological history. However, this is a serious problem only if the paleomagnetic pole for that time is poorly known like, for instance, in very old (Paleozoic) sequences or if the site underwent to intense tectonics.

4. – CONCLUSIONS

The history of the geomagnetic field reversals has found important applications in Stratigraphy. Since field reversals occur in a very short time compared to the periods of stable polarity, they can be precisely defined in the stratigraphic sequence. Unlike biostratigraphic events, field reversals occur simultaneously on the whole Earth, are not biased from varying environmental conditions and can correlate different environments such as volcanic, marine and continental sequences. Because of these reasons, magnetostratigraphy has been applied to a great variety of geochronological problems such as the example shown in Fig. 4.8. Nevertheless, since geomagnetic reversals have very little individuality, magnetostratigraphy is often used together with other methods such as biostratigraphy or radiometric dating.

The results that can be expected from magnetostratigraphy depend from several factors, the most important are:

– (1) the geomagnetic reversal rate during the considered epoch (this is, for instance, very high during the Miocene but there are no reversals during Turonian, Cenomanian and Albian);

– (2) the existence of a reference scale (GPTS), which is well established for the whole Cenozoic and part of the Mesozoic;

– (3) the rock magnetic properties and the preservation of the original NRM in the studied sequence.

Given favourable circumstances, magnetostratigraphy makes a unique stratigraphic tool that allows precise worldwide dating and correlation between sequences from different environments.

CHAPTER 5
BIOSTRATIGRAPHY
from taxon to biozones and biozonal schemes

J. Thierry & S. Galeotti

CONTENTS

1. – DEFINITION AND AIM 65
   1.1. Introduction 65
   1.2. The biostratigraphic procedures 65
   1.3. Index fossils, isochrony, dating and correlations 66
   1.4. Evolution of concepts and methods 66
   1.5. The biozone, the basic unit of biostratigraphy 66
   1.6. Definition and identification of biozones 66
2. – CONCEPTS, METHODS AND TERMS IN BIOSTRATIGRAPHY 67
   2.1. Classical biostratigraphy and biostratigraphic units 68
      2.1.1. Preliminary remarks 68
      2.1.2. Kinds of biostratigraphic units 69
   2.2. The logical biostratigraphy or Unitary Associations method 74
      2.2.1. Definitions 74
      2.2.2. The construction of biochronozones 74
   2.3. "Statistical biostratigraphy" 74
      2.3.1. Graphical methods 75
      2.3.2. Semi-empirical methods 76
      2.3.3. Multivariate analysis 76
      2.3.4. Probabilistic methods 77
   2.4. Hierarchy and sub-categories of biostratigraphic units 77
      2.4.1. Biohorizon, zonule and marker-bed 77
      2.4.2. Subbiozones and superbiozones 78
      2.4.3. Unitary associations and biochronozones 78
3. – PRACTICE 78
3.1. The steps of the procedure of biostratigraphy 78
   3.2. Results of the classical biostratigraphy 79
      3.2.1. Use of the classic biozones 79
      3.2.2. Possible "diachronism of bioevents" 82
      3.2.3. Significance, precision and reliability of the classical biozones 82
   3.3. Results of the logical biostratigraphy 83
   3.4. Results of the statistical biostratigraphy 83
   3.5. Relations between the classic biostratigraphic units, the units of the logical biostratigraphy and the units of the statistical biostratigraphy 85
   3.6. Relations between biochronology, geochronology and chronostratigraphy 85
      3.6.1. Biochronology and chronostratigraphy: biozone and stage boundary 85
      3.6.2. Biochronology and geochronology: the geochronologic calibration of biostratigraphic scales 86
   3.7. Recommendations in biostratigraphy 88
      3.7.1. Definition and denomination of biostratigraphic units 88
      3.7.2. Validation of biostratigraphic units 89
      3.7.3. Use of biozones and biostratigraphic scales 89

1. – DEFINITION AND AIM

1.1. INTRODUCTION

Biostratigraphy is the study of the stratigraphic distribution of fossils. The aim of biostratigraphy is to organize sedimentary rocks as units (**biozones**) with reference to their fossil content. The lateral and vertical extension of a biostratigraphic unit correspond, therefore, to the space and time interval where the fossil taxa used to define it occur. As a consequence, biozones can show large lateral variations in thickness and duration depending on the unequal distribution of fossils across different environments. In fact, biostratigraphic units can be defined only when their characteristic fossils are present whereas the fact that a barren sedimentary series has an age equivalent to that of a biostratigraphic unit does not justify to include the former into the latter.

Biostratigraphic units are distinct from any other kind of stratigraphic units in that the organisms whose fossil remains establish them show evolutionary changes through the geologic time that are not repeated in the stratigraphic record. This makes the fossil assemblages of any one age unique and distinctive from any other. The term biostratigraphy should be used in a purely descriptive sense since the observations made in the field can be presented independently from the establishment of formal biostratigraphic units (REMANE, 1991). This discipline must be clearly distinct from biochronology, which refers to the accurate dating of the evolutionary first appearance or extinction of a **taxon** using ages calibrated by various techniques and methodologies (e.g. radiometric methods or interpolation between radiometrically calibrated magnetic reversals). Hence, biochronology corresponds to the interpretation of the chronological position of biostratigraphic data.

The biostratigraphic units are called biostratigraphic zones, biozones, or simply zones. This chapter describes the meaning and the characteristics of the different kinds of biozones, and the use of these units as well as the results that can be obtained by applying different biostratighraphic procedures. It is, therefore, necessary to describe these different procedures prior to giving a detailed definition of the various kinds of biozone. We may however consider beforehand that the biozone is the basic unit of biostratigraphy corresponding to the vertical (time dimension) and lateral (geographical dimension) distribution of the fossil taxa used to define it; its fossil content makes it unequivocally identifiable by allowing to recognize its distinctive position in the irreversible process of evolution.

1.2. THE BIOSTRATIGRAPHIC PROCEDURES

In general terms, two procedures can be distinguished:
– the first, historically and methodologically older (d'ORBIGNY, 1850; OPPEL, 1856-58), is aimed at recognizing a hierarchy of units from greater to smaller scales, therefore

operating an ever-finer subdivision of the stratigraphic column. Accordingly, the stage – the fundamental unit recognized on the base of a stratotype – is subdivided into sub-stages and zones, which can be in turn subdivided into subzones. A similar subdivision of the stratigraphic column implies that the recognized units are perfectly contiguous and no overlaps or gaps occur across them;

– the second, more recent procedure (BUCKMAN, 1910), uses an opposite approach, i.e. from smaller to greater scales, by grouping local units into higher rank units. This method, which is based on the concept of "hemera" (BUCKMAN, 1893; 1898), is the most widely used nowadays. The hemera, as defined by BUCKMAN, corresponds to the time during which a particular species reaches its maximum abundance.

In both methods, however, these pioneers of biostratigraphy did not make any distinction, well established nowadays, between the biozone and the **chronozone**.

1.3. INDEX FOSSILS, ISOCHRONY, DATING AND CORRELATIONS

Any fossil shows an event of first appearance and an event of last occurrence in the fossil record which are potentially useful in establishing series of biohorizons and biostratigraphic scales. The most useful fossils for this purpose, the so-called index fossils, are those showing the following characteristics:

– limited distribution through time;

– wide (paleo)geographic distribution;

– abrupt (near-contemporaneous) evolution and extinction events;

– easily identifiable hard parts.

In this case, any biostratigraphic document is potentially useful in establishing series of **isochrons** (boundaries-time-lines or time-surfaces) allowing to define biostratigraphic units with the aim of correlation **(biochronological correlation)**. By identifying a series of successive biozones, each representing a time interval, biostratigraphy allows then the construction of **relative biochronologic scales** and gives indications on the relative age of rocks, i.e. their position in the geological record.

The **precision** of the biostratigraphic dating depends on the accuracy and (lateral) reproducibility of the biozonal subdivision. The reproducibility is related to the method used and the resulting kind of biostratigraphic unit. The accuracy is affected by the fossil group(s) used; each index-taxon (or group of taxa), in fact, exists over a more or less extended time interval whose duration is related to the **resolution power**. The obsolete definitions "stratigraphic fossil" (e.g. good marker) and "facies fossil" (e.g., bad stratigraphic marker) should no longer be used.

Nowadays, there is a tendency to use the expression "high resolution" (i.e. the subdivision of a specific interval in intervals of very short duration) instead of "precision of relative dating" (COPE, 1993), of a given fossil group. In this chapter, "resolution power" will be preferred to "high resolution".

1.4. EVOLUTION OF CONCEPTS AND METHODS

The increase of biostratigraphic precision derives from a series of factors including i) the progress of paleontology, notably the introduction of quantitative methods to study fossil populations replacing typological approaches on few specimens; ii) the emergence of new concepts and a better understanding of the evolutionary process; iii) the refinement of the methodology and technology; iv) investigation of previously unknown regions; v) analysis of more complete and expanded sedimentary sequences; vi) revision of the stratotypes and their substitution when unsuited, etc. It is since now evident that any geological study dealing with sedimentary sequences has to use biostratigraphy, which is a key tool of dating and correlation, in order to establish a stratigraphic succession that accurately represents the geological record.

With this goal, the use of different methods (either mathematical or not) to manage very large available data sets is often useful. Quantitative methods have, in fact, been gradually added to the naturalistic or conventional biostratigraphy. The two methods are not opposed. Rather, they represent developments of the knowledge which can provide similar or contradictory results.

In this chapter attempts a review of the classical biostratigraphic methods is presented also with the aid of case studies. Recently developed quantitative methodologies, terminology and nomenclature are also presented.

1.5. THE BIOZONE, THE BASIC UNIT OF BIOSTRATIGRAPHY

The different nature of the several kinds of biozones makes it difficult to give a simple definition of this term. Before giving a detailed description of these units, it is convenient to recall the following two major concepts:

– In a very general sense, a biozone is defined by its paleontological content and the resulting position within the irreversible process of biological evolution;

– On the other hand, following the concepts and methods developed during the last 150 years, a biozone can be defined either as a solid body, represented by rocks distinguished from contiguous strata on the base of their fossil content, or as a conceptual entity, with no "lithologic support", defined on the base of the presence, mutual exclusion and coexistence of a biological association.

1.6. DEFINITION AND IDENTIFICATION OF BIOZONES

Several basic rules must be observed in biostratigraphy:

– the identification and definition of biozones tend towards the most accurate reconstruction of the geological time based on the concepts of succession, duration, and simultaneity;

– the accuracy of the reconstruction depends on the quality of the biological and physical records (the fossils and the host sediments), which represent the only available documentation.

The criteria of construction of the biozones should therefore be only those actually related to these three concepts and their consequences on the presence of fossil material. It is then advisable to take into account every datums, even those which seldom appear in the paleontological record:

– the criteria of superposition that introduces the principle of succession and temporal (vertical) exclusion, and leads to the relation younger/older;

– the criteria of correlation that relies on the principle of simultaneity and co-occurrence (contemporaneity) and leads to the concept of lateral reproducibility.

5. Biostratigraphy

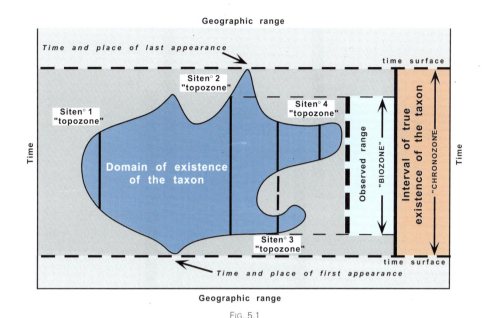

Fig. 5.1

Domain of existence of a taxon. The diagram illustrates the interaction between evolution and biogeography by representing the "space-time" range. The vertical bars represents the biochronological ranges of the taxon at different sites. For the sake of legibility, the second geographical axis of distribution was not illustrated.

In the application of the biostratigraphic methods, it should be always kept in mind that:

– the quality of fossil material affects the acquisition of data and, therefore, the interpretation;

– the biotic and sedimentary records are discontinuous in time and space;

– it always exists a more or less marked relation between the fossil content and the host sediment;

– this registration and this relation can entail an apparent **diachronism** in the observation of first appearance, disappearance, co-occurrence and exclusion of fossil rests, source data of the construction of biozones;

– this diachronism is related, on one hand to the real interval/duration of existence (**life span**) of the considered fossil species whose extension is never completely known (Fig. 5.1), on the other hand to the geographical distribution (paleobiogeography) that varies with time according to various factors controlling their area of distribution (biogeographic "provinces" or "domains" limited by climatic boundaries, transgressions/regressions, existence of gates and ways of exchanges, etc.);

– the taxonomic knowledge of fossil species (intra- and interpopulation variability) is very often incomplete and the interpretation of their taxonomy is influenced by each paleontologist's judgment, though this interpretation is made more reliable by the statistical methods.

Moreover, as already mentioned, biostratigraphy alone does not give indication on the absolute age of rocks since it allows to establish only the relative position of a succession of biozones in the geological record, without any information on the duration of each unit.

Only the **calibration** of zonal scale by radiochronologic data can give an idea of the duration of the biostratigraphic units. Indeed, the application of isotope geochronology to biostratigraphic scales allows an estimate of the absolute age of zonal boundaries and of the duration of certain units for certain time intervals. The preservation of orbital cycles of known duration (i.e. Milankovitch cycles) in continuous deep-sea sedimentary successions provides a further tool to calculate the duration of biostratigraphic units by simply counting the number of cycles contained within an individual biozone. However, orbital tuning counting back from the recent is limited to Neogene and possibly Paleogene and Cretaceous strata.

2. – CONCEPTS, METHODS AND TERMS IN BIOSTRATIGRAPHY

Since its beginnings, by the middle of the 19th century, until the Sixties, the definition of the biostratigraphic units had constantly progressed in the recognition and precision following the procedures of the so-called traditional biostratigraphy: the deductive naturalistic method, based on the rigorous observation of facts and the application of simple principles, followed by an interpretation.

In the last three decades, some biostratigraphic procedures have started to base their interpretations on logical methods – logical biostratigraphy – or on mathematical methods – statistical biostratigraphy; the construction of the so-called "automatic zonations" still rests on the observation, but it is also sustained by more rigorous investigative techniques thanks to the advances in information technology.

Although the two approaches are obviously complementary, differences in the results obtained are sometimes relevant. It is, therefore, useful to describe them separately and to stress the weaknesses and the strengths of both.

The organization of this chapter, which favours the practice, is in itself an expression of the successive developments in biostratigraphy which, though maintaining the basic concepts and methods, has recently achieved progresses inconceivable only some decades ago thanks to the progress in technology.

The logical and statistical techniques, still little used, are different faces of the data processing; they do not replace the traditional methods but implement them. In the future, their

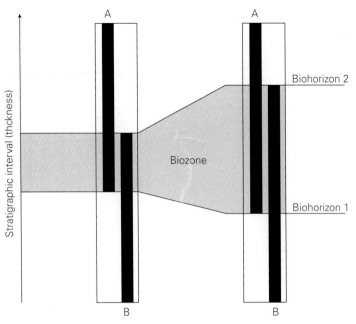

Fig. 5.2
Procedure of construction of a biozone (in this case concurrent range zone). Biohorizon 1 and Biohorizon 2 defined as the lowest occurrence of species A and the highest occurence of species B, respectively, subtend a biozone represented by the standard area. Note the lateral variability of biozone thickness.

rational use could bring to a better definition of the elements of the paleontological stratigraphy.

2.1. CLASSICAL BIOSTRATIGRAPHY AND BIOSTRATIGRAPHIC UNITS

2.1.1. *Preliminary remarks*

All the classical biostratigraphic units are represented by a body of rocks distinguished by a paleontological content which allows to distinguish it from adjacent rock bodies. The term biozone is used for all the different kinds of classical biostratigraphic units defined and identified on the base of the lateral and vertical distribution of their characteristic taxa (Fig. 5.2). As already mentioned (see paragraph 1.3.), index taxa should be chosen among species or subspecies characterized by a rapid evolution and a wide (paleo) geographic distribution.

This condition is necessary to obtain biostratigraphic scales allowing accurate subdivision of the stratigraphic successions and long distance correlations and depends on the fossil group considered.

The inappropriate use of terms as extension – distribution – range – duration – existence- life span, often rises to confusion between the duration of existence (or life span) and the observed **first occurrence** and **last occurence** of the index fossil(s). The two concepts are to be clearly distinguished since the former represents the invariable interval from the origination to the extinction of a taxon whereas the latter refers to the interval between the highest and lowest finds of the same taxon in a given stratigraphic succession and locality, which is obviously variable from site to site.

Indeed, although it is agreed that the biozone expresses only the idea of the paleontological content (presence/absence of one or more taxa assemblages, etc.), for many biostratigraphers it has a temporal meaning, even in the case of the partial range zone for which it is never certain to know the "total biozone" of a given taxon.

The nature of the classical biozones

Defined and recognized as bodies of strata, the classical biozones are closely related to the nature of the sediments, which makes their thickness and geographic extension largely variable. They can be closely related to or completely independent from the lithostratigraphic units of a given geological series within the same geographical **domain**. A biozone can range in thickness from a local thin bed to a unit of several thousands of meters recognizable over a vast geographical area and, theoretically, may have a global extension.

Therefore, the extension of a classical biozone is materially limited in space (marine or continental environments, biogeographic provinces, etc.) and time (first/last occurrence, coexistence/exclusion) by a series of factors controlling the distribution and the preservation of its characteristic fossil(s) across different depositional environment.

Biozones and expression of time

Due to the unavoidable discontinuities in the sedimentary processes and in the environmental conditions controlling the distribution of organisms, a biozone is very unlikely to have a global extension. Units based on the distribution of planktonic and nektonic organisms have better chances to have a wider extension within open marine settings at least at comparable latitudes. Being their geographic distribution less limited by environmental factors, in fact, taxa that live in the open water column, generally, have a wider geographical distribution. In contrast, benthic organisms tend to be less widespread, and are typically found only in particular environments. However, even based on pelagic organism, biohorizons (the biostratigraphic data bounding biostratigraphic units) are always diachronous because a species always originates at one particular time and place before reaching its maximum geographic dispersal. Similarly its final extinction will occur at a specific time and place corresponding to the last living specimen. For this reason and because biostratigraphic units exist only where their paleontological character can be observed, a biozone cannot be regarded as a time interval.

The improper use of the terms **biochron** (temporal entity) instead of biozone generates confusion about the nature of biostratigraphic units which remain stratal entities recognizable only in the presence of the characteristic fossil taxa. The concept of biochron and biochronology refer instead to "the organization of geologic time according to the irreversible process of evolution in the organic continuum" (BERGGREN & VAN COUVERING, 1978). Biochronostratigraphic units are "the sets of rock formed during biochrons, without reference to any particular stratigraphic section" (WALSH, 1998).

The confusion in the use of the terms biozone, biochron, biochronozone and chronozone explains – but certainly does not justify – why certain authors tend to give a universal (temporal) value to their biostratigraphic results; in this case they regard their biostratigraphic scales as continuous (contiguous or partially overlapping biostratigraphic units, without hiatus) and reflecting an uninterrupted time record, which is seldom or never the case, because of the essentially discontinuous character of the sedimentary processes.

Aiming at avoiding this confusion, it is here proposed that the concept of biozone maintains its stratal meaning and practical use, though apparently complex and ambiguous, to designate the basic unit of the classical biostratigraphy. This unit and its sub-categories represent the first step in the recognition of the temporal and geographical distribution of fossil taxa. Biostratigraphy, as any other naturalistic discipline, can progress only through a progressively wider spectrum of knowledge; the traditional biozone has a historical value that the progress of knowledge cannot completely cancel. It should be always kept in mind that the denomination and the identification of a classical biozone are based on the presence of one or more marker taxa in a body of strata representing a certain time interval. Thus, the classical biozone cannot be defined if its characteristic taxa are absent.

Similarly, the concept of chronozone maintains its original sense of "unit subdividing the stage"; it remains associated only to the concept of duration (not quantified) of a given time interval having a universal value, independently from the geographical distribution of the fossil (or fossil group) on which it is based.

As for the biochronozone, it is an "abstract unit" (by definition not represented by, though associated with, a body of rock) of the logical biostratigraphy. More related to time than to sediments, more objective than the traditional biozone, it is also more conceptual than the latter and tends towards the ideal zone which would be the chronozone.

2.1.2. Kinds of biostratigraphic units

Four major kinds of biostratigraphic units are commonly used (Fig. 5.3):

– the **assemblage biozone** is based on the presence of an assemblage of three or more taxa;

– the **range biozone** is based on the distribution of one or more organisms through time;

– the **abundance biozone** is based on the abundance of one fossil or fossil group;

– the **interval biozone** is based on the position of any (bio)event chosen in a biostratigraphic succession

Each of these categories can be named differently according to the different criteria of the various existing classifications. For instance, WHITTAKER et al. (1991) consider only the assemblage and the abundance biozones, grouping under the term "interval biozone" all the units based on the temporal distribution (duration of life or duration of existence) of the organisms; this simplification is not satisfactory because, from either a conceptual or a practical point of view, the definition of abundance and interval biozone – both considering evolutionary events – should form the base for **continuous biochronological scales**, whereas those based on exclusive assemblage allow the construction of **discontinuous biochronological scales**.

The assemblage biozone

The body of strata characterized by the coexistence (faunal zone and/or floral zone) of three or more fossil taxa which, taken together, distinguish it from adjacent strata, without considering the vertical distribution of the considered taxa. A biozone based on the association of only two taxa whose range overlap ("meeting zone" sensu SIGAL, 1984), can be considered as an intermediate term between the assemblage biozone and the distribution biozone, though classified in the latter category.

The term assemblage, here used following the Stratigraphic Guide of the International Commission on Stratigraphy, is often preferred to association, which can sometimes imply a causal relation (ecology, size, sorting, etc.) between the considered taxa. However, the term "assemblage" does not anticipate the reason, if any, for the grouping of the considered taxa which may derive from a fortuitous or causal relation. Indeed, assemblage may evoke a choice made by the biostratigrapher, which is not the case since biostratigraphy does not intervene on the "association" of taxa, which is rather the result of a "natural grouping".

The term cenozone, which is an alternative denomination for the assemblage biozone, derives from the Greek "koinos", i.e. common. Although having the advantage of deriving from a classical language, thus being well suitable to a translation in other languages, this term does not entirely express the meaning of an assemblage zone; for this reason, the term, "assemblage biozone" is preferable and the abandonment of the term cenozone is recommended.

The assemblage zone has been the first to be used in biostratigraphy; its meaning and definition has slightly changed from that time. An assemblage zone can be based on all the groups of fossil present, or restricted to certain.

The assemblage zones are sometimes associated to specific localities or regions, where the conditions favouring the development of their characteristic taxa occurred. Thus, they can be representative of these environments and, consequently, of a period of time characterized by such particular environments. This feature, giving them a marked paleoecologic character (**ecozone**), is one of the main obstacle to their use.

More generally, however, some assemblages, such those composed of marine planktonic organisms, can have a quasi-global distribution though in a restricted latitudinal interval and under conditions of slightly variable temperature; in this case they can also be representative of a time interval. Independently from any interpretation, the assemblage biozone is simply recognized by the coexistence of several taxa, which makes it a relatively objective unit of good value for local and regional correlations; out of the regional context where it has been defined, its correlation potential is generally low.

The boundaries correspond to surfaces marking the limits of occurrence of the fossil assemblage characteristic of

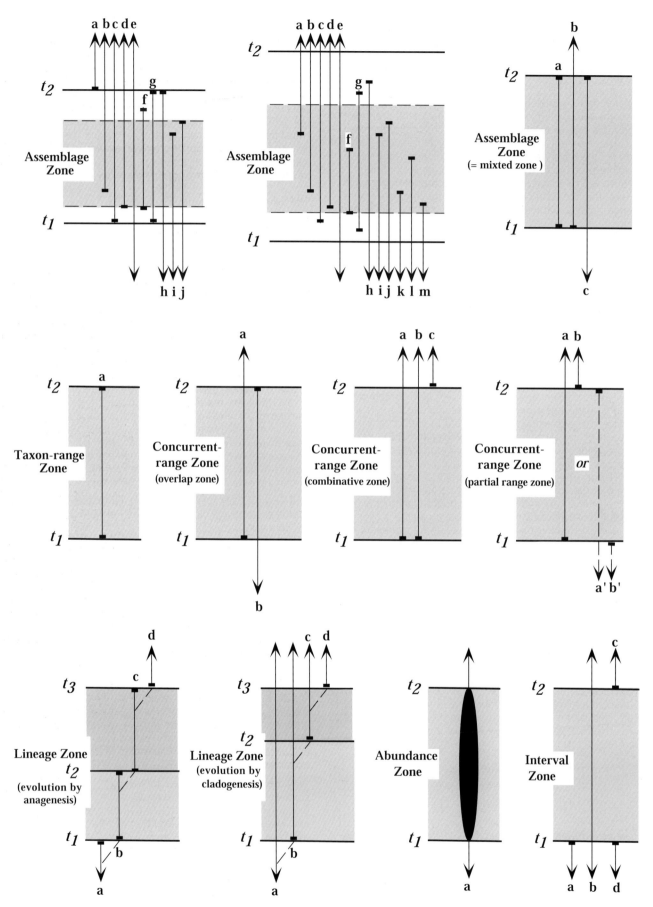

Fig. 5.3
Schematic representation of the kinds categories of biozones.
After Hedberg (1979), Salvador (1994), Sigal (1984) and Whittaker et al. (1991).

the unit. Not all the members of the specified assemblage have to occur in order for an interval to be assigned to an assemblage zone. Moreover, having each of the characteristic taxa a different vertical distribution, the range of some of the members may extend beyond the boundaries of the zone. The identification of the zone and its boundaries is, therefore, interpretative. Every time that an assemblage zone is established, its characteristic fossil assemblage and boundaries must be unequivocally described. However, if an assemblage zone is based on numerous taxa with a different stratigraphic distribution, the exact definition of its boundaries may be difficult.

An assemblage zone may be delimited by several events of first or last occurrences; the extreme case would be the "mixed zone" (SIGAL, 1984), a unit based on two categories of fossils selected on the base of their paleoecologic characteristics (planktic taxa within a set of benthic ones; spores and pollens in a marine series) or a zone characterized by one or more particular taxa (genus, family) intercalated in a series of zones based on the same fossil group.

The name of an assemblage zone is derived from the name of one taxa among the dominant and diagnostic constituents of the defining fossil assemblage. The suggestion by the "international Stratigraphic Guide" (HEDBERG, 1976) to derive the name from two taxa should be disregarded to avoid the possible confusion with the concurrent-range zone.

The range biozone

Several units based on different combinations of first/last occurrence of one or more taxa are included in this category. In its general meaning the range biozone is a body of strata corresponding to the known total distribution of an element or a set of elements within a fossil assemblage. The term distribution should be intended in a stratigraphic (vertical) sense. Indeed this criteria is most considered. In the practice, although, according to certain authors, the description of the range biozone takes sometimes into account also the geographical (lateral distribution) criteria. The term "occurrence biozone" has also been used.

The term "acrozone" – from the Greek "akros", the most elevated or extreme – has been proposed as synonym of "range zone" in order to use a word derived from a classic root because the English term "range" is not easily translated in all languages. However, the term "acrozone" should be avoided because it does not describe well this kind of biozone.

A range biozone can represent the stratigraphic range (interval of existence) of a certain taxon (species, genus, family, order, etc.) or group of taxa, or of any other paleontological element. Since a range zone is established, the data used for the definition of its limits must be clearly expressed. There exist two kinds of range zones: the "taxon-range zone" and the "concurrent-range zone".

To a certain extent, the Oppel zone, less rigorous on the criteria of vertical distribution of the considered taxa, and the lineage zone, based on a fragment of phyletic lineage of a taxon, could also be considered as range zones.

The terms "*teilzone*", "local range zone" or "topozones" have been used to indicate the local equivalents (i.e. based on the distribution of a taxon in a region or a particular locality) of the range biozone. However, the distribution in a local area is not meaningful unless the name of this area is given. The International Stratigraphic Guide (HEDBERG, 1976) suggests, to refer to the local distribution of a taxon as "range zone of the taxon A in the X section" or "in the Y well", or "in the Z region", without modification to the zonal term (SALVADOR, 1994).

Erroneously, "topozone" and "*teilzone*" are often associated to the concept of abundance zone. These units do not have any relation with the concept of abundance. They correspond, in fact, to partial zones referring to an association, often fortuitous, merely ecological, of taxa living in the same biotope; for this reason they have a limited chronological significance, more associated to the environment and relative limiting factors than time. They are locally, sometimes regionally, useful but their multiplication makes it complex the use of biostratigraphy on a regional or global scale.

Although the topozone and its synonyms remain very useful, these different units should no longer be used even temporarily for the risk of a definitive acceptance while waiting for a more complete documentation allowing to define one or more standard range zones.

It is recommended to use:

– (1) the "**total range zone** of a taxon"; based on the known total range of the stratigraphic and geographic occurrence of a particular taxon (species, genus, family, etc.). It represents the sum of the observed occurrences in all the individual sections and localities where the given taxon has been identified. It is sometimes termed "zone of total occurrence". Considering the "total range of a taxon" implies that a "partial range" of the same may exist. Therefore, not considering the criteria of coexistence and exclusion, its validity as a means of correlation is questionable, at least beyond a regional use where it can be very useful. Based on the entire vertical distribution of one or possibly more particular taxa, this unit differs from the concurrent-range zone.

Corresponding to the sedimentary unit containing the taxon on which it is based, the total range zone is often erroneously considered as an equivalent of the biofacies.

More than any other kind, the boundaries of a taxon-range zone can be constantly modified following new observations on the stratigraphic distribution and systematic revisions of the defining taxa. Moreover, due to the discontinuous character of the sedimentary processes and problems related to the preservation of organisms (e.g. dissolution, diagenesis or metamorphism) it is very likely that the taxon-range zone does not represent the actual distribution of the characteristic taxon. Finally, the vertical distribution of a taxon can vary as a function of different environmental factors (e.g., climate, water depth, sediments, etc.). The boundaries of a taxon-range zone are surfaces marking the outermost limits of known occurrence in every local section of the taxa whose range is to be represented by the zone. In any local section they correspond to the lowest and the highest stratigraphic occurrence of the specified taxon.

The taxon-range zone is named from the taxon whose range it expresses.

– (2) The "**concurrent-range zone**" for the body of strata including the overlapping parts of the range of at least two specified taxa selected among all the forms contained in a sedimentary succession from a particular locality. The other taxa can be regarded as characteristic elements of the zone, even marking the bottom or the top, but only two can be used to define its boundaries. Essentially, a biostratigraphic succession subdivided by means of successive concurrent-range zones presents "intervals without biostratigraphic zonation" or that do not overlap where the same layers are included in more than one zone.

In the definition of a concurrent-range zone it is important to select, among the taxa showing overlapping ranges, those having the most unambiguous chronological and geographical significance.

This kind of biozone considers an **interval of co-occurrence**; its potential of correlation is assured when respecting the criterion of exclusion of its characteristic taxa with respect to older or more recent taxa.

The lower boundary is defined by the lowest occurrence of the higher-ranging of the two defining taxa. The upper boundary is the highest stratigraphic occurrence of the lower-ranging of the two defining taxa.

The name derives from both the taxa that define and characterize the biozone by their concurrence.

The term concurrent-range zone is self explanatory. The term "overlap zone" (lower boundary coincident with the lowest occurrence of one taxon, upper boundary coincident with the highest occurrence of another taxon) and of "combinative zone" ("*zone combinatoire*" of SIGAL, 1984) may be used for certain kinds of biostratigraphic units when two or more than two taxa are considered. According to the latter author, a zone of partial distribution is characterized by the distribution of a taxon whose distribution does not overlap with that of two others taxa (i.e. the part of the range of the taxon above recognized disappearance of a given taxon, and below the known appearance of the other taxon); perfectly suitable for the definition of a zone of concomitant distribution, this type of biozone, has been sometimes regarded as a zone of interval.

The concurrent-range zone has also been termed with different names, such as concomitant-range zone, zone of coexistence, overlap zone or zone of overlap range by different authors.

Based on the occurrence of several characteristic taxa, the concept of concurrent-range zone had for a certain period led to the idea of biochronotype as an equivalent of the stratotype in the definition of chronostratigraphic units; this view seems to have been abandoned.

The concurrent-range zone is clearly distinct from the assemblage zone; in the former the vertical distribution of each taxon is taken into account independently whereas the latter considers only the coexistence or the general association of at least two fossil taxa. In other words, a concurrent-range zone can be recognized by observing the occurrence of only one taxon in a particular interval of a series provided that the vertical distribution of the other characteristic taxon is known in the underlying and/or overlying levels.

Remarks on the lineage biozone

Based on a segment of phyletic lineage (**chronospecies**), the **lineage biozone** – also called morphogenetic zone, evolutionary zone, phylozone, or phylogenetic zone) – can be defined as a particular kind of range zone. However, in agreement with the International Stratigraphic Guide, the lineage zone is discussed as a separate category because its definition and recognition require not only the identification of specific taxa but also that the taxa chosen for its definition represent successive segments of an evolutionary lineage. A sequence of range zones based on the successive evolutionary events of "taxa" on a lineage has a significance only if the succession of segments that defines it is recognizable on a large scale.

The lineage zone, which is sometimes called zone of consecutive existence, satisfies entirely the criterion of exclusion but not that of coexistence since, by definition, the recognized evolutionary stages do not overlap in time. On the other hand, if a transition stage is recognized, this can play the role of **interval of separation** and improve its temporal resolution.

Three kinds of lineage zone can be distinguished based on the evolutionary modes: the zone of phyletic derivation (evolutionary mode by **cladogenesis**), the zone of phyletic propagation (evolutionary mode by **anagenesis**) and the zone of "markers" (evolutionary mode by **punctuated equilibria**).

The phylozone, based on successions of chronological subspecies of a chrono-species (TINTANT, 1972 a, b), is among the best biostratigraphic units for a detailed and precise subdivision of time. However, its establishment requires a considerable work of fundamental paleontology (intra- and interspecific variability, systematics and taxonomy, evolutionary modes, paleoecology and paleobiogeography, etc.); besides that, the discovery of a number of chronological transients (small populations) sufficient to estimate the degree of evolution, is often aleatory.

Remarks on the "Oppel zone"

Named after the German biostratigrapher OPPEL, the "Oppel Zone", which has never been formally defined, has been considered either as a kind of assemblage zone or as a concurrent-range zone.

According to its author, there are two crucial points: the extension of the zone is independent from the petrographic nature of the strata; it is characterized by the continuous and exclusive occurrence of certain taxa of the same taxonomic level over a wide geographic area. This would result in an ideal succession of biozones having approximately the same age in all the regions where it is recognized. The "Oppel zone" has numerous features of the unitary associations and the biochronozones of the logical biostratigraphy.

The lower boundary of the Oppel zone is generally determined by the lowest occurrence of one diagnostic taxon and the upper boundary by the highest occurrence of another. Other characteristic taxa may be contained within the zone or extend to either side, although long-ranging and slowly evolving lineages are not usually included in the diagnostic assemblage. Not all the significant taxa are required to be present at all levels and in all places. The zone is named after one of the diagnostic species. Oppel zones are thus more flexible and subjective than concurrent range zones in the rigorously applied sense of that term, although the term "concurrent range zone" is commonly used for what is, in fact, an Oppel zone.

However, the Oppel zone does not correspond to any kind of biostratigraphic units currently used. For this reason and because since about thirty years it has progressively been abandoned, often to the profit of the interval zone, it is recommended not to use the Oppel zone (HEDBERG, 1976; SALVADOR, 1994).

The abundance biozone

It corresponds to the strata or the body of strata in which the abundance of a certain taxon or a specified group of taxa is significantly greater than that observed in the adjacent parts of the succession. The abundance zone has been given several different terms, the most used of which is acme zone.

Abundance is a relative and subjective idea which implies great number, frequency and richness; the term "bloom zone" has been used in cases when a taxon becomes suddenly and temporarily "more abundant", without flourishing. The acme is the maximal increase in abundance of a taxon, independently of its stratigraphic distribution. If the maximal abundance of a taxon or several taxa occurs in an interval always intercalated between other chronologically meaningful intervals, this can be used as a criterion for the definition of an abundance zone.

The boundaries of an interval of maximal abundance are not easily established and can be recognized only through a quantitative survey of the fossil assemblage. Moreover, their evaluation depends on the size of the survey sample.

Nevertheless, some authors consider the abundance zone as a valuable chronostratigraphic tool. The correlations based on pollen distribution charts are partly based on this notion. In this case, particularly for the Pleistocene, a classical biozonal scheme is only rarely developed and the fluctuations in relative abundance through time are used as for a "climatostratigraphic" interpretation.

Unusual abundances of one or more taxa may result from a number of processes of local extent (i.e. differential preservation and/or development of particularly favourable environmental conditions). Obviously, these conditions can be repeated in different places at different times. For this reason, a sound method to identify an abundance biozone is to trace it laterally. Moreover, the presence of each individual taxa must be evaluated in the light of the whole assemblage because the ephemeral development of unfavourable conditions may lead to a temporary decrease in its abundance or even disappearance. The abundance zone is thus only of local use.

The boundaries are defined by the level across which there is a notable change in the abundance of the specified taxon or taxa characterizing the zone.

The name is derived from the taxon or taxa whose significantly greater abundance it represents.

Generally, the correlation potential of an abundance biozone is limited by local factors even more than that of a total range zone. Its use is not recommended because it is based on rather vague and subjective notions. No evolutionary events are involved in its definition. Finally the abundance depend also on the relative abundance of other taxa within the same unit. Yet, this biozone preserves a weak chronological value in itself, because its biological content is linked to general changes in the evolutionary rate of the living world. Often, abundance zones are informally and temporarily defined during the geological mapping of a restricted region (i.e. at the scale of a sedimentary basin) for which they can be of great help.

The interval biozone

The body of fossiliferous strata between two specified bioevents of first or last occurrence of two different taxa. Such a zone (= "*zone intermédiaire*"; SIGAL, 1984) is not necessarily the range zone of a taxon or concurrent range zone of two taxa. It is defined and identified only on the basis of its bounding biohorizons independently from its fossiliferous content, which may be biostratigraphically insignificant (= "*zone blanche*"; SIGAL, 1984).

The interval zones are largely used for correlation, even though the rigorous determination of the species contained in it is sometimes difficult. The term "*zone intervallaire*" (SIGAL, 1984) should be refused for a possible confusion with the term "interval zone" previously used.

Barren intervals are frequent in the geological series and can occur either between two successive biozones (interzone) either within a biozone (intrazone). Consequently, these intervals should not be a subject of biostratigraphy and should not be named; however, they are informally identified and named by reference to the biozones that frame or contain them. Barren intervals have usually a local to regional extension, their occurrence depending on various local factors related to environmental parameters and preservational aspects.

The term interzone should be used with caution and only for truly barren intervals between perfectly characterized intervals. The experience suggests that sedimentary intervals referred to as azoic based on their macrofossil content can be very rich in microfossils. The preservation of organisms is largely influenced by their size and the chemistry of the test besides diagenetic processes following their fossilization.

The term intrazone should be used only in case of temporary disappearance of the index taxon or taxa within a definite zone. This non-fossiliferous interval whose importance is appreciable within a biozone can reflect a temporarily incomplete knowledge of their intervals of occurrence; in many cases, the preservational and/or environmental conditions leading to the temporary disappearance of the index taxa can be determined.

The boundaries of an interval zone are defined by the biohorizons selected for its definition.

The name may be derived from the names of the boundary biohorizons, with the name of the lower boundary preceding that of the upper boundary.

In the definition of an interval zone, it is desirable to specify the criteria for the selection of the bounding biohorizons, e.g. lowest occurrence, highest occurrence, etc. An alternative method of naming uses a single taxon name for the name of the zone. The taxon should be a usual component of the zone, although not necessarily confined to it. The choice of the name, therefore, can be completely independent from the definition of the zonal boundaries which can be based on the first/last occurrence of taxa different from that giving the name to the zone.

In subsurface stratigraphic work, where the section is penetrated from top to bottom and the paleontological identification is generally made from drill cuttings, often contaminated by recirculation of previously drilled sediments and material sloughed from the walls of the drill hole, interval zones defined as the stratigraphic sections comprised between the highest known occurrence (first occurrence downward) of two specified taxa are particularly useful. In this case, the term "retrozone" has been proposed by LIPSON-BENITAH (1992), but it is not used by oil company biostratigraphers.

2.2. THE LOGICAL BIOSTRATIGRAPHY OR UNITARY ASSOCIATIONS METHOD

Developed by the Swiss school (GUEX, 1977; 1979; 1987; 1991), it is known as the method of "**unitary associations**" or deterministic biostratigraphy. The units are defined by the occurrence of taxa recognized in a stratigraphic sequence, without a choice "*a priori*" of the taxa used to characterize a given interval. The presence of these taxa depends on the criteria of mutual exclusion (vertical constraint of superposition and distinction between older and younger taxa) and coexistence (lateral constraint of co-occurrence and relationship).

The unitary association method is based on the generation of a number of assemblage zones allowing a maximal stratigraphic resolution and a minimum number of superpositional contradictions.

Its application requires a software combining information on the coexistences and the chronological order of several taxa in a large number of geological sections, taking into account all the observed combinations, including the rare ones. This procedure is therefore fundamentally different from the so-called quantitative methods that, on the contrary, consider only the statistically most frequent cases.

Though this method leads to the construction of biochronozones, units defined by their authors as time intervals and, therefore, more related to chronostratigraphy, its place in the biostratigraphy chapter is justified by the use of fossil taxa assemblages.

2.2.1. Definitions

An unitary association is the maximal grouping of mutually compatible (contemporary) species that cannot be included into a larger grouping on one hand, corresponding to the maximal overlap of the **intervals of co-occurrence** of each of these species (smallest common taxonomic denominator), on the other hand.

One or more groups of unitary associations can be ordered in a sequence where each association is separated from the following and the previous ones by an **interval of separation**. Biochronologic scales based on unitary associations are, therefore, discontinuous (**discontinuous biochronologic scale**).

An unitary association or a sequence of unitary associations allows to define a **biochronozone**.

The search for taxa characteristic of the same zoological or botanical group within an unitary association grouping allows to determine its relative age through the biochronologic calibration; the term **integrated biochronology** is used when more taxonomic groups are considered.

2.2.2. The construction of biochronozones

A biochronozone is a time interval corresponding either to a unitary association whose "reproducibility" is demonstrated by comparing results from different sites, or to a continuous sequence of little reproducible successive unitary associations combined to increase its reproducibility. This points to consider as virtually associated the characteristic elements of the unitary associations that have been combined (SCHAAF, 1985).

The basic steps of the method are summarized in Fig. 5.4. In this figure stratigraphic sections with the local distribution of 7 species (labeled as 1 to 7) which may be present or absent in each individual section are taken into consideration. The observation of the coexistences of the various species under consideration allows to generate a species-species matrix which can be simplified by eliminating one or more species. The matrix can thereafter be rearranged by a permutation of its rows and columns to obtain various sets of mutually coexisting species. As a last step a series of maximal sets of intersecting species ranges are extracted from the rearranged matrix and represented in the Unitary Association range chart. This chart can be used to re-evaluate the ranges of the various species in the original biostratigraphic data set and assign relative ages to the stratigraphic intervals of the sections from which the latter was obtained.

Technically, a biochronozone is based on the reproducible subdivision of a **real reference system** or a **protoreferential**. It is characterized by the association of species/pairs of species characteristic of each of the unitary associations combined and by a set of new elements formed by the species/pairs of species deriving from the new combination. If this subdivision is recognizable over wide geographical areas it will be chronologically significant; if the temporal extent of one or more taxa is known, such a subdivision tends to an **ideal reference system** and a sequence of biochronozones constitutes a **zonal reference system**.

An unitary association marks a more or less extended time interval, separated from the preceding and the following ones in the time scale. Thus, unitary associations form a system of time intervals comparable to those defined by other tools such as magnetostratigraphy.

Hence, they are independent from the problem of the definition of precise boundaries often irresolvable due to the discontinuous nature of the sedimentary and fossil records. An unitary association is thus basically different from a traditional biozone since it is sometimes necessary to consider several regional unitary associations to obtain a "standard zone", particularly when the different areas studied have dissimilar faunal/floral assemblages.

According to LETHIERS (1987), a biochronozone is the part of a chronozone geographically delimited by vertical surfaces (temporal dimension) passing by points of maximal horizontal distribution (geographical dimension) of a given species. The identification of one or more taxa characteristic of a chronozone in a stratigraphic series allows to establish that the body of strata containing them falls in the relative biochronozone though not allowing to identify its boundaries. The complementary use of crosschecks and/or faunal associations ("cenozones based on biochronozones") increases the precision step by step; the maximal precision is reached when the vertical position of this unit will be no longer modified following the survey of new stratigraphic series.

2.3. "STATISTICAL BIOSTRATIGRAPHY"

We prefer the definition "statistical biostratigraphy" instead of "quantitative biostratigraphy" whose significance is too vague, since the techniques used "to quantify" the results derives essentially form the application of graphic, mathematical and statistical, probabilistic or deterministic techniques, univariate (comparison test), bivariate (linear regression and correlation coefficient) and multivariate (calculation of indices and distances, cluster analysis, principal components analysis, etc.). The statistical approach in biostratigraphy is particularly useful in the development of correlation methods.

Developed and refined in the last two decades thanks to the advances in information technology, these elaborate methods are used to obtain quantitative assessments of biostratigraphic units and time intervals corresponding to or separating them. They certainly allow to process quantitative data faster than the deductive reasoning of the biostratigrapher; however, if the mathematical treatment is objective in itself, the choice of the method is not, since it always depends on the biostratigrapher.

In fact, although the statistical methods allow to improve and refine correlations in specific cases, they rarely succeed in defining units and constructing biostratigraphic scales. They seem to work better when the index taxa and/or the specimens are abundant in the studied samples; for this reason they are better suited for micropaleontological rather than macropaleontologic studies.

The method of "grade-dating" (GOURINARD, 1983; 1984) is a tool of numerical dating which integrates isotope-derived ages and biometric statistical data from fossil populations (variability test) of a lineage characterized by gradual evolution (anagenesis). This method is not aimed at the definition of biostratigraphic units and scales. Being based on the evolutionary steps of successive populations (chrono-species recognition) the units obtained by this method are related to lineage biozone.

2.3.1. Graphical methods

Two methods have been developed during the 60s: the minimum distance (DE JEKHOWSKY, 1963) and the graphic correlation (SHAW, 1964).

Aimed at correlating stratigraphic logs, the method of the minimum distance take into account the differences between samples on the base of the abundance (frequency) of the

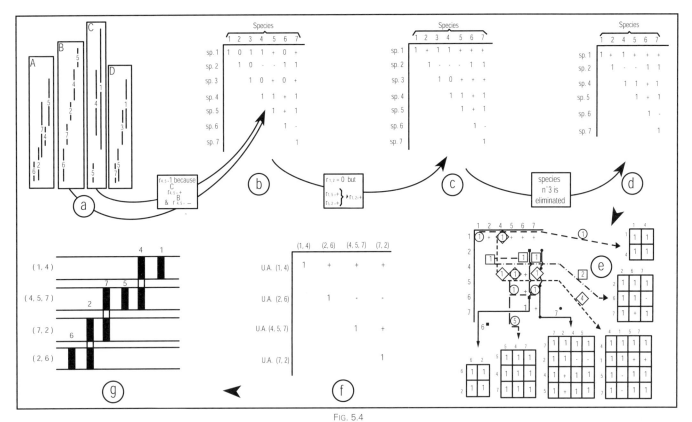

FIG. 5.4
Procedures for establishing unitary associations and biochronozones (after SCHAAF, 1985).

species recognized in each sample. Samples are considered synchronous when the statistical distance separating them is the lowest, independently from similarities or differences in their specific composition. Though based on criteria of abundance of species and individuals, this method does not allow to define units and biostratigraphic scales. It is rather controversial and rarely used; however, it allows to control the lateral changes within fossil assemblages by comparing contemporaneous stratigraphic sets from different depositional environments through time.

The method of graphic correlation (SHAW, 1964) applies a linear regression analysis to the vertical distribution of taxa (from the first appearance datum to the last appearance datum). The method is designed to separate biostratigraphically significant events of first and last occurrence from those with ecological (or other) significance, by comparing the stratigraphic order of bioevents at more than one site. The biostratigraphically significant events are then used for temporal analysis. It is one of the methods most widely used by oil company biostratigraphers (NEAL et al., 1994). As it considers not only the paleontological data, but all the available stratigraphic data and it aims at defining a reference (stratotypic) succession, this method will be discussed in chapter VIII.

2.3.2. Semi-empirical methods

This procedure, which adopts the principles of graphic correlations (SHAW, 1964), does not imply a statistical analysis; the evaluations are done by the biostratigrapher on an empiric base (best estimate on the diagrams built through the investigations) and affect the final result. Two methods have been proposed: the no-space graphs (EDWARDS, 1979) and the supplemented graphic correlation (EDWARDS, 1989).

These methods are prevalently used for well correlation during oil exploration in order to determine the maximal distribution of taxa in a given region by cumulating graphically their local distribution. The appearances/disappearances of taxa are regarded as synchronous and used as elements of correlation. Within wells, the depth of appearances/disappearances are not considered as absolute values; they serve only to establish the relative vertical relationships of the events observed.

The local thickness of intervals corresponding to the vertical distribution of each taxon in the resulting charts is not considered; only the relative position of the events is considered. Based on the succession of events of appearance and disappearance recognized in the surveyed sections an "hypothetical sequence" is constructed; this "sequence" is reviewed in the light of newly acquired data until it seems to be most in agreement with the starting assumption. No mathematical procedures are involved; these diagrams are therefore only a simple visualization facilitating the recognition of the vertical distribution of each taxa and designed to obtain total or concurrent "range zones".

Supplementary information such as chemostratigraphic, magnetostratigraphic data or geochronologic, are used besides the biostratigraphic data in the supplemented graphic correlation methods.

These methods are not aimed at defining formal units and biostratigraphic scales including a succession of "biozones". However they use the concept of range zone. All in all, they must be used with caution. Actually, if the initial correlation line between several points of observation (oil wells) is erroneously positioned, the resulting error will replicate twisting out the established chronological model beyond any possibility of control.

2.3.3. *Multivariate analysis*

This method is based on different mathematical techniques such as the calculation of indices, cluster analysis, principal component analysis, and factor analysis. The multivariate analysis is used to evidence similarities (or differences) in the specific composition of assemblages from different samples and sites; based on the application of rigorous investigations of quantitative data, its goal is to recognize elements (part of the analysed assemblage) that can be assimilated to "assemblage zones" by the user. The graphical representation used – dendrogram and projection on factorial axes – evidences groups of samples characterized by conforming compositions and behaviours in terms of presence/absence or abundance of each species, position of appearances or disappearances in the stratigraphic column, etc. with respect to a set of variables having a statistically meaningful weight. These methods, which are concerned with the combination of qualitative and quantitative paleontological data from a stratigraphic succession, also allow to associate some available physical data (complementary variables) such as lithology, geochemistry, magnetostratigraphy, geochronology, etc.

Poorly used in biostratigraphy, these methods are largely utilized in paleoecology and in the characterization of paleoenvironments (CUGNY, 1988), since they allow to evaluate data of different nature (e.g. biological and physical variables) from the same sample.

In terms of chronology, the interpretation of similarity and grouping, or differences and distance of the samples on the factorial axes or on a dendrogram, must be done with caution since the outcome have obviously not only a biochronologic meaning. This method is close to the **ecostratigraphy** an the units recognized to **ecozones** whose univocal chronological value is very questionable out of the limited area where they have been defined.

Although not bringing to the definition of units and biostratigraphic scales, the multivariate methods can be regarded as tools of correlation.

The cluster analysis evidences more or less homogeneous assemblages within different samples based on their affinity (similarity), either from the calculation of coefficients, or from the calculation of distances separating them. The "degree of similarity" between identified groups is expressed by classical indices such as the "Jaccard", "Dice", "Otsuka" and "Simpson", which were initially defined for ecological studies. The Jaccard index evidences differences between identified groups, whereas the Simpson index expresses similarities. Those of Dice and Otsuka are intermediate. The relationships of presence/absence of a species between two samples should not affect the calculation of their similarity, which may depend on many factors (sampling, preservation, environmental exclusion) other than time. Still, since these coefficients consider all the taxa occurring in the studied assemblages, each of them contributes to the calculation of similarity independently from their "relative biochronologic value". To improve the biochronologic use of these indices, several authors (Mc CAMMON, 1970; BROWER, 1984, 1985; BROWER *et al.*, 1978) have proposed indices that allow to weigh the influence of facies on the presence of taxa therefore allowing a better evaluation of the chronological value of the grouping obtained. The "lateral tracing" (MILLENDORF & HEFFNER, 1978; MILLENDORF & MILLENDORF, 1982) is a comparable, though less mathematical, method which consists in the evaluation of the correlation between samples from pairs of adjacent geological sections (spatial continuity) and between adjacent samples from the same section (vertical discontinuity). Finally, the distance between samples is used in the method of "ascending hierarchical classification" aimed at the construction of dendrograms (hierarchical trees, clustering samples and groups of samples). Here the choices of the user influence the results because it is necessary to take a series of decision that supposes the existence of groupings so that objects are well classified according to objectives to reach while respecting the starting data.

The "Principal Component Analysis" (PCA) is an essentially descriptive method that allows to delineate the interrelationships among a set of samples or among species within one or more samples by projecting them on two factorial axes. Though based on mathematical procedure, the PCA can be compared to the procedure used for the definition of assemblage zones in the classical biostratigraphy. The variables correspond to the sampling sites (surveyed stratigraphic successions) whereas the data are represented by the identified biostratigraphic events (first/last occurrences and relative abundance of taxa).

Multivariate analysis of biostratigraphic data is basically aimed at the identification of the most objective correlation between several sections, without definition of biostratigraphic units. The component with the maximum weight distributes the data on the first factorial axis which represents a "composite section" (HOHN, 1978, 1982) comparable to the "composite standard reference section" of the graphical methods (SHAW, 1964); the composite section obtained with the multivariate analysis is, however, more objective since it results from the comparison of all the sections considered and it is not constructed by successive comparisons with a reference section chosen *a priori*.

2.3.4. *Probabilistic method*

Based on the principle that the higher number of observations on a single taxon from numerous geological sections the better the knowledge on its vertical distribution, the probabilistic method aims at defining an ordered sequence of first/last occurrences, the so-called "optimal sequence". The events observed are ranked in the most probable order and the distance separating them is calculated on a relative time scale (inter-event statistical distance and gradual optimal sequence). The lowest and highest occurrences are not taken into account. Only the midpoint, which represents the level of most probable occurrence, is evaluated and verified in terms of confidence intervals and significance test. This technique is aimed at constructing biostratigraphic scales including zones of most likely vertical extension of single species.

The concept of probability is, therefore, involved at two different levels: in the definition of mathematical models allowing to describe the optimal sequence (binomial, trinomial and normal laws) and in the evaluation of the confidence level of the model (statistical test).

Complex computations are carried out to define an optimal sequence and a matrix representation as a support to the biostratigraphic information. There are two kinds of procedure:

– the "ranking procedure" (HAY, 1972; BLANK, 1979; BLANK & ELLIS, 1982) determines the most likely sequence of first and last appearances of taxa by estimating the probability of the stratigraphic relationships for pairs of events. A statistical test is run to check if the stratigraphic relation between each pair of events is accidental or not;

– the "ranking and scaling method" (AGTERBERG & NEL, 1982 a, b; GRADSTEIN & AGTERBERG, 1999) is aimed at the evaluation of the average distances between consecutive events; in practice, it gives additional insights to the ranking procedure and is supported by statistical tests. Unlike graphic correlation, the ranking and scaling ('RASC' for short) method takes into account the stratigraphic order of all (pairs of) fossil events in all wells simultaneously, and calculates the most likely sequence of events. In the optimum sequence obtained, the position of each event represents an average of all individual positions observed in the wells. Scaling of the optimum sequence in relative time is a function of the frequency with which events in each pair in the optimum sequence cross-over their relative positions (observed records) from well to well; the more often any two events cross-over from well to well, the smaller their interfossil distance. Final distance estimates are expressed in dendrogram format, where tightness of clustering is a measure of nearness of events along a stratigraphic scale. The scaled version of the optimum sequence features time successive clusters, each of which bundles distinctive events. Individual bundles of events are assigned zonal status. The process of zone assignment in the scaled optimum sequence is somewhat subjective, as guided by the stratigraphic experience of the users. Large interfossil distances between successive dendrogram clusters agree with zonal boundaries, reflecting breaks in the fossil record due average grouping of event extinctions. Such extinctions occur for a variety of reasons, and may reflect sequence boundaries. From a practical point of view, it suffices to say that taxa in a zone on average top close together in relative time.

Theoretically, the probabilistic model of scaling allows to integrate data on non biologic events which can be regarded as synchronous "marker levels" (body of rocks recording a certain event: magnetic anomaly, volcaniclastic beds, etc.).

2.4. HIERARCHY AND SUB-CATEGORIES OF BIOSTRATIGRAPHIC UNITS

Theoretically, a stratigraphic interval can be indifferently subdivided into range zones, interval zones, assemblage zones, abundance zones, unitary associations, biochronozones, etc. based on a single fossil group or different groups; similarly to the units of the logical biostratigraphy, the various traditional biozones and their subdivisions (sub-biozones) do not have a hierarchical value, and do not exclude each other.

In classical biostratigraphy, depending on the paleontological groups and the kind of biozone, it is possible to have an overlap between biozones, and between biozones and stage boundaries.

In the logical biostratigraphy, a biochronozone can be represented by one unitary association or it can group several of them. Theoretically, the chronostratigraphic nature of biochronozones should lead to the exact coincidence between stage boundaries and the intervals of separation between two successive biochronozones; in practice, the examples are still too few to form a basis for discussion.

This not a matter of debate in statistical biostratigraphy whose methods are not intended to define standard biostratigraphic units; it is never considered, *a fortiori*, neither to subdivide the obtained units, nor to group them in higher rank units, though their authors term them "total range zone" (statistical or semi-empirical method), "concurrent range zone" (semi-empirical method), " probable range zone" (probabilistic method), and "assemblage zone" (multivariate method).

2.4.1. *Biohorizon, zonule and marker-bed*

Biohorizon

According to the International Stratigraphic Guide (HEDBERG, 1976; SALVADOR, 1994), a **biohorizon** is a boundary, a surface or an interface in correspondence of which a remarkable change in the biostratigraphic data occurs. It can correspond to the surface delimiting two biozones or occur within a biozone. The criterion (**biomarkers**) on the base of which the biohorizon is identified within a stratigraphic succession are generally the lowest occurrence, the highest occurrence, the changes in the abundance, or a notable morphological change of a given taxon. The **First Appearance Datum** ("FAD") biomarker and the **Last Appearance Datum** ("LAD") biomarker are frequently used biohorizons; these biomarkers have, therefore, the status of (bio)event. According to BERGGREN & VAN COUVERING (1978) FAD and LAD, are "features in the paleontologic record which mark the most widespread, easily identified, and rapidly propagated events". This concept was first developed and named "datum event" by BANDY (1963, 1964) for biostratigraphic markers in planktonic foraminifera identified in marine deposits in southwestern France and the Philippine Islands. BANDY characterized these datum events as significant biostratigraphic markers with widespread (and assumed near-contemporaneous) distribution in deep-water deposits. The concept of FAD and LAD implies, therefore, a choice, among the many observed biostratigraphic data, of those shown to be regionally (or globally) synchronous.

In this meaning of "time-surfaces" or "interfaces", biohorizons have been called in different ways including surfaces, horizons, markers, datums, reference marks, reference layers, key horizons, etc.

Commonly used by macropaleontologists and, particularly, ammonite specialists, the term " biohorizon" has been used to designate a thin stratigraphic unit characterized by a distinctive fossil association having a high potential of correlation, within which no finer subdivision is possible and whose base and top must be clearly defined in a reference section (GABILLY, 1976). However, this definition is different from that given by the International Stratigraphic Guide, which defines the biostratigraphic horizon or biohorizon as "a stratigraphic boundary, surface, or interface across which there is a significant change in biostratigraphic character. A biohorizon has no thickness and should not be used to describe very thin stratigraphic units that are especially distinctive". Therefore, a discrepancy exists between the two definitions. For French biostratigraphers, a biohorizon, or more currently "horizon", corresponds to the subdivision without hiatus of a subzone (sequence of faunal association) whereas according to the International Stratigraphic Guide, a "faunal horizon" is a sequence of short-lived and discontinuous faunal events within a subzone (CALLOMON, 1984a, 1994; PAGE, 1995).

Zonule

Created by FENTON & FENTON (1928) and used by HEDBERG (1976) to replace the term biohorizon, the term **zonule** is generally used to subdivide a biozone or a subzone (SALVADOR, 1994; PAGE, 1995). However, this term has received different meanings and its use is discouraged by the International Stratigraphic Guide.

In its original meaning, the zonule is defined by the presence of a fossil assemblage in a stratigraphic succession (chronological distribution) in an area of occurrence (geographical dispersal); being repetitive through time under favourable conditions, this assemblage (faunal or floral) corresponds, therefore, to an ecological unit.

Following a different interpretation, some specialists of Jurassic ammonites have attempted to use the zonule as the finest possible subdivision defined by a reference stratotype within an organized succession of chronostratigraphic units (PHELPS, 1985). This definition implies subtle, not easily appreciable, differences between biohorizon as used by some biostratigraphers and zonule (PAGE, 1995).

Marker-bed

Relatively thin stratigraphical intervals which can be easily distinguished from adjacent ones on the base of the abundance or peculiarity of their fossil content are named "marker bed". Characterized by a species or a species assemblage, they are regarded as isochronous units within a basin or part of a sedimentary basin, notably in continental environment. The procedure based on marker-beds seems more effective than the biozonal subdivision as a tool for correlation within a restricted geographic area. It does not make it possible to date a sedimentary sequence, but rather to characterize the observed events from a biostratigraphic point of view. Condensation may, however, adversely affect this procedure by including extended intervals in a single marker-bed.

In spite of this problem and the absence of an international agreement, one can speak of "event biostratigraphy", as shown by many examples in marine and continental settings.

Brachiopod marker-beds, coinciding with maximum flooding events on the Jurassic carbonate platforms of the Paris basin (GARCIA & LAURIN, 1996), provide an accessory biostratigraphic tool for intervals without ammonites. The events of first/last occurrence of foraminiferal taxa are associated to eustatic events and used as reference mark to delimit "biozones", in the Cretaceous (HART, 1993).

The mammal biochronologic scales in the continental Paleogene (VIANNEY-LIAUD et al., 1993) are based on the identification of marker-beds. Using phylogenetic assumptions, the polarity of the morphological modifications (teeth, elements of the skeleton) observed in the lineages are defined, in order to identify intermediate evolutionary steps (chronospecies). The convergence of polarity between different lineages confirms the identified successions and leads to the establishment of a biochronologic scale. The latter is consolidated by the addition of the traditional events of appearance/disappearance and the criteria of stratigraphic superposition.

Effective, experimental and perfectible through the contribution of new data filling the observational gaps (basically discontinuous character of the sedimentary processes in continental environments), the calibration of these scales is however difficult; this possibility is provided, however, by the integration with magnetostratigraphic data.

2.4.2. *Subbiozones and superbiozones*

The term subzone is more often used than subbiozone. Most subzones are defined using the same taxonomic group (genus, sub-genus, family) on which the biozone containing them are based. Though not obligatory, this approach ("homo-biochronology") is strongly recommended to obtain a continuity or a contiguity of subzones. The use of different groups is, in fact, likely to produce overlaps between each defined unit.

A biozone can be entirely or only partially subdivided into subzones. The superzone is a grouping of more biozones. Contrarily to the subzones, which are routinely employed in biostratigraphy, the grouping of more biozones into a superzone is rarely used.

2.4.3. *Unitary associations and biochronozones*

A single **unitary association** or a group of them forms a biochronozone; in the latter case the unitary associations are sub-units within such a biochronozone and give the possibility of a more precise subdivision of time. The unitary association, which represents the finest subdivision of the logical biostratigraphy, can be, in certain cases, assimilated either to a subzone or to a biozone of the classical biostratigraphy.

3. – PRACTICE

3.1. THE STEPS OF THE PROCEDURE OF BIOSTRATIGRAPHY

As any discipline of stratigraphy, biostratigraphy proceeds by successive procedural steps. Whatever the method used, the following steps are followed:

– a local analytical and static phase aimed at the definition of the most precise and reliable markers;

– a regional and dynamic analytical stage, calling on the principle of correlation and aimed at the establishment of a biostratigraphic succession (biozonation or zonal scale) ranked and defined on the base of the vertical evolution of the fossil content within biostratigraphic units;

– an interpretative step, when the regional scheme obtained is framed into a geographically wider context aiming at establishing a firm regional biostratigraphy (Tab. 5.1), comparable with the global biostratigraphy and eventually leading to the elaboration of a global chronostratigraphy.

The first two stages call on the basic procedures and operative techniques of biostratigraphy (identification of first/last occurrences, concomitances, abundance; selection of index taxa; denomination of recognized units; etc.). Each of these stages is intended to the classification, subdivision and organization of the biostratigraphic successions into units named after their fossil content.

The classical biostratigraphy and the logical biostratigraphy are very valuable for this purposes, though both showing advantages and disadvantages. The statistical biostratigraphy, instead, is not very effective in the construction of biozones except for some rare examples. It is however, a useful supplementary tool of correlation allowing to quantify similarities and differences between assemblages from different sites especially in the presence of many variables (taxa) and locations.

3.2. RESULTS OF THE CLASSICAL BIOSTRATIGRAPHY

Used since the beginnings of stratigraphy, then refined little by little by a permanent contribution of new data, the classical biozonations, based on numerous fossil groups, have been developed for the entire Phanerozoic. It is beyond the scope of this chapter to discuss the various biozonal schemes used for the different periods and based on different fossil groups. Yet we can try to recall the most common tendencies in the practical use of biostratigraphy, such as the most used kinds of unit, the advantages and disadvantages of using the ones or the others. In different circumstances the use of each kind of biostratigraphic units has different limitations related to the possible diachronism of bioevents,

5. Biostratigraphy

TABLE 5.1

Example of an ammonite-based zonation (biozones, subbiozones and biohorizons) for the Callovien stage (Middle Jurassic) of western Europe; the existence of two biogeographic provinces requires the establishment of two comparable scales (J. THIERRY et al., in Collectif Groupe Français d'étude du Jurassique, 1991).

CALLOVIAN		SUB-BOREAL PROVINCE (CALLOMON, 1955, 1964; MARCHAND, 1979; CALLOMON & SYKES, 1980; CALLOMON, 1984, 1985; CALLOMON et al., 1988, 1989, 1992; PAGE, 1988, 1989; CALLOMON & DIETL, 1990; DIETL, 1991, 1993)						SUBMEDITERRANEAN PROVINCE (CARIOU, 1969; CARIOU et al., 1971 a & b; CARIOU, 1971 a & b, 1980, 1984; CARIOU et al., 1988 a & b, 1990; THIERRY et al., 1991)					
	UPPER	LAMBERTI	Lamberti		Paucicostatum			Paucicostatum	Lamberti	LAMBERTI	UPPER		
					Lamberti	Lamberti		Lamberti					
					Praelamberti	Distractum		Praelamberti					
			Henrici		Henrici	Flexispinatum		Athletoides	Poculum				
						Entospina		Subtense / Carioui					
					Messieani	Megaloglobulus		Nodulosum (Angustilobata)					
		"ATHLETA"	Spinosum		Spinosum	Punctulatum		Collotiformis	Collotiformis	ATHLETA			
						Prorsosinuatum		Piveteaui (Odysseus)					
						Fraasi							
			Proniae		Proniae	"Evexa" II		Trezeense / Athleta	Trezeense				
						"Evexa" I		Leckenbyi					
						Complanatoides							
			Phaeinum		Phaeinum (Interpositum)	Berkhemeri		"Pseudopeltoceras"	Rota				
						Aculeatum/ Balticum		Rota / Regulare					
	MIDDLE	"CORONATUM"	Grossouvrei		Grossouvrei	Doliforme		Waageni	Leuthardti	CORONATUM	MIDDLE		
					Obductum Posterior			Leuthardti					
			Obductum		Obductum	Coronatum		Baylei	Baylei				
								Villanyensis					
		JASON	Jason		Jason b	Jason b		Richei	Tyranniformis	ANCEPS			
					Jason a	Jason a		Blyensis					
			Medea		Medea	?		Turgidum	Stuebeli				
						Medea		Bannense					
	LOWER	CALLO-VIENSE	Enodatum (Planicerclus)	Calloviense	Enodatum c	Enodatum c	Calloviense	Posterius (Kiliani)	Patina (Oxyptyca)	GRACILIS	LOWER		
					Enodatum b	Enodatum b		Pamprouxensis					
					Enodatum a	?		Boginense (Oxyptica/Proximum)					
			Calloviense		Micans			Michalskii	Michalskii (Ardescicum)				
					Calloviense	Calloviense							
			Galilaei		Galilaeii	?							
		KOENIGI	Curtilobus		Trichophorus	Bullatocephalus Subcostarius		Laugieri	Laugieri				
					Tolype					Voultensis			
					Curtilobus	?		Tyranna / Pictava	Pictava				
			Gowerianus		Gowerianus	Macrocephalus		Grossouvrei	Grossouvrei				
					Metorchus	Megalocephalus b Megalocephalus a		Prahecquense	Prahecquense				
					?	Toricelli							
			Kamptus		Kamptus c	Kamptus		Moorei					
					Kamptus b								
					Kamptus a / Herveyi	?							
		HERVEYI	Terebratus		Terebratus b	Terebratus		Leptus	Bullatus	BULLATUS ("MACROCEPHALUS")			
					Terebratus a	Suevicum b		Furculus					
			Keppleri		Verus	Suevicum a Quenstedti							
					Keppleri (jacquot 1)			Demariae					

the resolution power and precision, the potential of correlation depending on the characters of the studied fossils.

3.2.1. Use of the classical biozones

The assemblage biozone

Implicit in the concept of assemblage biozone, the criteria of coexistence and exclusion give to this kind of biostratigraphic unit a correlation potential which favours its extensive use in spite of the clear relationship between faunal/floral assemblages and environment. Examples are reported by several authors (Archaeocyaths: DEBRENNE et al., 1990; DEBRENNE & ZHURAVLEV, 1992 – Graptolites: LEGRAND, 1996 – Chitinozoa: PARIS et al., 1995 – Benthic foraminifera: MOULLADE, 1974, 1984 – Ostracods: LETHIERS, 1984, 1993; etc.). Yet, this inconveniency is limited by the use of pelagic organisms whose assemblage zone can usually be traced and correlated confidently at least within the same biogeographic domain (e.g. the Paleozoic north Gondwana or the Mesozoic Tethys and Boreal domains). Similar bioprovinces exist also in respect to benthic organisms distribution (e.g. the Cambrian North American or Perigondwana Trilobite provinces).

Some biostratigraphers tend to avoid the use of the assemblage zones since they are likely to hide the presence of hiatuses by a process of globalization (natural continuous zonation); on the other hand, because of the imprecision in the definition of the lower and upper boundaries of these units they could be diachronous.

Once faunal assemblage data are grouped to establish a zonal scheme, the ecological and biogeographic factors controlling species distribution make the stratigraphic range of taxa often trespassing within adjacent zones; in case of barren intervals, the extension of the zone is considered to continue until the successive fossiliferous level. For these reasons a zonal scheme based on successive assemblage zones is constantly updated in the light of new data on the distribution of the characteristic taxa.

The concurrent-range biozone

Based on the interval of coexistence in the vertical distribution of taxa, the good correlation potential of the concurrent-range zone is reinforced if the criterion of exclusion of its characteristic taxa, in relation to other older and younger taxa, is achieved. Being this criterion respected well enough in the practice, this kind of biozone is widely use and examples exist for many fossil groups (Conodonts: PERRET & WEYANT, 1994 (Tab. 5.2) – Graptolites: LEGRAND, 1996, COOPER, 1999- Ostracods: LETHIERS, 1984 (Tab. 5.3); 1993 – Brachiopods: ALMERAS et al., 1990; etc.).

TABLE 5.2
Example of a conodont biozonation of the Carboniferous of the Pyrenees (from Perret & Weyant, 1994).

		PERRET 1989	STOPPEL 1976 in CRILAT 1981	BUCHROITHNER 1979	BOYER, KRYLATOV & STOPPEL 1974	BOERSMA 1983	MARKS & WENSINK 1970	WIRTH 1967	PERRET 1977 1985 & 1988	
		I. delicatus								PENNSYLVANIAN
	Bachkir.	Id. sulcanus parvus						présente ?	Id. sulcatus parvus	
		Id. sinuatus		noduliferus		Gn. macer	Gn. macer		Id. sinuatus	
	Serpouk.	D. noduliferus	?		?			?	Id. noduliferus	
VISEAN		Pa. nodosus-		Gn. bil. bollandensis		Gn. commutatus - nodosus	Gn. commutatus nodosus		Gn. bil. bollandensis Gn. bil. bilineatus Gn. nodosus	MISSISSIPIAN
		Gn. bilineatus		Gn. commutatus	Gn. bilineatus bilineatus ?			présente		
		Gn. bilineatus - Pa. commutatus	Gn. bilineatus bilineatus ?	Gn. commutatus commutatus		Gn. commutatus	Gn. commutatus commutatus	présente ?	Gn. commutatus - Gn. bil. bilineatus	
		Pseudognathodus homopunctatus	interregnum Sc. anchoralis - Gn. bilineatus	Gn. typicus	interregnum Sc. anchoralis - Gn. bilineatus	Gn. typicus	Gn. typicus	anchoralis - bilineatus interregnum	Gn. symmutatus homopunctatus	
TOURNAISIAN		Sc. anchoralis		Sc. anchoralis	Sc. anchoralis	Sc. anchoralis	Sc. anchoralis	Sc. anchoralis ?	Sc. anchoralis	
		Gn. punctatus - Siphonodella	Si. crenulata	Si. crenulata ?	Si. crenulata	crenulata		crenulata ?	Do. bouckaerti Gn. semiglaber P. c. carina	
		Si. cooperi	Ps. tr. triangulus Ps. tr. inaequalis	Ps. tr. triangulus Ps. tr. inaequalis	Siphonodella et Ps. tr. triangulus	Si. tr. triangulus			Siphonodella - Ps. tr. triangulus	
		Si. duplicata			Si. duplicata Ps. tr. inaequalis	Si. triangulus inaequalis				
		Si. sulcata - Pr. kockeli	Si. sulcata	Pr. kockeli Ps. dentilineatus	Si. sulcata et Pr. kockeli	Pr. kockeli -	?	?	Si. sulcata - Pr. kockeli	
	Fam.	Protognathodus - Si. praesulcata	Protognathodus	Protognathodus	Faune à Protognathodus	Si sulcata			Protognathodus - Si. praesulcata	DEVONIAN

A biostratigraphic scheme based on successive concurrent-range zone is, by definition, discontinuous. Influencing the occurrence of organisms, the palaeoenvironmental conditions may largely change the timing of biohorizons (first/last occurrences) from place to place.

It is therefore useful to distinguish the "true" interval of existence, from the "observed" interval of distribution of species in a given locality, the later being only part of the former. The assessment of the "real" distribution interval of the index-taxa in a considered domain can be attempted by calibrating the observed events of first and last appearance to a scale based on independent data (e.g. magnetostratigraphy).

The total range biozone

Not including the concepts of coexistence and mutual exclusion the definition of a total range zone is likely or have a low correlation potential beyond a restricted area in view of the mode of life (benthic or planktonic) of the fossil group used (Foraminifera: Pélissié et al., 1984 – Chitinozoa: Paris, 1990). The total range biozone of a taxon, in fact, merely corresponds to the stratigraphic intervals and areas favourable to its preservation, which are very unlikely to include always its entire "interval of existence" when different localities are considered.

Acquisition of a large number of observations at different sites and crosscheck with the distribution of other taxa, however, may allow a good control of the real extension of this kind of biostratigraphic unit.

The lineage biozone

The lineage biozone possesses a great potential of correlation and precise relative dating. However, the establishment of this kind of biozone may prove to be difficult because it requires an accurate assessment based on morphological studies of large populations of the evolutionary relationships within a phyletic lineage. The massive preliminary work of fundamental paleontology may sometimes discourage its usage (Ammonites: Tintant, 1963; 1972a, b).

The abundance biozone

This kind of unit is easily recognized in the stratigraphic successions. However due to its obvious dependence on local conditions and the resulting low correlation potential, the abundance biozone is often used only at a preliminary stage in stratigraphy.

The interval biozone

The interval biozone is delimited by two bioevents of two different taxa. They can be two events of first or last occurrence or of last occurrence and first occurrence at the lower and upper boundary, respectively. In the case the lower boundary is a first occurrence bioevent and the upper boundary is a last occurrence bioevent, they will define a concurrent-range zone. When the lower and upper boundaries consider only one taxon, they obviously define a total range zone.

5. Biostratigraphy

TABLE 5.3

Example of an ostracod-based biozonation of the upper Devonian of the Boulonnais region (LETHIERS, 1984).

CHRONO-STRATIGRAPHY				Cephalopods Zones (1)	Marker beds Namur, 1974 (M.g.m.) (2)	Conodonts Zones (3)	Ostracods Entomozoids Zones (4)	Spores Zones (5)	Foraminifers Zones (6)	Index (Dinant synclin.) (7)	Ostracods Zones (8)
CARBONIFEROUS (pars)	DINANTIAN (pars)	TOURNAISIAN (pars)	HASTARIAN (pars)	Cu II (pars) / Cu I — Pericyclus / Gattendorfia	52 - 57	Siphonodella sulcata z. ZIEGLER, 1969 and Siphonodella plus volues	R. (R.) latior zone RABIEN, 1960	TE	D. trivialis N. in K / H. explanatus K.	CF 1 Chernyshinella zone	Tn2 (pars) / Tn1b (α,β,γ)	CTO 2 / CTO 1
DEVONIAN (pars)	UPPER DEVONIAN	STRUNIAN (sensu Conil, al 1976)		do VI — Kalloclymenia Wocklumeria	50 - 51	"Protognathodus fauna" ZIEGLER 1969	hemispherica / latior interzone	PLs V. pusillites (K.)			DSO 8 upper / lower
					44 - 49	Bispathodus costatus z. ZIEGLER SANDBERG et AUSTIN, 1974	R. (M.) hemispherica (RICHTER, 1948) and R. (M.) dichotoma (PAECKELMANN, 1913) zone	PLm D. &N. / S. lepidophytus (K.) ST.	ε δ	Tn1a (δ) / Tn1a a,b Fa2d	
				do V — Oxyclymeniae/Gonioclymenia	40 à 43	Polygnatus styriacus z. ZIEGLER, 1962		VUs H. versabilis K. / VUi S. cf. uncatus H.	γ β	Quasiendothyra zone DF 3	Fa2c
		FAMENNIAN	UPPER	do IV — Prolobites Platyclymenia	39		R. (F.) intercostata (MATERN, 1929)	GMs Ar. gracilis K.		Fa 2 Fa2b	
					38	Scaphignathus velliferus z. ZIEGLER, 1962		GMm E. gr. minutus	α		DSO 7
					36 à 37	Palmatolepis m. marginifera z. ZIEGLER, 1962	Serratostriata / intercostata interzone	GMi H., S. & M.			
					32 à 35	Palmatolepis rhomboidea z. ZIEGLER 1962 SANDBERG & ZIEGLER, 1973	E. (R.) serratostriata (SANDBERGER, 1845) and E. (N.) nehdensis (MATERN 1929) zone	GHs Ar. gracilis K. cf. Ac. hirtus N.		Fa2a	
			LOWER	do III — Cheiloceras	31			GHi		Fa1b	
					29 - 30	Palmatolepis crepida z. ZIEGLER, 1962		GP4			DSO 6
					28		U. sigmoidale zone (MULLER-STEFFEN, 1964)	GP3 P. greggsii (M.)	DF 2 ? Septatournayelle zone	Fa1 Fa1a	DSO 5
					27						
				post do Iδ	26	Palmatolepis triangularis z. ZIEGLER 1967	E. (E.) splendens zone (WALDSCHMIDT, 1885) interzone	GP2			
					25						
					24						
		FRASNIAN	UPPER	do Iδ — Crickites holzapfeli	23	Palmatolepis gigas z. ZIEGLER, 1962, 1971	reichi, schmidti, volki, materni, barrandei zone / cicatricosa-barrandei interzone	GP1 R. planus D. & N.	DF 1 ? Naticella zone	F 3 upper / lower	DSO 4
					22		Z a E. (E.) varcostrata (CLARKE, 1884)				
			LOWER	do I — Manticoceras cordatum / Manticoceras β/γ	21	Ancyrognathus triangularis z. ZIEGLER, 1962 / Polygnatus asymmetricus zones ZIEGLER, 1962, 1971	B. (R.) cicatricosa (MATERN, 1929)	LTs A. langi (T.L.) A.	F2	j / i / h	DSO 3
					20		cicotricosa / torleyi interzone	LTi S. triangulatus A.		g / f / e	DSO 2
					18 - 19					d / c	DSO 1
				do I α — Pharciceras lunulicosta	17	Schmidtognathus hermanni-Polygnathus cristatus z. ZIEGLER, 1966	U. torleyi zone (MATERN, 1929)			b	

Interval biozones are largely used by micropaleontogists (Chitinozoa: VERNIERS et al., 1995 – Ostracods: LETHIERS, 1984 (Tab. 5.3); 1993 – Foraminifera: MAGNIEZ-JANNIN, 1995; ROBASZYNSKI & CARON, 1995; etc.).

Since it combines the criterion of presence, coexistence and exclusion, the interval zone should theoretically be the most valuable kind of biozone in terms of resolution and correlation potential. Its use is strongly recommended.

Remarks on ecobiostratigraphy and climatostratigraphy

As already mentioned in other parts of this chapter (see also the ensuing paragraph 3.2.2. – Possible "diachronism of bioevents") the control that environmental factors exert on the distribution of both planktonic and benthic organisms is one of the main limitation in the application of biostratigraphy. Changes in environmental conditions across different domains (e.g. across different latitudes when considering planktonic organisms or water depths when considering benthic taxa) can, in fact, result in large differences in the temporal range of taxa from place to place. This may result in large temporal differences between the events of true evolutionary origination of a taxon, ideally the appearance of the first specimen in a given locality, and the first occurrence in a different locality which will coincide with the first event of immigration during the phase of dispersal.

Yet, environmental changes induced by global factors (e.g. climate-related changes) can provide a sequence of regionally to globally recognizable bioevents (ecobiostratigraphic events). The resulting biozones (ecobiozones) are defined on the basis of the temporary (non evolutionary) appearance or disappearance of specific taxa or their changes in the relative abundance within fossil assemblages. Therefore, ecobiozones often correspond in the procedure of construction to the abundance biozone since they can be based on the recognition of changes in the relative abundance of taxa within fossil assemblages. However, the definition and validation of ecobiozones requires an integrated stratigraphic approach combining lithologic, biological, isotopic data allowing a comprehensive understanding of the timing and causes leading to the environmental changes that is ultimately recorded by ecobiostratigraphic horizons.

The observation of quantitative changes in the fossil assemblage and the calibration to climatic events of known absolute age (climatostratigraphy) is a method widely used in the study of the Quaternary interval for which a detailed record of past climate changes is available. The application of this method has led to the definition of several ecozone based on fluctuations in abundance among calcareous plankton assemblages which are near contemporaneous over wide regions. As an example, ecobiostratigraphic scales have been developed for the Quaternary of the

Mediterranean region (JORISSEN et al., 1993; CAPOTONDI et al., 1999, ASIOLI et al., 1999, SPROVIERI et al., 2003). The use of independent data (radiometric age, secular paleomagnetism, tephro-chronology), shows that the temporal resolution of this biozonal schemes is very high (millennial – to centennial- scale). Moreover, the results of these studies show that the marine ecozonations developed reflect the global climatic oscillations on which the continental Quaternary stratigraphy is based, providing a link between marine and terrestrial biostratigraphy.

3.2.2. *Possible "diachronism of bioevents"*

The succession of biozones is a record of both evolutionary and sedimentary processes; in principle a sequence of biozone can provide indications on the relative position of stratigraphic intervals and should allow to correlate them from place to place.

Among the different kinds of classical biostratigraphic units, those including the concepts of occurrence, coexistence and exclusion (i.e. the concurrent range zone and the interval zone) seem to be best suited for this purpose.

Often, the classical biozonations are presented as not containing gaps; in certain cases however, they are discontinuous. Actually, by defining a boundary on the base of a first or last occurrence datum, only the interval where the index taxa occur are included in the biozone. Indeed, the level of any biostratigraphic events (or biohorizon), can be geometrically figured out as linked by surfaces running through the rocks from site to site and subtending intervals that represent the biozones. However, as anticipated in other parts of this chapter, it cannot be assumed that a biohorizon is a level in time. Likewise, the term biostratigraphic event does not necessarily imply that the considered "event" occurred simultaneously everywhere. Rather, there are several examples showing that this is often not the case for both planktic- or benthic taxa-based bioevents.

By using the same, globally synchronous, upper Eocene iridium anomaly as an age-control point in the Massignano section (central Italy) and the Antarctic ODP Site 689B MONTANARI et al. (1993), have shown that the events of first and last occurrence of several planktic foraminiferal and calcareous nannofossil index species, in particular *Globigerinatheka index* and *Istmolithus recurvus*, have largely different absolute ages in the two regions (Fig. 5.5). A geochronologic calibration with an uncertainty of ±0.4 Ma for this interval, showed that this "biological diachronism" amount to about 1 My.

The influence of water temperature on the distribution of marine organisms has also been evidenced by with regards to the position of the *Ericsonia subdisticha* biozone in the late Priabonian (CAVELIER, 1972) and the consequences of this on the definition of the Eocene/Oligocene boundary in its type area.

Other paleoenvironmental factors controlling the distribution of fossil species may affect the procedure of biozonation. It is the case of the oceanic anoxic event occurring in the latest Cenomanian boundary which had a large influence on the distribution and evolution of planktic foraminifera (CARON et al., 1993; BEAUDOIN & CARON, 1995). Across this interval, the boundaries of the *Whiteinella archeocretacea* total range zone are well recognizable in areas where anoxic conditions developed, whereas they are hardly identified and apparently diachronic in areas characterized by less marked anoxic conditions.

These examples show that bioevents of first and last occurrence must be used with caution and calibrated to other scales when establishing an interval biozone. To the aid of progress in stratigraphy we should make choices that can be re-discussed in the light of integrated biostratigraphy.

3.2.3. *Significance, precision and reliability of the classical biozones*

Biostratigraphers have tried to obtain ever higher resolutions on the base of the distribution of the classical biomarkers.

Looking at the different biostratigraphic scales based on different groups and for different periods, it is clear that the stratigraphic resolution of the classical biozones varies considerably depending on the fossil group considered. Obviously, an integrated biostratigraphic approach is crucial in the obtaining a satisfactory resolution and a control on diachronism of bioevents; the use of integrated scales clearly increases their precision, also in the case of organisms characterized by low evolutionary rates or whose distribution is largely controlled by environmental factors.

In conclusion, the construction of a reference relative chronology that includes classical biozones and its precision depends largely on the "biostratigraphic qualities" of the paleontological group used. However, a good understanding of the paleobiology of the group and the taxonomic interpretation of the biomarkers are also very important factors.

Any biostratigraphic interpretation should clearly refer to the biozonal scale used (i.e. reference to the fossil group, to the biogeographic province, to the author and the year of publication); the accuracy of the dating should be discussed (subzone, zone, first/last occurrence, etc.) and a margin of error proposed.

Finally, we must stress the importance of integrating the biostratigraphic data with non-paleontological results (i.e. calibration with magnetostratigraphy, chemostatigraphy, sequence stratigraphy, etc.), which is an undeniable approach in modern stratigraphy (DE WEVER et al., 1986; ROBASZYNSKI et al., 1980; 1990…).

3.3. RESULTS OF THE LOGICAL BIOSTRATIGRAPHY

The examples developed (GUEX, 1987; 1991; DE WEVER, 1982; SCHAAF, 1985; BOULARD, 1993) clearly show that the successions of biochronozones are discontinuous (i.e. separated by "intervals of separation" where no associations are described), not related to body of rock, and characterized by assemblages distinctive and mutually exclusive. Thus, they have the characteristics of the concurrent range zone. Though giving indications on the coexistence of certain taxa they don't provide information on the level of first or last occurrence of them, which lies between two unitary associations (within the intervals of separation).

This method seems to be free from the inconveniences of recognizing the appearance of taxa, which can coincide with the evolutionary origination of a taxon but very often is locally controlled by multiple environmental factors, preservation, etc.

Since an unitary association is defined and characterized by the interval of occurrence of one species and/or the co-occurrence or the exclusion of several species, the corresponding interval is univocally defined. Independent from defining the endpoints of taxon range, the unitary associations and associated biochronozones can be correlated on the base of the species that characterize them; originally

5. Biostratigraphy

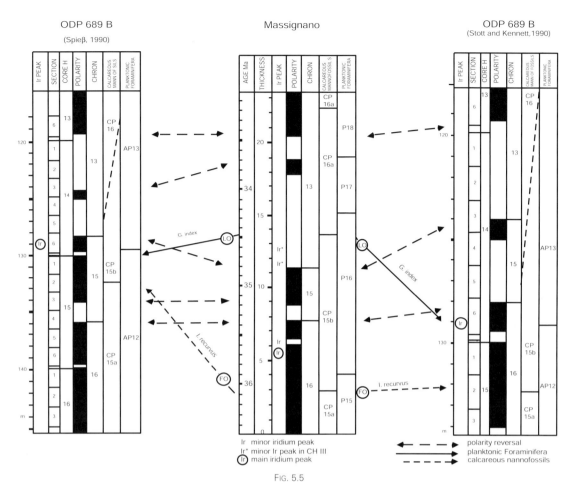

FIG. 5.5
Possible diachronism of bioevents. Alternative possible biostratigraphic and magnetostratigraphic correlations between the Massignano Quarry (Italy) and ODP Leg 113 Site 689B (Antarctica) across the Eocene-Oligocene boundary (MONTANARI et al., 1993).

defined as a theoretical unit, it can turn out to be a body of rock.

An unitary association seems therefore to have a chronological value greater than an interval biozone or any other classical biozone because it considers a greater number of data (presence/absence and co-occurrence/exclusion of taxa).

However, the two methods are not at odds. Neither the fundamental phase of knowledge derived from the classical biostratigraphy approach should be disregarded, nor it should be argued against the "conceptual nature" of the unitary associations and biochronozones. Rather the integration of the results obtained by one and the other method is a means to improve our knowledge and the biostratigraphic tool.

Still little developed, the logical biostratigraphy would probably deserve a wider use in the light of the encouraging results obtained on certain fossil groups where the classical biostratigraphic method was little effective. However, as with the classical methods, a correct taxonomic interpretation is fundamental for the reproducibility of results.

3.4. RESULTS OF THE STATISTICAL BIOSTRATIGRAPHY

In spite of the large number of excellent examples and studies on this topic (SCHAAF, 1985; GRADSTEIN et al., 1992 (Tab. 5.4); 1994; BOULARD, 1993; BERGGREN et al., 1995), the statistical methods seem to be not yet adapted to biochronology.

Apparently due to its "reductionist approach" (Fig. 5.6), the method of ranking and scaling does not allow the construction of a coherent sequence of biostratigraphic events that respect the relation of superposition of the original data.

These methods are statistically satisfactory since they propose the likeliest solution; however they are not well adapted to biochronology since they do not consider all the data and exclude the rare cases. A fossil record of exceptional quality is a prerequisite for their use; however, this is rarely the case.

3.5. RELATIONS BETWEEN THE CLASSICAL BIOSTRATIGRAPHIC UNITS, THE UNITS OF THE LOGICAL BIOSTRATIGRAPHY AND THE UNITS OF THE STATISTICAL BIOSTRATIGRAPHY

Due to their different natures the units created following these three methods cannot be directly compared and correlated.

Though having identical names, the units of the classical biostratigraphy and those of the statistical biostratigraphy are not related. Based on the likeliest distribution of one or more taxa, the latter, are constructed using a reductionist approach and do not take into account the cases of coexistence or

TABLE 5.4
Cenozoic foraminiferal and dinoflagellate cyst biostratigraphy from the Labrador and North Sea based on the ranking and scaling method (RASC analysis) (GRADSTEIN et al., 1992).

AGE Ma BERGGREN et al. 1985	EPOCH	STAGE	BLOW (1969)	MARTINI (1971)	Foraminifera	Dino-flagellates	COSTA & MANUM (1988) Dino-flagell.	KING (1989) North Sea Planktonic forams	KING (1989) North Sea Calcareous Benthic forams	KING (1989) North Sea Aggl. forams
	QUARTERNARY	Calabrian	N23	NN20/NN19					NSB 14-16	NSA12
	PLIO-CENE L	Piacenzian	N22/N21	NN18/NN16	C. grossa			NSP16		
	PLIO-CENE E	Zanclean	N19	NN15/NN13	N. atlantica			NSP15	NSB13	
		Messinian	N18/N17	NN12/NN11	?		D20	NSP 12-14	NSB 11-12	NSA11
10	MIOCENE L	Tortonian	N16	NN10/NN9/NN8	B. metzmacheri		D19			
	MIOCENE M	Serravalian	N15/N14/N13/N12/N11/N10	NN7/NN6	?		D18	NSP11	NSB10	
		Langhian	N9/N8	NN5/NN4	G. praescitula		D17			
20	MIOCENE E	Burdigalian	N7/N6	NN3			D16	NSP10	NSB9	NSA10
		Aquitanian	N5/N4	NN2/NN1	?				NSB8	NSA9
	OLIGOCENE L	Chattian	N4-P22	NP25/NP24	G. ex gr. officinalis	T7	D15	NSP9		NSA8
30	OLIGOCENE E	Rupelian	P21/P19/20	NP23	R. bulimoides	T6B/C	D14		NSB7	NSA7
			P18	NP22/NP21	D. seigliei	T6A	D13			
	EOCENE L	Priabonian	P17/P16/P15	NP19/20/NP18			D12		NSB6	NSA6
40		Bartonian	P14/P13	NP17	R. amplectens	T4C/D	D11	NSP8		
	EOCENE M	Lutetian	P12/P11	NP16/NP15	A. aubertae		D10	NSP7	NSB5	NSA5
50			P10	NP14		T4A/B	D9	NSP6	NSB4	NSA4
	EOCENE E	Ypresian	P9/P8/P7	NP13/NP12/NP11	S. patagonica	T3C/T3B	D8/D7	NSP5	NSB3	NSA3
			P6	NP10	Coscinodiscus spp.	T3A	D6			
	PALEOCENE L	Selan-dian (Thanet.)	P5/P4	NP9/NP8/NP7	R. paupera	T2C/T2B	D5/D4	NSP4	NSB2	NSA2
60			P3	NP6/NP5	T. ruthven murrayi	T2A	D3	NSP3/NSP2		NSA1
	PALEOCENE E	Danian	P2/P1	NP4/NP3/NP2/NP1	S. pseudobulloides	T1B/T1A	D2/D1	NSP1	NSB1	

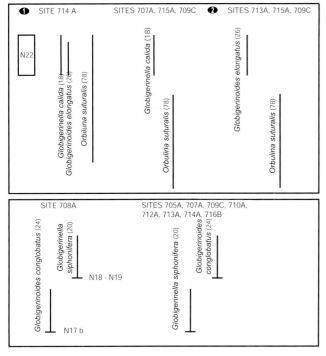

FIG. 5.6
Exemplification of the "reductionist effet" of statistical biostratigraphy on Neogene foraminiferal data from various ODP sites (from BOULARD, 1993). Upper figure: assessment of the stratigraphic relations between *Orbulina suturalis* (78), *Globigerinella calida* (18) and *Globigerinoides elongans* (26) at sites 714A(1), 707A, 709C, 713A and 715A(2); lower figure: assessment of the stratigraphic relations between the first appearances datum (FAD) of *Globigerinella siphonifera* (20) and *Globigerinoides conglobatus* (24) at sites 705A, 707A, 708A, 709C, 710A, 712A, 713A, 714A and 716B.

exclusion of rare taxa. The cross-check of the identified succession of events with those observed in a different fossil group or with geochronologic data is a necessary step for its validation. Following this first step, it is necessary to come back to the original data set and reconsider the distribution of each taxon; however, such a method has not yet been developed.

The two methods have different aims and use different types of data and the choice of method will depend on the purpose of the investigation. The RASC method is likely to produce higher resolution scales. However, the potential of correlation obtained will be somehow affected because some zonal boundaries will be based on facies-controlled or geographically constrained events rather than global (evolutionary) originations and extinctions. The approach of Unitary Associations, on the other hand, will likely produce lower resolution scales but is theoretically more robust for long-distance correlations.

Accordingly, there is no equivalence between the classical biozones and the unitary associations-biochronozones. However, for certain fossil groups and stratigraphic intervals, similar results can be achieved independently by the two methods.

Two examples are represented by the study of lower Jurassic (DOMMERGUES & MEISTER, 1987; Tab. 5.5) and upper Jurassic (BOULARD, 1993) ammonites. In these cases, the use of the two methods has shown a good agreement between the results of the "classical zonation" and the "automated zonation". This examples show that when applied to the distribution of a well-known group such as the ammonites, the unitary association method confirms and possibly refines the results of the classical biostratigraphy.

Results of similar studies applied to microfossils have been promising (DE WEVER, 1982; LAHM, 1984; CARBONNEL, 1986; SCHAAF, 1985, Fig. 5.7; BOULARD, 1993) but sometimes contradictory (GUEX, 1987). The latter authors stressed the possibility of a "diachronism of datum levels" (FADs et LADs), which would imply a profound rediscussion of the entire "classical biozonation" previously established. However, the basic data treated by the automated zonation derive from different biostratigraphic works that do not necessarily use the same taxonomic concepts. The use of this technique on such data set should be preceded by a systematic revision guarantee a rigorous control of taxonomy and a better reliability of the results.

3.6. RELATIONS BETWEEN BIOCHRONOLOGY, GEOCHRONOLOGY AND CHRONOSTRATIGRAPHY

3.6.1. *Biochronology and chronostratigraphy: biozone and stage boundary*

For a given time interval the boundary of the biozones are positioned according to the evolutionary events within the fossil group; in certain cases they can correspond to a stage boundary; in some other cases the biozones can extend across these limits.

Is it necessary to avoid the possibility of overlap? Should biozones and stages be intimately related? Is it necessary that the zones are strictly a subdivision of the stage or that the stage represents a strict grouping of biozones? Can stages and biozones be defined independently from each other and their boundaries differ? The answer to this questions assumes a great importance in chronostratigraphy since it influences the definition of chronostratigraphic units: in particular, should the stage be defined prior to identify its subdivision into ("chrono") zones? Or should the stage represent the sum of previously established, contiguous or not, ("chrono") zones? These questions can be referred to two different stratigraphic methodologies: on one hand the successive subdivisions of the stratigraphic column in progressively thinner and more precise units; on the other hand the progressive grouping of local units into units of higher rank.

This practice has some fundamental consequences on the recent tendency of definition of stage boundary stratotypes, that hold the stratigraphically concrete nature of a point in a rock succession of a given locality and of a fossiliferous biohorizon (in the sense of a rock body) containing one or more biomarkers above (or underneath) this surface. A boundary defined in this way is precise and non ambiguous since it represents a unique stratigraphic surface and time; the remaining difficulty is the recognition of this boundary in the space, when the criterion to recognize it is missing in other localities. This problem raises some discussion.

On one hand, according to the International Commission on Stratigraphy (COWIE et al., 1996) a stage boundary is defined as a point (GSSP – *Global Stratotype Section and Point*) in a reference section on the base of several stratigraphic data including biostratigraphy. On the other hand the biozones are defined independently and only on the base of paleontological criteria. It is obviously not indispensable that the limits of the two categories of stratigraphic units correspond. The same obviously, the "ideal" GSSP would coincide with a simultaneous change in any detectable stratigraphic record (e.g. evolution in a given fossil group, remanent magnetization, chemical composition, etc.), which would make it easily and univocally identified using different stratigraphic techniques.

TABLE 5.5

Example of a good correspondence between traditional ammonite biozones and unitary associations-biochronozones established for the Carixian (Jurassic, middle Lias) of Western Europe (data taken from DOMMERGUES & MEISTER, 1987, 1991 et DOMMERGUES et al., in Collectif Groupe Français d'étude du Jurassique, 1991).

"AUTOMATIC" ZONATION BASED ON UNITARY ASSOCIATIONS (Dommergues et Meister, 1987)		"EMPIRIC" TRADITIONAL ZONATION (Dommergues, 1979, 1984; Schlatter, 1980; Comas Rengifo, 1985; Cubaynes et al., 1984; Phelps, 1985; Meister, 1986; Dommergues et Meister, 1991)			STAGE	SUB-STAGE	SYSTEM and SERIES
Unitary Associations	Biochronozones	Horizons	Subzones	Zones			
UA 1	BCZ 1 = "DEPRESSUM"	Depressum	SUBNODOSUS	MARGARITATUS	PLIENSBACHIAN	DOMERIAN	JURASSIC - MIDDLE LIAS
UA 2		Celebratum					
UA 3	BCZ 2 = "NITESCENS"	Nitescens	STOKESI				
UA 4	BCZ 3 = "MONESTIERI"	Monestieri					
UA 5	BCZ 4 = "OCCIDENTALE"	Occidentale					
UA 6	BCZ 5 = "FIGULINUM"	Figulinum	FIGULINUM	DAVOEI		CARIXIAN	
UA 7		Angulatum					
UA 8	BCZ 6 = "CAPRICORNUS"	Crescens / Davoei	CAPRICORNUS				
UA 9		Capricornus					
UA 10		Lataecosta					
UA 11							
UA 12	BCZ 7 = "MACULATUM"	Maculatum / Sparcicosta / Fimbriatum	MACULATUM				
UA 13	BCZ 8 = "CRASSUM"	Luridum	LURIDUM	IBEX			
UA 14		Crassum					
UA 15		Rotundum					
UA 16	BCZ 9 = "LEPIDUM"	Alisiense	VALDANI				
UA 17							
UA 18	BCZ 10 = "ACTEONI"	Centaurus (Venarense/Acteoni)					
UA 19							
UA 20							
UA 21	BCZ 11 = "VALDANI"	Valdani					
UA 22	BCZ 12 = "MAUGENESTI"	Maugesteni					
UA 23	BCZ 13 = "ARIETIFORME"	Arietiforme					
UA 24	BCZ 14 = "MASSEANUM"	Masseanum	MASSEANUM	JAMESONI			
UA 25							
UA 26	BCZ 15 = "JAMESONI"	Bronni (Evolutus + Pettos)	JAMESONI				
UA 27							
UA 28	BCZ 16 = "BREVISPINA"	Brevispina / Submuticum/Tenuilobus	BREVISPINA				
UA 29	BCZ 17 = "NODOSOFISSUM -BIRUGA"	Polymorphus/Braga	POLYMORPHUS				
UA 30		Taylori	TAYLORI				
UA 31	BCZ 18 = "NODOGIGAS"	Nodogigas					

It seems that it is the order of acquisition of knowledge and its progress which guides the stratigrapher, and that the main concern of the working groups on the stage boundaries is the development of "zonations allowing to establish the distribution of different fossil groups in correspondence of the studied boundary and the correlation of these zones based on the first appearance of certain species" (LEGRAND, 1996). The biostratigraphic framework is fundamental and impossible to circumvent, and with rare exceptions, the stage boundaries (and chronostratigraphic higher categories) are constructed on biozonal scales.

3.6.2. *Biochronology and geochronology: the geochronologic calibration of biostratigraphic scales*

The correlation of two or more scales based on different fossil groups and the calibration to the radiochronologic scale, allows a precise assessment of the geologic time. For this goal a set of radiochronologic data that can be interpolated to at least one of the biostratigraphic scale is necessary.

Since the publication of the first geochronologic scales in the 50s, the time interval represented by stratigraphic units has been estimated by considering the numerical ages of the stage boundaries and calculating its average duration taking into account the uncertainty in the numerical age determination; then, this duration was divided by the number of biozones recognized in this interval. This method, subject of many criticisms related to the fact that the numeric data were being gathered only very gradually and often with very different values for a given chronostratigraphic limit, has been the only available for several decades.

Sometimes even, this method suffered from circular reasoning, when in the absence of absolute age determination, the average duration of a chronostratigraphic unit was calculated, starting from the number of biozones recognized in this interval, the average duration of a biozone being estimated from another interval. As an example, the ages of stage boundaries of the Lias, a subsystem for which no reliable radiometric data were available, was calculated on the base of the average duration of ammonite zones estimated at 1 Ma for the stages of the Middle and Late Jurassic.

Identical examples could be exposed on Silurian and Cenozoic biozones based on graptolite and microfossil distribution, respectively.

If geochronologists had been at the same stage of knowledge as biostratigraphers, it would have been possible earlier to locate a group of biozones in a time interval therefore avoiding sometimes hazardous extrapolations.

Yet, taking the "likely" age of stage boundary published on geochronologic scales as a basis, many stratigraphers have

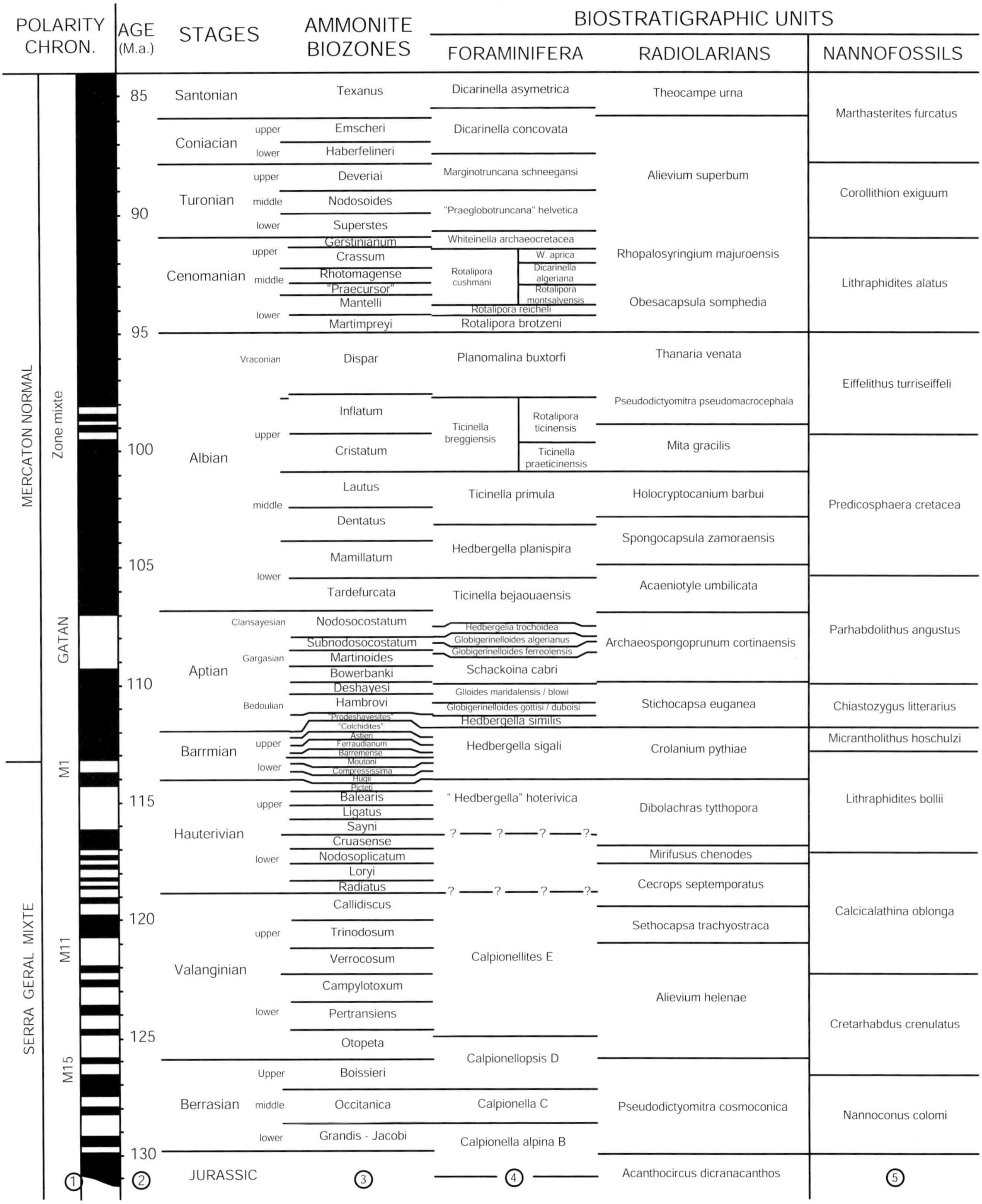

5. Biostratigraphy

POLARITY CHRON.	AGE (M.a.)	STAGES		AMMONITE BIOZONES	BIOSTRATIGRAPHIC UNITS		
					FORAMINIFERA	RADIOLARIANS	NANNOFOSSILS
MERCATON NORMAL / Zone mixte	85	Santonian		Texanus	Dicarinella asymetrica	Theocampe urna	Marthasterites furcatus
		Coniacian	upper	Emscheri	Dicarinella concovata	Alievium superbum	
			lower	Haberfelineri			
		Turonian	upper	Deveriai	Marginotruncana schneegansi		Corollithion exiguum
	90		middle	Nodosoides			
			lower	Superstes	"Praeglobotruncana" helvetica		
		Cenomanian	upper	Gerstinianum	Whiteinella archaeocretacea	Rhopalosyringium majuroensis	Lithraphidites alatus
				Crassum	Rotalipora cushmani / W. aprica		
			middle	Rhotomagense	Dicarinella algeriana		
				"Praecursor"	Rotalipora montsalvensis		
				Mantelli		Obesacapsula somphedia	
			lower	Martimpreyi	Rotalipora reicheli		
	95				Rotalipora brotzeni		
		Albian	Vraconian	Dispar	Planomalina buxtorfi	Thanaria venata	Eiffelithus turriseiffeli
			upper	Inflatum	Ticinella breggiensis / Rotalipora ticinensis	Pseudodictyomitra pseudomacrocephala	
	100			Cristatum	Ticinella praeticinensis	Mita gracilis	
			middle	Lautus	Ticinella primula	Holocryptocanium barbui	Predicosphaera cretacea
				Dentatus		Spongocapsula zamoraensis	
	105			Mamillatum	Hedbergella planispira		
			lower	Tardefurcata	Ticinella bejaouaensis	Acaeniotyle umbilicata	
GATAN		Aptian	Clansayesian	Nodosocostatum	Hedbergella trochoidea	Archaeospongoprunum cortinaensis	Parhabdolithus angustus
			Gargasian	Subnodosocostatum	Globigerinelloides algerianus		
				Martinoides	Globigerinelloides ferreolensis		
	110			Bowerbanki	Schackoina cabri		
			Bedoulian	Deshayesi	Glloides maridalensis / blowi	Stichocapsa euganea	Chiastozygus litterarius
				Hambrovi	Globigerinelloides gottisi / duboisi		
				"Prodeshayesites"	Hedbergella similis		
M1		Barrmian	upper	"Colchidites"	Hedbergella sigali	Crolanium pythiae	Micrantholithus hoschulzi
				Astieri			
				Ferraudianum			
				Barremense			
			lower	Moutoni			
				Compressissima			
	115			Hugii			
		Hauterivian	upper	Picteti	"Hedbergella" hoterivica	Dibolachras tytthopora	Lithraphidites bollii
				Balearis			
				Ligatus			
				Sayni	? — ? — ? — ?—		
				Cruasense		Mirifusus chenodes	
			lower	Nodosoplicatum			
				Loryi		Cecrops septemporatus	Calcicalathina oblonga
				Radiatus	? — ? — ? — ?—		
	120			Callidiscus		Sethocapsa trachyostraca	
M11		Valanginian	upper	Trinodosum	Calpionellites E		
				Verrocosum			
				Campylotoxum		Alievium helenae	Cretarhabdus crenulatus
			lower	Pertransiens			
	125			Otopeta	Calpionellopsis D		
M15		Berrasian	Upper	Boissieri		Pseudodictyomitra cosmoconica	Nannoconus colomi
			middle	Occitanica	Calpionella C		
			lower	Grandis - Jacobi	Calpionella alpina B		
SERRA GERAL MIXTE	130	JURASSIC				Acanthocircus dicranacanthos	

① ② ③ ④ ⑤

1: VAN HINTE (1976), LOWRIE & ALVAREZ (1979) ; 2: ODIN et al. (1982) ; 3: PORTHAULT (1974), BUSNARDO (comm. pers.) ; 4: VAN HINTE (1976), PREMOLI-SILVA & BOERSMA (1977), SIGAL (1977) ; 5: THIERSTEIN (1971, 1973, 1976), ROTH (1973).

FIG. 5.7

Integrated biostratigraphy based on ammonite, foraminiferal, radiolarian and calcareous nannofossil distribution established on the basis of different methods (traditional biozones and unitary associations – biochronozones) for the Cretaceous of Europe (*from* SCHAAF, 1985).

calculated the average length of the different zonal units within a system or a stage (CALLOMON, 1984b; 1994), sometimes even "correcting" the geochronologic scales on the base of the calculated duration of the biozones (WESTERMANN, 1984).

Based on a duration of the 70 My for the Jurassic (ODIN & ODIN, 1990) the average duration of a biozone in the ammonite scales available for this system (Groupe Français d'Etudes du Jurassique, 1991), would be 0.9 Ma, whereas those of the subzones and the horizons[1] would be 0.43 Ma and 0.26 Ma, respectively (HANTZPERGUE, 1993). The same calculation made by taking into account the duration of each stage shows important variations within the Jurassic: the longest average duration of a biozone (1.5 Ma) is reached in the Bajocian and the shortest (0.35 Ma) in Tithonian; for the subzones the average values range from 0.67 Ma (Hettangian) to 0.30 Ma (Tithonian) and for horizons[1] from 0.175 Ma (Pliensbachian) to 0.475 Ma (Bajocian).

For comparison with other geological intervals, the average duration of an Archaeocyathid zone in the Cambrian is 10 Ma (the entire Tommotian), whereas that of Trilobite zones of the same system is estimated to 2-3 Ma. The Ordovician Graptolite classical biozones have an average duration longer than 4 Ma; their relative precision seems higher in the Silurian with an average value of 0.8 Ma. However, it must be stressed the hazard of this estimates: for only the Arenig, the interval for which the best subdivision based on Trilobite distribution is available, the average precision would be 3 Ma; however, the Arenig is a part of the Ordovician whose chronostratigraphy is still debated due to the presence of hiatuses. As for Devonian Conodont biozones, an assessment based on the system chronostratigraphy gives 1.7 Ma as an average duration, but this value can decrease to 1 Ma for the Emsian and event less for Late Devonian conodont subzones.

These large variations demonstrate the unreliability of calculating average durations of biochronologic units, due to the still too large imprecision of the radiometric determination of stage boundary ages and the lack of a firm biostratigraphic subdivision of certain intervals because of the poor knowledge of the fossil material.

A representative case is the Mesozoic ammonite zonation for which the interpolation with numerical ages has led to a re-evaluation of the duration of the biozones (middle Triassic: HELLMANN & LIPPOLT, 1981 – Anisian-Ladinian-Carnian: BRACK & RIEBER, 1993 – early-middle Oxfordian: FISHER & GYGI, 1989 – Bajocian-Bathonian: ODIN et al., 1993 – Late Aptian: KEMPER & ZIMMERLEE, 1978 – Late Cenomanian – Early Maastrichtian: Mc ARTHUR et al., 1994; OBRADOVICH, 1993).

The method consists in calculating the average duration of biostratigraphic units between two "reference biozones" whose likeliest numerical age is known. Accordingly, the average duration of the ammonite zones would be 3.2 Ma in the lower Bajocian, 1.2 to 1.8 Ma in the upper Bajocian-lower Bathonian, and 2.2 Ma in the lower Oxfordian. In the lower Turonian the average duration is 0.2 Ma, 0.7 Ma in the middle Turonian, 0.30 Ma in the upper Turonian; 0.60 to 0.75 Ma in the Coniacian, the Santonian and the lower Campanian. In the upper Campanian-lower Maastrichtian interval ammonite biozones have a duration ranging from 0.55 to 0.75 Ma.

---

1. The term horizon is here used to define a small stratigraphic interval characterized by a distinctive faunal assemblage. We recall that, according to the International Stratigraphic Guide, this term is instead used to define a surface.

3.7. RECOMMENDATIONS IN BIOSTRATIGRAPHY

3.7.1. *Definition and denomination of biostratigraphic units*

Problems of terminology and writing

The term "biozone" is routinely used as an abbreviation for "biostratigraphic zone"; moreover, the term "biozone" is frequently reduced to "zone". According to the Stratigraphic Guide of the International Commission of Stratigraphy a zone is a "minor body of rock in many different categories of stratigraphic classification. The type of zone indicated is made clear by a prefix, e.g., lithozone, biozone, chronozone". It is recommended that the prefix "bio" is used before the term "zone" to distinguish a biostratigraphic zone from any other kind of stratigraphic unit, to avoid any confusion, particularly between biozone and chronozone. Indeed, both can be named according to one or several fossils though representing fundamentally different concepts. The chronozone, in fact, can be identified even in the absence of its index fossil(s). This confusion is avoided when chronozones are designated by clean names, such as done for the Paleozoic: Whitwell chronozone or Gleedon chronozone of the Homerian stage of the Wenlock series (Silurian).

For both practical and historical reasons it is recommended to maintain the use of the abbreviated term "zone" (and "subzone") to indicate the biozones resulting from the traditional biostratigraphy, though this practice is only rarely accompanied by an explicit explanation for its use. This would avoid confusions and indicate *ipso facto* the type of procedure used: the "Agnostus pisiformis Zone", would indicate the biostratigraphic zone conventionally defined by the "assemblage" dominated by the species *Agnostus pisiformis* at the base of the upper Cambrian of Northern Europe. In phrasing the name of this biostratigraphic unit, the name of the index taxa should be written as normal text, in order to distinguish it from the designation of taxa on which, according to the rules of scientific publication should be in italic.

A second convention must be applied to the name of traditional biozones. Once this has been clearly defined, either by its author, or by the users, it is not necessary to report the entire name of the biozone every time it is mentioned. Thus, the classical biozones and associated categories can be designated by the name of the defining taxa, writing in capital the first letter of the species name, in order to not confuse it with the formal taxonomic writing recommended by the zoological and botanical nomenclature codes. Accordingly, the "Helvetica Zone" is the "Helvetoglobotruncana helvetica" range zone; the "Davoei Zone" is the *Prodactylioceras davoei* abundance zone of the upper Carixian (lower Jurassic) of West Europe (Euroboreal province); the "Medea Subzone" is the lineage biozone of a "chronospecies" within the *Kosmoceras (Zugokosmoceras) enodatum, K.(Z.) medea et K.(Z.) jason* lineage denoting the Lower-Middle Callovian boundary (Middle Jurassic) of north-western Europe (sub-Boreal province).

The above mentioned rules have the advantage of reducing the length of the biostratigraphic text; their use is submitted to the initiative of the authors. However, the style must remain consistent throughout the entire text in the same publication. In order to avoid any confusion, it is useful to point out at least once in the text, which genus the index species used belongs to.

In logical biostratigraphy (unitary association and biochronozones) and less frequently in the classical biostratigraphy (biozones or biohorizons), the units recognized are sometimes named by numbers or letters followed by numbers, such as for the Cenozoic calcareous nannofossil zonations.

A code instead is frequently used when multiple taxon names should be used to nominate the zone (e.g. assemblage biozones or concurrent range zones). Although practical, a succession of biostratigraphic units indicated by letters or numbers is much less informative than a succession of units denominated by taxon names, even if those are reduced to only the name of species. Moreover, a biozone-interval associated to a code may change its defining taxon without changing its code arising confusion. On the other hand, the use of codes instead of long taxon names has the advantage of a more immediate use for the non-specialists.

Denomination, boundaries and definition

The definition and the denomination of the biozones and their subdivisions, as well as the selection of criteria to establish their boundaries, were treated in the preceding paragraphs on the different kinds of biozones. Avoiding repetitions, it is important to insist on a fundamental point: the choice of an index taxa should not only satisfy the standard of the various kinds of biozone; it must also attend the paleobiologic aspect including its intraspecific variability, systematic position, environmental preferences, geographical distribution, etc.

Too often, biostratigraphic units are based on the distribution of poorly known taxa whose taxonomic revision has obvious consequences on the definition of biozones. As already discussed, a critical factor influencing the definition of any biostratigraphic unit is the personal taxonomic interpretation of the paleontologist/biostratigrapher.

Rule of priority

Biostratigraphic units are based on the distribution of index-taxa whose taxonomic name is conform to the rules of the international zoological and botanical codes, including genus, species (subspecies), author and year of publication. Following the international nomenclature, when a change of taxonomic rank intervenes in view of a systematic revision (either elevation of a subspecies to the rank of species or regrouping of subspecies in only one species; either subdivision of a genus into sub-genus or elevation of one sub-genus to the rank of genus, etc.), the rule of priority is applied.

This rule must be applied to biostratigraphic units also. The publication of any biostratigraphic scale must be accompanied by a list of the biostratigraphic units, including criterion for definition and kind, definition of their boundaries and description of the paleontological content. The possible changes of rank must be plainly clarified; their equivalence with the old units must be expressed similarly to the synonymy lists used for taxa.

It is then agreed that when a subbiozone is elevated to the rank of biozone, the latter maintains the name of its original author and the year of publication.

If its definition or its paleontological content is modified, changes must be clearly defined; the name of the unit (as in the case of a modification in the definition of a taxon) will be followed by the term "*emend*", the name of the reviser and the year of emendation. If, for systematic reasons, the name of the index taxon is changed (synonymy and application of the rule of priority), this change must be clearly circumstantiated. The same way, if several subzones are grouped into a new biozone the authors must give a fully description of the newly created unit.

When a zone is subdivided into subzones without changing the name of the zone, each new unit takes the name of its author and the year of the publication. If one of the units of a lower rank holds the same name as the unit of higher rank, it maintains the name of the original author and year of publication.

3.7.2. *Validation of biostratigraphic units*

The validation of the biostratigraphic units in specific regions should fall within the competence of the various national committees on stratigraphy. The committees evaluate only the respect of the rules of nomenclature (definition, boundaries and denomination) and the criterion of priority, the scientific responsibility remaining on the authors.

A judgment is made upon request of the authors or following a scientific review by qualified working groups, emanations of the Committee.

The procedure should start after the deposit of a scientific publication containing the definition or the emendation of the proposed units. These, must be substantiated in a review or a report with sufficient distribution. After a review of the group in charge of the judgment, the Committee votes the official approval. The list of the units approved is published each year and transmitted for information to the International Commission on Stratigraphy of the International Union of Geological Sciences.

3.7.3. *Use of biozones and biostratigraphic scales*

Any publication using one or more biostratigraphic scales must refer to their author in an explicit way. In the first place, this reference makes it possible to judge the reliability and the degree of accuracy of the scale used and the relative units. Secondly, it permits comparisons and correlations between scales proposed by different authors on the base of the same or different fossil groups.

The assertion without discussion and reference of a biochronologic age can only involve the doubt and discredit for work not yielding with this fundamental rule of any research in geosciences.

Based on the numerical age of the studied material, the "biochronologic age" of the recognized events must take into account the "confidence interval" of the radiometric determination and the relative "margin of error" reported. The determination of a biochronologic age, if not accompanied by such a discussion, can only arise doubts and discredit for a work not respecting this fundamental rule of any research in the field of geosciences.

It is not a matter of giving the biochronologic age an absolute value, which would be in conflict with the rules of definition of biostratigraphic units. Rather, according to the paleontological material and the scale(s) used, it is necessary to specify the interval covered by one or more biostratigraphic units, in which the established relative dating has the greatest possibility of being identified.

CHAPTER 6
ISOTOPE GEOCHRONOLOGY

N. Clauer & A. Cocherie

CONTENTS

1. – INTRODUCTION 91
2. – DESCRIPTION OF METHODS 92
   2.1. Potassium-argon methods: K-Ar, $^{40}Ar/^{39}Ar$ 92
   2.2. Isochron type methods 93
   2.3. U-Th-Pb method on separated minerals 93
3. – TECHNICAL ASPECTS 94
   3.1. Which geochronometers? 94
   3.2. Specific characteristics of sedimentary geochronometers 94
   3.3. Initial isotopic homogenization 95
4. – DIRECT ISOTOPIC DATING OF SEDIMENTARY ROCKS 96
   4.1. K-Ar, Rb-Sr and Sm-Nd dating of Precambrian sediments 96
   4.2. Other methods of stratigraphic dating of sediments 97
5. – INDIRECT ISOTOPIC DATING OF SEDIMENTARY ROCKS 97
6. – CONCLUSIONS 98

1. – INTRODUCTION

Isotope geochronology and ***radiochronology*** are geological disciplines dealing with age determinations based on the natural decay of unstable **radioactive isotopes** and on the production of **cosmogenic isotopes** by cosmic radiation, respectively. The stratigraphic application of ^{10}Be, ^{14}C, ^{26}Al, ^{32}Si, ^{36}Cl, ^{53}Mn, ^{59}Ni and ^{81}Kr isotopes that result from nuclear spallation remains limited, at the moment, to the study of polar ices, oxides, carbonates and young organic materials. The method based on ^{14}C radioactivity in materials originally in contact with atmospheric carbon dioxide is the most widely used, with applications to carbonates of various origins and wood (dendrochronology). These methods, as well as that based on the observation of the spontaneous fission tracks of ^{238}U, correspond to a specific field of radiochronology that will not be discussed here. Most of the methods currently used are based on the principle of the decay of radioactive isotopes trapped in minerals of plutonic, volcanic, metamorphic and sedimentary rocks. Only methods based on natural decay will be discussed here, since they can be applied to time intervals covering the entire history of the Earth, whereas those based on the production of cosmogenic isotopes are best suited for dating the limited time interval from the Quaternary to the Present.

Geochronology became a recognized discipline in Earth Sciences in the late 1940s following major scientific advances in nuclear physics with a very fruitful period in the 1980s due to deterministic improvements in mass spectrometry and a better understanding of crystal chemistry. The first attempts to date directly sedimentary rocks and minerals were only published at the end of the 1950s, whereas indirect dating of sedimentary rocks by studying plutonic and volcanic rocks intruded or interbedded in sedimentary rocks started earlier. The use of new technologies during the 1980s, such as infrared spectroscopy, Mössbauer spectroscopy, X-ray energy dispersive analysis, electron spin resonance and varied coupled-plasma spectrometry was also a key factor because of improved mineralogical and geochemical understanding allowing more "realistic" applications based on better knowledge and thus on more rigorous concepts. The last and most spectacular step was *in situ* dating by microprobe and laser equipment.

It may be remembered here that the principles of the use and application of isotope geochronology depend on physical parameters such as natural decay, radiation, physical conditions of mineral formation, mineral retention properties, isotope diffusion coefficients, etc., as well as on chemical parameters relating to the isotopes considered and the materials analysed. Details of these aspects can be found in the latest edition of "Isotopes" (FAURE & MENSING, 2005). As a consequence, results are sometimes difficult to interpret, because a geological interpretation of isotopic analyses needs complementary physical and chemical information on the rocks or on the separated minerals, as well as a geodynamic knowledge of the region where the samples were collected.

Isotope geochronology is a complementary approach to the various stratigraphic disciplines such as chemiostratigraphy, magnetostratigraphy, biostratigraphy, lithostratigraphy, chronostratigraphy, etc. ODIN (1995) gave a detailed description of "stratigraphic geochronology" as well as the relative terminology and applications. Therefore, these aspects are not detailed in the present chapter, which is rather intended to give an overall evaluation of the methods that might be used for direct and indirect isotopic dating of sedimentary successions. Basically, isotope geochronology has the unique potential of determining the age and the timing of mineral formations and of identifying and dating geological events, which none of the other tools and disciplines mentioned above is able to do. It is especially obvious that the contribution of isotope geochronology becomes particularly crucial when no biostratigraphic references are available in sedimentary successions, such as in Precambrian periods.

Whatever the method used, each isotopic determination firstly provides an **apparent age** (a data) associated with an analytical uncertainty, which needs to be interpreted in the light of independent criteria before being considered as a geologically meaningful age. The most complete knowledge of the selected minerals (the so-called "***geochronometers***" – ODIN, 1982) is crucial, as will be seen hereafter, in isotopic dating of sedimentary rocks.

The analytical uncertainty generally provided in publications refers to the analytical part of the work done in the laboratory. It is generally expressed at the 2σ level (95% confidence level). This uncertainty only takes into account the probability of obtaining the same result from a second independent aliquot of the same sample within the calculated uncertainty. Such uncertainty does not take into account the sampling effect. For instance, a single zircon grain cannot be analysed twice, nor can an area of a μm scale be analysed twice by "secondary ion mass spectrometry" (SIMS) spot analysis. In fact, the analytical uncertainty is part of three kinds of uncertainties in isotope geochronology for stratigraphic application, which were formalized by ODIN (1982). They include the stratigraphic uncertainty relating to the precise sampling location in a sedimentary succession and to the geological uncertainty depending on the specific characteristics of the selected geochronometers. These characteristics, which can add to the overall uncertainty, include the duration of the crystallization process of the geochronometers, the moment of the resetting of their isotopic system and their behaviour in changing geochemical contexts during post-formation history. Actually, it may be agreed on the fact that the overall uncertainty depends on the analytical methods and on the type and number of geochronometers analysed. In summary, a good geological background is needed to precisely identify the uncertainties related to the studied material. In addition, the combination of the use of more than one method on more than one single geochronometer is the best way to ensure the geochronological meaning of **isotopic ages** (e.g. U-Pb and $^{40}Ar/^{39}Ar$ on zircon and mica grains of volcanic rocks).

2. – DESCRIPTION OF METHODS

Unlike cosmogenic isotopes produced by cosmic radiation, the radiogenic isotopes that are discussed herein result from the decay of radioactive isotopes. They are combined in the K-Ar, K-Ca, $^{40}Ar/^{39}Ar$, Rb-Sr, Sm-Nd, Lu-Hf, Re-Os, U-Th-Pb methods and that based on the U-Th-Ra disequilibrium series.

Isotope geochronology is based on the measurement of the parent (**radioactive**) and daughter (**radiogenic**) **isotopes** and at least one stable isotope taken as reference. When isotope geochronology first began, the most widely used equipment was the thermo-ionization mass spectrometer (TIMS), the working principle of which is an acceleration of thermally emitted ions by an electrical field and separation of the various isotopes in a magnetic field according to their mass. This method needs a prior chemical purification of the element to be analysed, as the analysed masses need to correspond to the selected element. In the early 1980s, it became possible to simultaneously detect the currents of various ions on multiple Faraday cups, which resulted in a very high degree of accuracy in the isotopic ratios – for instance, it is now possible to measure the $^{87}Sr/^{86}Sr$ ratio with a relative uncertainty of 0.002%. At the same time, the development of high-resolution ion microprobes (highly sensitive SIMS with high-resolution such as "sensitive high-resolution ion microprobes": SHRIMP) enabled *in situ* determination of isotopic ratios on rock chips or thin sections without preliminary purification, with investigation sizes as small as 15 μm. A recent alternative to the use of the very costly SIMS is the coupling of a laser ablation system to a multiple-collector (Faraday cups and ion counters) inductively coupled plasma mass spectrometer (MC-ICP-MS). Finally, among the improvements in electron microprobe performance in the 1990s, a 'chemical' (i.e. non-isotopic) monazite geochronology became reliable because i) the quantity of daughter isotopes initially present in the studied mineral is completely negligible compared to the abundance following decay of the parent isotopes (^{232}Th, ^{235}U and ^{238}U), and ii) ^{232}Th is an abundant stoichiometric constituent of the mineral.

The K-Ar, $^{40}Ar/^{39}Ar$, Rb-Sr and Sm-Nd methods are already routinely applied to plutonic and metamorphic rocks, micas and hornblendes, to volcanic rocks and feldspars and to specific sedimentary minerals. The U-Th-Pb method is generally used to date igneous or metamorphic accessory minerals, such as zircon, monazite, titanite and apatite. In addition to zircon, the Pb-Pb method allows volcanic rocks and sedimentary carbonates to be dated. Alternatively, the K-Ca method has a limited application because it can only be applied to minerals not containing any Ca, such as K-feldspars, some micas and salts (e.g. sylvinite and langbeinite). The Lu-Hf method is generally used to characterize magmatic activities or mantle evolution and the use of the Re-Os geochronological method is mainly restricted to molybdenite, while the U-Th-Ra disequilibrium series are applied to very young materials, especially biogenic and inorganic carbonates. It has, for instance, been used to date coral skeletons and mollusks, but also young volcanic rocks.

2.1. POTASSIUM-ARGON METHODS: K-AR, $^{40}AR/^{39}AR$

K-Ar and $^{40}Ar/^{39}Ar$ isotopic systems are used routinely to date whole rocks and separated mineral species. The decay pattern of ^{40}K is slightly complex: 88.8% decay to ^{40}Ca (β^-) and 11.2% decay to ^{40}Ar in three ways, but mainly (11%) through electron capture with g emission during the return to the ground state of the excited ^{40}Ar. The additional ways are through electron capture without g emission (0.16%) and through a very rare $^{40}K \rightarrow {^{40}Ar} + \beta^+$ reaction (0.001%). The formula for the age calculation is:

$$t = 1/(\lambda_K + \lambda_\beta) \, \text{Log} \, [1 + (\lambda_K + \lambda_\beta) \times (^{40}Ar/^{40}K)]$$

where $\lambda_K = 0.581 \, 10^{-10}$ year^{-1} and $\lambda_\beta = 4.962 \, 10^{-10}$ year^{-1}. The $^{40}Ar/^{40}K$ ratio is obtained by measuring the radiogenic ^{40}Ar content by isotopic dilution and the K content by a physical-chemical method (e.g. atomic absorption) and then calculating the ^{40}K amount, the value of which, 0.0117%, is constant in nature except for a few exceptional samples.

The main problem of the method is related to the overwhelming occurrence of Ar in the atmosphere (99% ^{40}Ar). However, corrections for this atmospheric Ar can be made because air has a constant $^{40}Ar/^{36}Ar$ ratio. Also to be remembered are two scenarios that can complicate the interpretation of the obtained ages: i) a loss of Ar by diffusion due to the effects of thermal events, even of short duration and low intensity, which is common in surficial environments, and ii) an initial trapping of radiogenic ^{40}Ar associated with fluid circulation before closure of the geochronometer, which can result in the calculated ages being too old. In as much as the closing temperatures of most minerals are known and that some of them are efficient 'traps' of radiogenic ^{40}Ar, they can be selected to date ultimate thermal events affecting a given geological unit. This can be the case, for instance, for K-feldspar (150-200°C), biotite (300-350°C), muscovite and hornblende. For whole-rock determinations, the application of the K-Ar method is relevant for young (1 – 10 Ma) rapidly-cooled rocks (of volcanic type) because of the decay constants and the opening of the component minerals at relatively low temperature. Rocks with a vitrified matrix shall be discarded because Ar behaviour in glass is very variable.

The $^{40}Ar/^{39}Ar$ method is a technically interesting variation of the preceding method as the K content can be determined on the same aliquot as the radiogenic ^{40}Ar. To do so, neutron activation is used as it induces the following reaction:

$$^{39}K + \text{neutron} \rightarrow {^{39}Ar} + \text{proton (n, p).}$$

One can therefore measure ^{39}Ar by spectrometry in the same way as the $^{40}Ar/^{36}Ar$ ratio. The concentration of ^{39}Ar is directly proportional to ^{39}K with an abundance of 93.258%, and thus to ^{40}K in the rock. This means that with a single measurement one can determine the parent and daughter isotope fractions by comparison to an irradiated reference sample. This facility makes it possible to analyse a sample at increasing temperature steps until fusion. The radiogenic ^{40}Ar released at low temperatures is associated with fragile structural sites in the grains such as mechanical defects or chemically weathered areas. With increasing temperatures, Ar is extracted from more retentive sites of the analysed samples or minerals with the $^{40}Ar/^{36}Ar$ ages remaining constant and closest to the true crystallization age. The shape of the obtained age spectrum (flat, stair-case or saddle shaped) is indicative of the events that might have or not affected the mineral.

The two $^{40}Ar/^{39}Ar$ and conventional K-Ar methods coexist according to their applications: briefly, the conventional K-Ar method is used for rocks and minerals (micas, amphiboles) that have a simple evolutionary history and/or are fairly young (a few million years), whereas the $^{40}Ar/^{39}Ar$ method is better suited to minerals with a more complex history in the age range of a few to a thousand million years.

2.2. ISOCHRON TYPE METHODS

Isochron methods are based on the general equation expressed as follows for the Rb-Sr method (λ: radioactive decay constant = $1.42 \cdot 10^{-11}$ year $^{-1}$):

$$(^{87}Sr)_{t, mes} = (^{87}Sr)_o + (^{87}Rb)_{t, mes} (e^{\lambda t} - 1)$$

in which ^{87}Sr and ^{87}Rb are the total number of atoms at the present time for the isotopes, while $(^{87}Sr)_o$ is the number of atoms initially present at the time of closure of the system. Because isotope equipments are designed to record isotopic ratios, a stable isotope, ^{86}Sr in the case of the Rb-Sr method, is taken as a reference isotope:

$$(^{87}Sr/^{86}Sr)_{t, mes} = (^{87}Sr/^{86}Sr)_o + (^{87}Rb/^{86}Sr)_{t, mes} (e^{\lambda t} - 1)$$

If the isotopic constitution of the Sr is initially the same for all the analysed samples (rocks and/or minerals) and if the system has remained closed, the data points define a line, also known as an isochron, when the age is geologically meaningful. Its slope $(e^{\lambda t} - 1)$ allows the age "t" to be calculated, while the intercept gives the initial isotopic ratio of the daughter element. However, it is essential to remember that an excellent alignment with a correlation coefficient close to 1 is not automatically an isochron. It may also be a mixing line between two end-members of different composition giving rise to intermediate compositions, initially leading to a mixing line.

The methods are applicable to both whole rocks and separated minerals. A minimum of five or six analyses is necessary to define properly the isochron slope, but the calculated age is only valid if the various samples crystallized from a single, isotopically homogeneous, magmatic or volcanic system, or were entirely homogenized during a high-grade metamorphic event. In the case of whole-rock dating, the magmatic liquid needs to have been differentiated during crystallization in order to obtain a spread of points in the isochron diagram reflecting variations in the parent/daughter elemental ratio. In the case of separated-mineral analyses, the data-point spread is often wide and analytically favourable, but the obtained age will only be significant if closure in each of the various mineral systems was simultaneous. The following isochron type methods might be useful for stratigraphic purposes in dating magmatic, volcanic or metamorphic rocks: i) Rb-Sr with a range cover from 10 Ma until the Earth's beginning (4.6 Ga); ii) Sm-Nd with a very similar geochemical behaviour for the two rare-earth elements, which should incline to use only separated minerals, because only they can provide a spread of the analytical points; iii) Re-Os with an interesting application for dating molybdenite, which is Re-enriched and Os-depleted; and iv) the Lu-Hf method.

2.3. U-TH-PB METHOD ON SEPARATED MINERALS

The U-Th-Pb method is based on three decay chains of ^{238}U, ^{235}U and ^{232}Th, which might be considered in secular equilibrium for ages greater than 1 Ma. This approach is based on the following simplified nuclear reactions:

$$^{238}U \rightarrow {}^{206}Pb \ (\lambda = 1.55125 \cdot 10^{-10} \text{ year}^{-1}),$$
$$^{235}U \rightarrow {}^{207}Pb \ (\lambda = 9.8485 \cdot 10^{-10} \text{ year}^{-1}) \text{ and}$$
$$^{232}Th \rightarrow {}^{206}Pb \ (\lambda = 0.49475 \cdot 10^{-10} \text{ year}^{-1}).$$

The association of the first two chronometers is the most accurate and the most widely used for obtaining relatively old ages (>10 Ma). The selected minerals must contain U and Th and, if possible, be devoid of any initial Pb. The daughter/parent ratios (i.e. $^{206}Pb/^{238}U$ and $^{207}Pb/^{235}U$) are corrected by measuring the $^{206}Pb/^{204}Pb$ isotopic ratio where the non-radiogenic ^{204}Pb isotope is taken as reference. The correction assumes knowledge of the $^{206}Pb/^{204}Pb$ and $^{207}Pb/^{204}Pb$ crystallization ratios of the mineral to be dated. These are approximate ratios, referred to as "common Pb", and are given by a trend model of the Pb isotope in the Earth's crust (STACEY & KRAMERS, 1975). The data corrected for initial Pb can then be plotted in a $^{206}Pb^*/^{238}U = f(^{207}Pb^*/^{235}U)$ Concordia diagram (WETHERILL, 1956). The line fitted through the data refers to samples for which the $^{206}Pb^*/^{238}U$ age is identical to the $^{207}Pb^*/^{235}U$ age. The most frequently used mineral is zircon, known to be highly refractory to any further alteration and containing very little initial Pb. Minerals with a simple history plot directly on the Concordia and consequently provide directly the age of their crystallization. The data plots of larger aliquots often deviate from the Concordia curve as they represent mixtures of two mineral generations, because partial reopening of the initial state can be expected during later metamorphism. Specific loss of radiogenic Pb could also have occurred in weathered/altered and/or in U-enriched minerals; a loss of radiogenic Pb by diffusion is often envisaged for the latter and the age analysis is then considered to be "discordant".

Two techniques are basically used to obtain useful data in a Concordia diagram: i) according to the conventional method developed by KROGH (1973), by dissolving a mono-mineral in an acid solution to which artificial Pb and U isotopic tracers (spikes) are added, and by measuring the Pb and U isotopic ratios by TIMS; the U and Pb concentrations are determined by isotopic dilution with the same equipment, and ii) according to a more recently developed method based on the design and engineering of a high-resolution (M/ΔM ~5000) ion microprobe that makes it possible to carry out the same type of measurement but *in situ* at a 15 to 20 µm scale (COMPSTON et al., 1982). In the case of routine analyses (COCHERIE et al., 2005a), the absolute uncertainty is between 1 and 3 Ma for the first method and between 3 and 5 Ma for the second method, regardless of the considered age range.

A recent study by COLEMAN (2002) showed that a very high degree of accuracy can be attained with the conventional U-Pb method: analysis of zircon grains from different parts of the same granite pluton (Tuolumne, USA) provided ages between 92.6 ± 0.2 and 88.1 ± 0.2 Ma, inclining the author to suggest that the granite emplacement lasted over a period of 5 Ma. This case outlines the fact that an accuracy of <1 Ma is necessary to evaluate accurately the duration of the continuous filling of a magma chamber that could have lasted several millions of years. Also, a recent improvement to the conventional zircon U-Pb chemical abrasion method (MATTINSON, 2005) makes the analyses more concordant in a Concordia diagram and increases the precision to 0.1%.

The recent development of a laser coupled to a MC-ICP-MS and a multi-ion counting system outlines the potential of reaching the same resolution and analytical precision as an ion microprobe in saving time and money (COCHERIE et al., 2005b).

The electron probe microanalyser (EPMA) is also potentially very suited for determining *in situ* ages on U and Th-enriched minerals (Suzuki & Adachi, 1991). After a decade of tests, it is now demonstrated that this method can provide reliable ages on monazite at a 1-2 µm scale, making it both a complementary and an alternative method to the isotopic U-Pb (ID-TIMS) method (Montel et al., 1994, 1996; Cocherie et al., 1998, Williams et al., 1999). The grains can be dated using EPMA either directly on thin sections or on separated grains mounted as polished sections. Thin sections are preferable because they allow small grains that are difficult to separate (~10 µm) to be analysed and because they can retain the textural constraints associated with the calculated ages (Williams et al., 1999; Be Mezème et al., 2005). The second procedure tends to statistically overestimate the proportion of grains larger than 50 µm, but it almost avoids analysis of grain borders with an unknown shape beneath the surface and consequently erases the related potential matrix effect (Pyle et al., 2005). It allows the study of complex zoning in large monazite grains (Cocherie et al., 1998, 2005a). Using a suitable isochron diagram, this "chemical" method is able to provide reliable ages on monazite with a precision of about ± 5 Ma on a set of analyses performed at a µm scale.

Additional information about these methodological and technical aspects of the isotope methods can be found in Faure & Mensing (2005).

3. – TECHNICAL ASPECTS

The geochronometers suitable for stratigraphic applications can be whole rocks, minerals or organic components of volcanic, plutonic or sedimentary origin. In some cases, metamorphic materials might also be helpful to solve stratigraphic questions. Very often, the choice of the method depends initially on the material available and the purpose of the analysis. The intrinsic qualities of the material (i.e. its resistance to post-formation alteration induced by weathering or temperature increase) needs to be taken into account. Also, the geological context in which the material has evolved as well as the potentials and limits of the analytical methods must be carefully evaluated. For instance, volcanic zircons are very resistant to meteoric weathering, but the U-Pb method used to date them cannot be applied to relatively young materials because of the very long half-life of the radioactive isotopes. Furthermore, the Sm-Nd method, which relies on the analysis of rare-earth elements, may be among the best methods for dating minerals formed in superficial settings, as rare-earth elements are among the least mobile elements in these settings. However, because of analytical precision, the half-life of the radioactive isotope does not allow any application to materials younger than 100 Ma.

These different aspects point to the fact that the rules in the application of isotope geochronology to stratigraphic purposes cannot be rigid. They also explain why published attempts have not always properly achieved the planned goals. This highlights the necessity of a critical analysis in the available studies dealing with stratigraphic application to balance the best compromise between goals, methods and appropriate analysed material. To do so, we have selected several studies dealing with direct and indirect isotopic dating of sedimentary rocks, in which a proper identification of the best-suited material and a realistic evaluation of the homogenization processes potentially affecting this material during crystallization are discussed.

3.1. WHICH GEOCHRONOMETERS?

The available experience in isotopic dating of sedimentary rocks (Clauer & Chaudhuri, 1995) shows that reliable geochronometers are not numerous, assuming that isotopic dating of bulk rocks is expected to have a stratigraphic application only if its components are rigorously homogeneous, such as for volcanic lavas. In fact, sedimentary rocks are usually heterogeneous, which means in turn that the components that potentially formed during sedimentation need to be extracted and isolated. The basic quality of a good geochronometer is its precise location in a stratigraphic succession and its capability "to record" the evolution of its host succession, that is to say its capability of registering thermal and chemical changes that might have occurred after deposition. This is generally not the case for minerals of plutonic rocks, in which cooling may obscure or even erase their crystallization age compared to the time of the intrusion. For this reason, any isotopic dating of plutonic material with a stratigraphic application has to be considered carefully in favouring the combination of several isotopic methods applied to varied cogenetic or successive minerals.

Alternatively, volcanic lavas and associated pyroclastic deposits contain excellent geochronometers. These include zircons for which, however, the most reliable ages must be calculated from concordant U-Pb data. Volcanic lavas also contain biotite, which is a basic geochronometer for the K-Ar, $^{40}Ar/^{39}Ar$ and Rb-Sr methods, amphibole, which is precisely datable by $^{40}Ar/^{39}Ar$ and less by K-Ar, and feldspars. Widely used in stratigraphic applications using volcanics, feldspars require some attention as they are quite sensitive to external processes. The selection of volcanic geochronometers is often relatively simple: in addition to an extreme purity of the fractions, the individual crystals need to be free of any meteoric weathering.

In sedimentary rocks, reliable geochronometers are few, as stated above, because of their sensitivity not only to meteoric weathering but also to post-sedimentary processes, such as burial or hydrothermal temperature increase. The inventory of suitable minerals is limited to some clay minerals, salts, alunite and Ca-carbonates. The latter can be dated indirectly but precisely (justifying the use of the term "high resolution stratigraphic dating"), which is a method that does not follow the classical principle of radioactive decay in a closed mineral system (Faure in Odin, 1982) and the potential of which can be evaluated in the publication by Stille et al. (1994).

Among clay minerals, glauconites, for which a complete review of the various aspects of their use was published by Odin and collaborators (1982), are among the most studied. However, despite very impressive results such as those of Odin & Hunziker (1982), Rousset et al. (2004) (Fig. 6.1) and others, recent studies of Smith et al. (1993) and Clauer et al. (2005) question some aspects of mineral homogeneity of glauconite fractions even when considered pure. Illite and the mixed layers illite-smectite have also been largely used, especially in Precambrian sediments (e.g. review in Clauer & Chaudhuri, 1995). Attempts also exist with smectites and palygorskite. These sedimentary minerals can be dated by the traditional K-Ar, Rb-Sr and Sm-Nd methods.

3.2. SPECIFIC CHARACTERISTICS OF SEDIMENTARY GEOCHRONOMETERS

Stratigraphic dating has been applied successfully for a long time to minerals of volcanic origin. On the contrary, its application to sedimentary minerals has often been questioned, because of their theoretical inability to retain

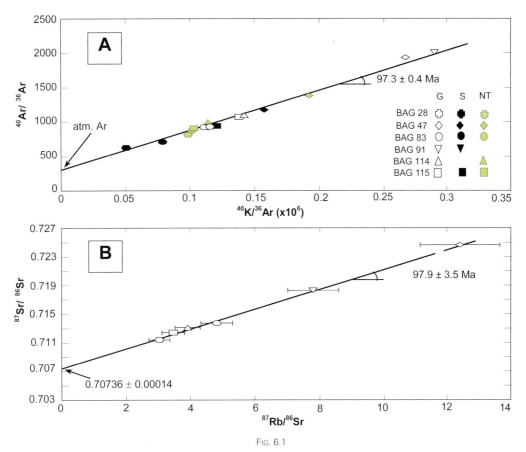

FIG. 6.1

K-Ar and Rb-Sr data of glauconite separates from Albian siltstones of southeastern France. (A) K-Ar isochron plot for varied glauconite fractions (NT = non-treated glauconite separates; S = suspended matter after ultrasonic treatment; G = glauconite grains after ultrasonic treatment. (B) The same G samples plotted in a Rb-Sr isochron diagram (after ROUSSET et al., 2004).

radiogenic isotopes, especially Ar, during post-sedimentary evolution. Recent studies on the smallest technically separable clay particles (width 20-30 nanometers and thickness 1-2 nanometers) have shown that clay particles of illite type are retaining the whole amount of radiogenic ^{40}Ar produced from decay of radioactive ^{40}K (CLAUER et al., 1997). In addition, this diffusion aspect is far less determining than the effects of meteoric weathering, which can be avoided through careful sampling. Thermal experiments in the laboratory and analysis of material collected in sedimentary rocks subjected to progressive temperature increase concur in suggesting that a significant thermally-induced Ar diffusion requires significant burial depths (several thousands of meters) during tens of millions of years.

Until about twenty years ago, the sample preparation and the characterization of the separated material might have contributed to maintaining a suspicion against the stratigraphic dating of sedimentary minerals. Indeed, when isotopic methods were first applied to clay minerals, the procedure consisted in crushing the rocks, often of pelite type, separating the fraction <2 µm enriched in clay minerals by sedimentation in distilled water and characterizing the separated fraction by X-ray diffraction. This method, which was not applied to glauconite-type separates purified by magnetic enrichment, was gradually improved:

– by replacing the crushing technique, which reduces millimeter-sized detrital components to the size of the clay fraction, by cryostatic disintegration, which better preserves the natural size of the different components of the rocks;

– by refining the separation of the clay fraction with ultra-centrifugation, which makes it possible to decrease the size range of the particles to 0.2 µm and even to 0.02 µm in some cases;

– by supplementing the mineralogical characterization through scanning and transmission electron microscopic observations.

3.3. INITIAL ISOTOPIC HOMOGENIZATION

Besides the need for any mineral to represent an isotopically closed system, another aspect that must be taken into account in stratigraphic isotope dating is the initial isotopic homogenization among the same geochronometers. If initial isotopic homogenization can be proven for most volcanic and even plutonic materials, this is not the case for sedimentary geochronometers. Thus, ODIN & MATTER (1981) and ODIN et al. (1982) worked on the isotopic dating of glauconitic minerals, which certainly have the best potentials for stratigraphic isotopic dating among the materials of sedimentary rocks. In this context, CLAUER et al. (1992) showed that the process of glauconitization requires a specific microenvironment in which Fe-bearing smectites evolve into glauconies, then into glauconites by progressive incorporation of K. They showed that the process of glauconitization can be subdivided into two steps: i) a progressive dissolution-crystallization of detrital clay-type material embedded in coprolite grains, followed by ii) a crystal growth in a sea water-influenced environment, tending to an isotopic homogenization with sea water. Finally, only the glauconitic material containing at least 6 to 6.5% K_2O attains isotope equilibrium in the formational environment, so that only this latter material can be used for stratigraphic dating.

A similar behaviour was observed in mixed layered illite-smectite of bentonite, which makes it possible to consider

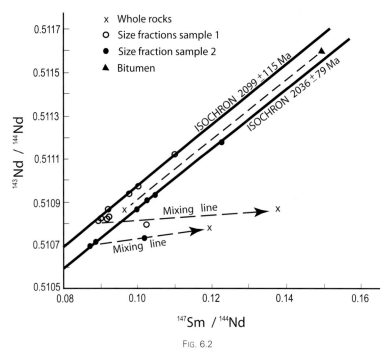

Fig. 6.2

Sm-Nd isochron plot showing two isochrones, consisting in the data points of varied size fractions of two different samples (3002 and 3004) as well as the plot of the organic poles (the bitumen and the organic-rich whole rock 3005) (after Bros et al., 1992).

that ***isotopic homogenization*** of clay minerals can occur shortly after the time of deposition, even when dispersed in a restricted environment that could have been isolated from a marine environment by the low permeability of the rock (Clauer et al., 1997; Rousset & Clauer, 2003). A consequence of this observation on the isotopic dating of clay minerals is that one cannot systematically postulate by convention that the Sr, Nd, Pb and Ar isotopic ratios of the environment of formation yield a marine signature, even if the depositional setting is marine. However, the difference between the isotopic compositions of the environment of formation and the marine environment is likely to be small. But the difference can lead to an underestimation of the mineral age when the isotopic ratios of the environment of formation are higher than the normal marine values. This effect is larger on young material characterized by lower Rb/Sr, Sm/Nd or K/Ar ratios.

4. – DIRECT ISOTOPIC DATING OF SEDIMENTARY ROCKS

Many studies of stratigraphic isotopic dating performed on different types of sedimentary rocks, or of other types of rocks, have been reported by Odin (1982). These aspects will, therefore, not be presented again as they are available elsewhere. However, it shall be recalled that dating glauconite-type separates does not systematically provide data that are useful for stratigraphic purposes. The further evolution of the host rocks is to be considered and even the formation process of the glauconite pellets has to be evaluated carefully as suggested recently by Smith et al. (1993) and Clauer et al. (2005), who both show that detrital "impurities" in the glauconite pellets significantly bias the results. A few examples are reported hereunder, to evaluate some of the recent developments of the methods or the new fields of application.

4.1. K-AR, RB-SR AND SM-ND DATING OF PRECAMBRIAN SEDIMENTS

As stated at the beginning of this chapter, isotopic dating for stratigraphic purposes is particularly appropriate for Precambrian rocks, in which biostratigraphy cannot be applied. Among recent work, it certainly is interesting to report the few Sm-Nd isochron studies made on Precambrian sedimentary rocks, as this method was theoretically inapplicable to these rocks because of the very limited mobility of the elements in supergene conditions.

The first Sm-Nd dating of Proterozoic iron-bearing sedimentary rocks ("*banded-iron formation*") from Canada provided an isochron Sm-Nd age of 2076 ± 248 Ma (Stille & Clauer, 1986). The K-Ar and Rb-Sr apparent ages of the same samples were significantly younger (about 1470 Ma). The Sm-Nd age was considered to approach closely the time of deposition of the formation, while the K-Ar and Rb-Sr values were interpreted as resulting from further recrystallizations initiated by diagenetic events occurring about 600 Ma later in the formation. These results also outlined the differential behaviour of the three methods with a much more resistant Sm-Nd system to late alterations than the two others. Bros et al. (1992) obtained a similar result by studying the Sm-Nd, Rb-Sr and K-Ar systems of the clay fractions from Lower Proterozoic pelites of Gabon. The clay fractions of two samples were separated in varied sizes, which all together provided two Sm-Nd isochron ages of 2099 ± 115 and 2035 ± 79 Ma. These ages were interpreted as resulting from an early isotopic homogenization of the clay material. The K-Ar and Rb-Sr apparent ages appeared again to be significantly younger at about 1800 Ma. Again, the Sm-Nd system of the clay-type material appears more resistant to any further alteration than the K-Ar and Rb-Sr systems, which have a similar behaviour as they provide similar data. Bros et al. (1992) reported a third Sm-Nd isochron with a whole-rock enriched in organics and a sample of bitumen parallel to the two other isochrons. This feature points to the existence of a Nd isotope homogenization among the organic matter and the clay material. In addition, the Nd isotope homogenization is only achieved at a small scale, which is that of each sample size, confirming the limited mobility of these rare-earth elements in surficial conditions (Fig. 6.2).

In summary, the new information is about the potential of the Sm-Nd method, which appears to be of interest for stratigraphic dating of Proterozoic sedimentary rocks. It is able to remain close even during diagenetic to hydrothermal alteration that is common in such old material. For additional information about this contrasted behaviour of the different isotopic clay systems and also about the oldest isotopic age obtained on clay material at the Earth surface, the reader is directed to the publications by TOULKERIDIS et al. (1994; 1998).

4.2. OTHER METHODS OF STRATIGRAPHIC DATING OF SEDIMENTS

Among other recent developments in isotopic methods likely to provide stratigraphic data on sedimentary materials is the $^{40}Ar/^{39}Ar$ method, which has been tested on different clay minerals (FOLAND et al., 1984; KUNK & BUSEWIITZ, 1987; ZWINGMANN, 1995). Its applicability to sedimentary materials remains controversial: experimental work on irradiation is a critical aspect and it is essential that its effects on the mineral structures are perfectly known and understood before any routine application. However, despite the fact that many results of $^{40}Ar/^{39}Ar$ dating of glauconitic minerals (e.g. FOLAND et al., 1984; YORK et al., 1981; BELL, 1985) provide rather negative conclusions because of the secondary effects induced by the necessary irradiation of the samples, a recent study of SMITH et al. (1993), as already mentioned above, opens possible applications. The major step in the application of this method is still in the understanding of the effects of the necessary radiation on the mineral structures. In particular, there is the "retention" of the ^{39}Ar produced by neutron activation from ^{40}K present in the measured separates, and the diffusion of ^{40}Ar due to the temperature of the reactor, which is often higher than that during crystallization of the analysed minerals.

Besides the clay minerals and the methods described above, other geochronometers and methods are potentially applicable to the isotopic dating of sedimentary material. Among these is that based on the comparison of the isotopic composition of Sr in marine carbonates, sulfates, oxides and phosphates to the contemporaneous marine reference, which is known to change through time making it possible to "date" these mineral phases by comparison with a reference (e.g. STILLE et al., 1994). This method is described in chapter III (Chemiostratigraphy).

The widespread deposition of carbonate rocks during the Precambrian has also encouraged their isotope dating for stratigraphic use, especially by the Pb-Pb method. However, the first attempts of U-Pb and Pb-Pb datings were rather disappointing and only in the mid-1980s did Pb-Pb dating of metamorphic limestones renew interest in this application (MOORBATH et al., 1987). Since then, many applications to Paleozoic or Proterozoic carbonates have been published, because the isochron ages could be attributed to known sedimentation or diagenetic events.

5. – INDIRECT ISOTOPE DATING OF SEDIMENTARY ROCKS

Once the principles of plate tectonics had been established, the next step was dating the major phases of the various plate movements from creation to resorption. This implied dating the associated thermal events, which occur in various ways: emplacement of plutonic bodies (granite, etc.), extrusion of lava flows, metamorphism with total or partial recrystallization of given minerals, etc. Some of these activities may occur in sedimentary rocks and therefore their occurrence can be useful from a stratigraphic point of view. As already stated, the general principle of isotope geochronology is to measure the time "t" since the closure of the considered geological system, i.e. since the solidification of minerals and/or rocks. In the case of metamorphic activity, measuring the age of partial reopening of isotopic systems, such as during diagenetic or hydrothermal activities in sedimentary rocks (by using separated minerals for instance) can sometimes be of interest, especially in the application discussed here.

The evolution of refractory minerals is often quite complex, with several events being recorded by the same grain. Consequently, a global analysis, even of a single grain, can yield an analytically precise age but of no geological significance because it reflects a mixture of several crystallization or recrystallization stages. However, by grinding a polished section to the crystal equator level of a mineral (a zircon in this case), preliminary observations by a scanning-electron microscope equipped with cathodoluminescence imaging and a transmitted-light optical microscope, help to determine the poorly crystallized grains and areas (probably releasing radiogenic Pb). This knowledge ensures that the in situ ion-microprobe analysis, if working on small areas (<20 µm), will avoid the crystal defects, fractures and inclusions likely to contain initial Pb, and it makes it possible to focus on the fresh homogeneous zones able to provide easily interpreted concordant analyses.

A case study is given by zircons extracted from a migmatitic gneiss in French Guiana close to the Brazilian border. The grains were divided into two populations: those being elongate and relatively pale by cathodoluminescence observation, and those being smaller, round and very dark. It could also be observed that the elongate grains were surrounded by a dark overgrowth. The analyses were carried out using the SHRIMP II ion microprobe (SIMS) of the Australian National University (Canberra, Australia). The concordant $^{207}Pb*/^{206}Pb*$ spot ages corrected for initial Pb of the elongated zircons gave ages around 2170 Ma, whereas those of the round grains and the borders of some elongated grains gave significantly younger ages of 2100 Ma. Except for two analyses, all plot near the Concordia in two distinct populations: one at 2094 ± 7 Ma and the other at 2173 ± 9 Ma (Fig. 6.3) calculated at 95% confidence (2σ). Monazite grains extracted from the same rock yield a single U-Th-Pb "chemical" age (electron microprobe) of 2168 ± 3 Ma (Fig. 6.4; COCHERIE et al., 2005c). The spot dating of monazite and zircon provides the context of the evolutionary history of the migmatite rock body. The first elongated zircon grains and monazite initially crystallized concurrently around 2170 Ma, dating the emplacement of the initial (the protolith) migmatitic rock. About 70-80 Ma later, an intense metamorphic event induced partial melting of the protolith and favoured the formation of migmatitic gneiss (or migmatite). Some of the protolith zircon grains resisted the metamorphic event, while new zircon formed, both as crystals and as overgrowths on the earlier grains. This study also outlined the very refractory nature of monazite giving the age of the protolith and not the age of the migmatization, which is as one would expect considering the closure temperature of this mineral often assumed to be less refractory than zircon. This study shows the key contribution of in situ spot age determinations by making possible the analysis of very small homogeneous areas of high crystallographic quality, even if the available minerals are rare, small and inhomogeneous from the geochronological point of view.

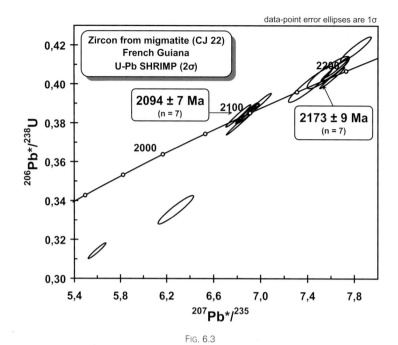

Fig. 6.3
U-Pb Concordia diagram for the spot analyses carried out on zircons from migmatitic gneiss of southern French Guiana. To facilitate the reading of the diagram, the error ellipses are shown at ± 1σ, although the calculations were done at ± 2σ.

One common stratigraphic application of isotope geochronology is to establish the geochronologic time scale of chronostratigraphic units. For instance, Jurassic chronostratigraphic units were defined on ammonite biochronology based on a previously established zonal standard from Europe. In this respect, PÁLFY et al. (2000) provided a high precision isotopic age data base considering a selection of U-Pb and $^{40}Ar/^{39}Ar$ ages of volcanoclastic rocks with precisely known stratigraphic age. They estimated the boundary of Jurassic stages with a precision of 2 to 5 Ma and sometimes even better.

6. – CONCLUSIONS

Many isotopic methods can potentially be applied to sedimentary material for direct stratigraphic isotope dating as well as to plutonic, volcanic and even metamorphic material for indirect stratigraphic application. However, these geochronologic studies are only successful if based i) on a rigorous approach implying careful location and identification of the selected rocks, ii) on a strict selection of the geochronometers (mainly, if not only, mineral separates) in

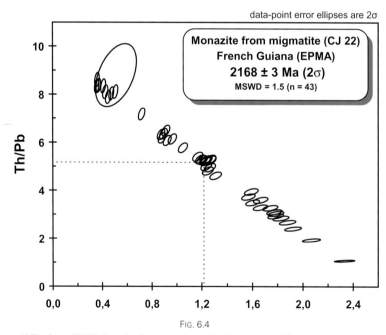

Fig. 6.4
Plot of Th/Pb vs. U/Pb, from EPMA data, for five monazite grains from migmatitic gneiss of southern French Guiana. The regression line fits well with the theoretical isochron (dashed line) at 2168 Ma, allowing the mean age at the centroid of the population to be calculated, where it is the most precise. Ellipse errors and error calculations are done at ± 2σ.

most applications, iii) a detailed study of the geochronometers (mineralogy, morphology and geochemistry), and iv) an integration of complementary information (sedimentologic, geodynamic and structural). The overall approach is also strengthened when several isotopic methods are applied to varied mineral separates.

Beyond the traditional application of isotopic methods for stratigraphic application, the novelties that have been tested and applied in the last decade or so include the application of Sm-Nd systematics on clay material of Precambrian sedimentary rocks, as well as new technological developments allowing *in situ* analysis by microprobe equipment or *in situ* laser ablation coupled to ICP-MS. These latter applications should now be tested on sedimentary material. Future research in this field will probably focus more on improvements to equipment rather than on a wider application of the existing methods.

CHAPTER 7
SPECIFIC STRATIGRAPHIES

P. Lebret, R. Capdevila, M. Campy, M. Isambert, J.P. Lautridou,
J.J. Macaire, F. Menillet, R. Meyer & A. De Goër de Herve

CONTENTS

1. – STRATIGRAPHY OF THE METAMORPHIC
AND PLUTONIC TERRAINS
(R. Capdevila) 101

   1.1. The study of lithology, space-relations
and chronology of plutonic and metamorphic
bodies is pertaining to stratigraphy 101

   1.2. How should plutonic and metamorphic bodies
be considered? as lithostratigraphic units
or as units of lithostratigraphic class or
as different units of class? 101

   1.3. Stratigraphic classification of bodies of igneous
and metamorphic rocks 102

      1.3.1. Bodies of volcanic rocks and low-grade
metavolcanic and metasedimentary rocks 102

      1.3.2. Bodies of unstratified plutonic
and metamorphic rocks 102

      1.3.3. Layered intrusions 103

2. – GEOCHRONOMETRY OF PRECAMBRIAN TIME
(R. Capdevila) 103

   2.1. Principles of geochronometric subdivision
of the Precambrian 104

   2.2. Subdivision method 104

   2.3. Geochronometric units of the Precambrian 104

      2.3.1. Archean 104

      2.3.2. Protérozoic 106

3. – STRATIGRAPHY OF SURFICIAL FORMATIONS
(P. Lebret (coord.), M. Campy, M. Isambert,
J.-P. Lautridou, J.-J. Macaire, F. Menillet & R. Meyer) 106

   3.1. Définition 106

   3.2. Vocabulary: surficial formations ; study concepts
and units 107

   3.3. Study methods 107

4. – QUATERNARY STRATIGRAPHY
(P. Lebret (coord.), M. Campy, M. Isambert,
J.-P. Lautridou, J.-J. Macaire, F. Menillet & R. Meyer) 108

   4.1. Définition 108

   4.2. Study methods 108

      4.2.1. The Quaternary in an oceanic environment 108

      4.2.2. The Quaternary in a continental environment 108

   4.3. Practice: stratigraphic scales for the Quaternary 109

5. – STRATIGRAPHY OF VOLCANIC TERRAINS
(A. de Goër de Herve) 109

   5.1. Presentation 109

   5.2. Vocabulary 109

      5.2.1. Volcanic and volcanogenic materials 109

      5.2.2. Facies associations and lithostratigraphic
units 111

      5.2.3. Questionable vocabulary 112

   5.3. Practice 113

      5.3.1. Periodicity 113

      5.3.2. Dating 113

1. – STRATIGRAPHY OF THE METAMORPHIC AND PLUTONIC TERRAINS

1.1. THE STUDY OF LITHOLOGY, SPACE-RELATIONS AND CHRONOLOGY OF THE PLUTONIC AND METAMORPHIC BODIES IS PERTAINING TO STRATIGRAPHY

Stratigraphy is generally considered as the science of sedimentary terrains. However, very early stratigraphy has integrated the study of volcanic rocks interstratified with the sediments and it also considered plutonic intrusions and the metamorphic basement as geochronological markers. The recognition of volcanic rocks belonging to different magmatic series has become very important in the reconstruction of palaeoenvironmental and palaeogeographical models. In addition, for many years batholits and high grade metamorphic terrains have been subdivided into lithological units whose space-time relations are studied as the "inside stratigraphy". Nowadays the contribution of the study of plutonic and metamorphic terrains in the framework of the earth crust history has become as important as that of sedimentary rocks. Stratigraphy must therefore integrate all kind of rocks. This is what stated for example by the *International Subcommission on Stratigraphic Classification* (1987) and the *North American Commission on Stratigraphic Nomenclature* (1983), which widely deal with the stratigraphy of the metamorphic and igneous bodies (Fakundini & Longrace, 1989; Salvador, 1987). The present book is in agreement with the philosophy of the above commissions.

1.2. HOW SHOULD PLUTONIC AND METAMORPHIC BODIES BE CONSIDERED? AS UNITS OF LITHOSTRATIGRAPHIC CLASS OR AS DIFFERENT UNITS OF CLASS?

Principles and methods governing the stratigraphic study of plutonic and metamorphic terrains partly differ from those at the base of studies of the sedimentary terrains. Therefore a question is: must plutonic and metamorphic bodies, in particular when non-stratified, be considered to be lithostratigraphic units? Presently, the ISSC recommends to consider them as lithostratigraphic units whereas the NACSN define plutonic and metamorphic bodies as units of different class, so called lithodemic. This opinion was followed by the Norsk Stratigrafiska Komitè (Norvay) and the lithodemic nomenclature was integrated in the *Glossary of Geology* of the American Geological Institute. Finally, the stratigraphic nomenclature used for the study of layered intrusions is even different and codified by Irvine (1987).

The different systems of stratigraphic classification of bodies of plutonic and metamorphic rocks (by ISSC and NACSN; Hedberg, 1976; Laajoki, 1989; Hattin, 1991) all found on valid scientific arguments and therefore is difficult to predict wether the recommendation will be largely accepted in the future. Geologists who study non stratified crystalline terrains are inclined to apply the lithodemic classification whereas petrologists of the layered intrusions continue to use the Irvine's nomenclature. The present work made the choice of the pragmatism. We maintain the lithostratigraphic classification for the sedimentary terrains, we recommend the use of lithodemic nomenclature for the stratigraphic study of non stratified terrains and the Irvine's classification, sligthly modified, for the layered intrusions.

1.3. STRATIGRAPHIC CLASSIFICATION OF BODIES OF IGNEOUS AND METAMORPHIC ROCKS

Many cases have to be considered, following the nature of the studied bodies and the employed stratigraphic principles.

1.3.1. Bodies of volcanic rocks and low-grade metavolcanic and metasedimentary rocks

Bodies of volcanic rocks interstratified with sedimentary rocks have been always considered as lithostratigraphic units (e.g.: Marsac Formation, rhyodacites, Armorican Massif). Bodies of metavolcanic and metasedimentary rocks where the pristine stratification is still well recognized are considered litostratigraphic units as well (e.g. Armorique-Trédrez Formation, metabasalts and metagraywackes metamorphosed to green schists facies, Armorican Massif). The general rule is to consider the bodies of metamorphic rocks which suffered negligible transformations as lithostratigraphic units so that principles of stratigraphy can be applied and their relative chronology can be established on the field.

1.3.2. Bodies of unstratified plutonic and metamorphic rocks

Classical principles of stratigraphy can not be applicable to non stratified bodies. The latter are therefore considered as lithodemic rather than litostratigraphic. In order to show the use of the lithodemic classification, we are going to deal with the case of large granitic batholiths and that of the high grade metamorphic rocks of crystalline terrains. This classification is applied to all the non stratified plutonic and metamorphic bodies.

Batholiths

A granitic pluton is a massive intrusive body, consisting of one or more magmatic intrusions with an area in outcrop in the order of 100 – 200 km^2. A batholith is a linear group of plutons whose length can exceed a thousand of km and width that can reach a hundred of km. It is constituted by hundreds of differents superimposed and conformable plutons and its complete cooling may exceed a time interval of 50 Ma. For example, the oldest plutons of the Coastal Batholith of Peru are 100 Ma old whereas the youngest are 35 Ma old. The relative chronology among plutons is based on the principles of cutting and enclosing (replacing in this case the principle of superimposition): a pluton B intruding a pluton A, or cutting a pluton A, and/or enclosing inclusions of A, is younger than the latter. A pluton is rarely homogeneous and constituted by only one lithostratigraphic unit. In certain places, it can show gradual zoning from core to rim, which can represent an *in situ* evolution of the magma (facies variation). Generally plutons consist of different lithostratigraphic units (tonalites, granodiorites, granites) corresponding to injections of previously differentiated magma batches. The study of the framework of the pluton, the structures of each unit, and most of all of the contact between adjacent units, allow establishing different timing of intrusion or their contemporaneous injection (synplutonism).

If applied to batholiths, the lithodemic hierarchy is the following: the fundamental unit is the **lithodem** (intrusive), which corresponds to the formation among the lithostratigraphic nomenclature. It is an intrusive body which can be mapped, homogeneous, showing gradual zoning, and often constituted by small and separated magmatic intrusions, generally cogenetic and coheval. Practically a pluton is an intrusive lithodem. With respect the lithodem (sheet, septa, zone) the units of lower rank are presently considered as informal. With respect to the lithodem the unit of higher rank is the **suite** (corresponding to the **group**). An intrusive series is a group of akin and contemporaneous plutons, showing the same lithological sequence, chemical trends, inclusions, synplutonic veins and roughly of the same age. The term **supersuite** (corresponding to supergroup) is often employed when the series are of regional importance, constituted by tens of different plutons, separated and distributed over hundreds of km of distance. For example, the Lima segment of the Coastal Batholith of Peru was subdivided into tens of superseries having different ages and nature.

This nomenclature can be used when the plutons are not associated in large batholiths of west America type, for example in the European hercynian chain.

High grade metamorphic crystalline terrains

These crystalline terrains are basically constituted by high grade metamorphic rocks, of ortho, para or undetermined type, and by migmatites. All these rocks are often associated with granitoids. The term "non stratified", applied to high grade gneissic rocks, indicates that, on the field, criteria allowing to establish the relative chronology of their components can not be observed anymore. The fabric of the gneissic rocks is the result of tectonic and metamorphic processes and generally it does not reflect the original stratification of the sedimentary or volcanic protoliths. In order to establish the relative chronology of the crystalline terrains, particular methodological approaches are requested. Swarms of basic veins emplaced between two orogenic phases are frequently used as chronological markers. For example, among the Archean rocks of Greenland, the basic veins of Ameralik, which are strongly deformed, dismembered, metamorphosed and re-injected by the products of partial melting of their country rocks (Sederholm effect), are allowed to divide the gneiss of the Godthab district into the old Amîtsoq gneiss containing the veins and the younger Nûk gneiss never cut by these veins. The relative chronology of the crystalline terrains is also based on the study of the sequence of metamorphic and strain phases suffered by the protoliths. The discrimination between polymetamorphic and monometamorphic units belonging to the same terrain is fundamental. Finally, the application of the geochronological methods is a very useful tool to reconstruct the history of the crystalline terrains. The distinction between lithostratigraphy and lithodemic stratigraphy is supported by the fact that the relative chronology of the crystalline terrains is the result of laboratory analyses whereas in sedimentary terrains the relative chronology is already defined in the field, at the beginning of the stratigraphic study.

The lithodemic nomenclature recommended for the intrusive bodies is also applied to metamorphic and migmatitic rocks. The lithodem is the practical unit which can be mapped in order to study and describe a crystalline terrain. The metamorphic lithodems whose protoliths are sedimentary (para) or volcanic rocks can be generally distinguished in informal units. The metamorphic lithodems whose protoliths are plutonic, generally show a massive structure and therefore they are rarely divisible. The migmatitic lithodems can not be divisible because the observed lithological differences are representative of different degrees of migmatization (metatexites, diatexites). The higher rank of the lithodem is the series which is made up of two or more associated lithodems of the same lithological class. A supersuite consists of two or more suite, laterally or vertically associated.

A group of rocks belonging to two or more genetic classes (metamorphic, plutonic, migmatitic, sedimentary or of undetermined origin) is called a **complex**. This term is a very useful tool when chronological relations among the different components are uncertain, or when it is impossible to map and separate each body of rocks. Complex can be also used to define wide portions of metamorphic rocks of different origin, partially melted (migmatites) and cut by plutons. The hierarchic position of a complex is undetermined, but commonly compared to that of the suite or supersuite.

1.3.3. Layered intrusions

Some basic and ultrabasic plutonic bodies with tholeiitic or alkaline affinity are generally found in layered intrusions, mainly consisting of a piling up of lots of km of plutonic rocks in cumulate layers, roughly similar in appearance to sedimentary rocks. The thicknesses of the layered intrusions are generally on the order of some km, exceptionally exceeding 8 km. In these intrusions, the magmatic cumulate rocks form horizontally by gravity settling, from bottom to top, showing some typical structures of sedimentation: various kinds of grain sorting, current marks, cross bedding, channels, convolutions, sequences (rhythmic units). Being the principle of superposition commonly respected in the cumulate layers of the magma chambers, the stratified units of these intrusions should be considered as lithostratigraphic units. However layered intrusions are not entirely made of stratified units, as commonly shown by veins and marginal units. In order to establish the relative chronology of layered intrusions, the principle of superposition, intrusive cross cutting and structural criteria have to be used together. In this way the whole units of layered intrusions represent a specific stratigraphic class which is different from the lithostratigraphic and lithodemic classes: the plutonostratigraphic class, using peculiar stratigraphic nomenclature proposed by IRVINE (1982).

However layered intrusions can be divided in first order groups on the basis of structural or "petrostratigraphic" criteria (appearing and disappearing of cumulus minerals considered as useful keys to unravel the stratigraphic columns of the layered intrusions) and named "**plutonic divisions**". These are represented by the series, the zone and the sub-zone in decreasing order of importance. The **plutonic series** is the most important division of the intrusion itself. It is defined on the basis of structural criteria; it groups a stratigraphic succession of cumulate rocks and describes the general structure of the intrusion. In the famous layered intrusion of Skaergaard, the Layered Series, the Marginal Border Group and the Upper Border Group are the most important divisions of the intrusion. The **plutonic zones** and sub-zones are respectively the first and second order stratigraphic subdivisions of the series. In most of the tholeiitic stratified series the zones are identified by the appearance (lower limit) and disappearance (upper limit) of the cumulus olivine. The sub-zones are marked by the first appearance of other cumulus minerals (augite, magnetite, apatite). However the precise criteria to distinguish zones and sub-zones will depend on the nature of the intrusion. All the zones are not divisible in sub-zones (e.g. the Middle Zone of Skaergaard) and a zone has to be at least mapped on a common scale.

On the basis of petrographic criteria the layered intrusions are therefore separated in different parts which are denominated "**plutonic units**" and are smaller then the divisions. Since different criteria govern divisions and units there is no direct correspondence between the second order divisions and the units. In particular the units are not sub-sets of the divisions. A single unit, for example a rhythmic unit, can start in the upper part of a zone and end in the lower part of the successive zone.

Among the units can be distinguished groups, members, strate and rhythmic units. A **plutonic bed** is a stratified cumulate unit consisting of a distinct unit through its petrographic features (composition, texture) and confining towards the top and the bottom by discontinuities or by sharp petrographic variations (for example a stratum of gabbro). A horizon is a plain of reference mark within a stratum or separating two strata of contrasting nature. It has no thickness and can not be used as synonim of thin stratum. A **plutonic member** is a distinct stratified unit made of some levels naturally associated. For example, it can be formed by two associated levels, one leucocratic and the other melanocratic or by levels of different nature but having a common key mineral (olivine-bearing member). A **plutonic group** is an important set of rocks constituted by similar units (frequently members) which are close stratigraphically but not necessary consecutive. For example the Triple Group of Skaergaard consists of non consecutive three members, each one made of an anorthosite level followed by an overlying gabbro level. A **plutonic rhythmic unit** is defined as a regular succession of levels of the same nature (for example the succession of levels with the same grain sorting, similar composition and thickness) or as a succession of levels showing the same regular stratigraphic repetition (for example some levels made of alternated gabbro/anorthosite). A **plutonic cyclic unit** is a rhythmic unit characterized by the first appearance of cumul minerals corresponding to the same order of fractional crystallization of the parental magma.

Formal names can be attributed only to well defined and most important divisions and units of the layered intrusions. Each division or unit of an intrusion is generally marked by an alphabetical or alphanumerical sign on the basis of its position along the stratigraphic column.

In order to avoid misunderstanding with the lithostratigraphic, biostratigraphic and chronostratigraphic terms and to join international rules, we recommend to ad the prefix "plutono" to the Irvine's divisions and units (plutono-zone, plutono-series) in the stratigraphic study of the layered intrusions.

2. – GEOCHRONOMETRY OF PRECAMBRIAN TIME

Subdivision of the Phanerozoic time scale is generally based on **geochronologic units** that were defined on the basis of **chronostratigraphic units**. However, the subdivision of the Precambrian as proposed by the Subcommission on Precambrian Stratigraphy (SPS) and largely accepted by the ICS and IUGS, is based on **geochronometric units** (NORTH AMERICAN STRATIGRAPHIC CODE, 1983) with conventional boundaries expressed in years.

These two methods of dividing time are thus based on different principles. A geochronologic unit is the time interval during which a reference rock unit with a well-determined position in the Earth's history (chronostratigraphic unit) was deposited. A geochronometric unit is a conventionally defined time interval that does not refer to material units. Its boundaries correspond as much as possible to periods with minimal geologic activity, or to world-wide hiatuses in the recording of such activity thus bracketing major events in the Earth's history.

These two methods of subdividing geologic time illustrate quite different concepts of stratigraphy and application of the chronometric method for dividing Precambrian time has caused vivid discussions (HOFMANN, 1990; 1991; 1992). Proponents of the geochronologic method reproach geochronometry that it does not conform to the rule of defining a reference standard in rock (HEDBERG, 1976), in short, that it confuses calendar with history and thus has nothing in common with stratigraphy and historical geology (PREISS, 1982; TRENDALL, 1984; CLOUD, 1987; NISBET, 1982; 1991). Defenders of geochronometry point out that, today, it is easier to measure the age of a rock, especially in the Precambrian, than to determine its age by correlation with a chronostratigraphic unit. They add that the Phanerozoic time scale, too, is to some degree a conventional one and that the formal subdivisions and nomenclature are especially useful for grouping rock units on small-scale maps and for facilitating scientific communication (PLUMB & JAMES, 1986; PLUMB, 1992). Without wishing to enter into these discussions, it is clear that the geochronometric approach followed by the SPS (VIDAL, 1987) is convenient in view of the difficulty to establish precise stratigraphic correlations in the crystalline rocks that make up most Precambrian terranes.

2.1. PRINCIPLES OF GEOCHRONOMETRIC SUBDIVISION OF THE PRECAMBRIAN

1) Geochronometric units are determined so as to reflect the succession of major events in the Earth's history. For Precambrian history, these are tectonomagmatic cycles and other global events of a sedimentologic, biologic, or climatic nature.

2) As many major geologic events are diachronous on a world-wide scale, the boundaries of geochronometric units should bracket such events, rather than corresponding to their apogee.

3) Resulting from the two preceding points, the boundaries of geochronometric units correspond to periods of minimal geologic activity or to hiatuses in the recording of such activity, bracketing major events.

4) Precambrian stratigraphy mostly being based on the radiometric dating of rock units, the boundaries of geochronometric units are defined in years (e.g. 2.5 Ga) rather than by referring to material units whose age can be uncertain. As these boundaries are defined in terms of global geologic events, they are conventional and, once defined, no more reference is made to the events that were used. Because of their conventional nature, these boundaries are expressed in years without an error margin.

2.2. SUBDIVISION METHOD

The approach adopted by the SPS initially was based on the work by Regional Working Groups that drew up synthetic stratigraphic columns, integrating all Precambrian geologic data in terms of time for each region under consideration (PLUMB & JAMES, 1986; 1987). Composite columns were drawn up for the main Precambrian domains in North America, the Baltic Shield, the Ural Mountains, the Ukrainian Shield, China, India, South America, South Africa and Australia (see excerpt of Fig. 7.1). The columns show tectono-metamorphic and/or magmatic events, specific rock units, unconformities and hiatuses. Particular attention was paid to tectonomagmatic cycles, whose four classic stages are shown in the columns:

– (1) Pre-orogenic sedimentation and volcanism stage;

– (2) Orogenic stage, characterized by intense deformation, metamorphism and syn-orogenic intrusions;

– (3) Late- to post-orogenic stage, characterized by moderate deformation, abundant magmatism and molasse deposition;

– (4) Platform-sediment deposition stage.

Rock units in the columns include greenstone belts, highly metamorphic gneiss complexes, basic intrusions that generally were emplaced during periods of tectonic quiet, glacial deposits, etc. The constituting elements of each column (events, rock units, etc.) and the hiatuses represented therein are arranged according to time, younging upward, their thickness being proportional to their duration. The ages and durations shown result from a critical compilation of isotopic, paleontologic, stratigraphic and paleomagnetic data. Graphic codes indicate good, probable, estimated or uncertain precision of the durations.

The columns, drawn at the same scale, are then arranged parallel in a table, with their geographic location shown on the abscissa and the ordinate showing time in years (Fig. 7.1). The trace of a horizontal line (isochron) in this table, intersecting the vertical columns, visualizes the geologic events that took place at the same time on Earth. Despite the variability of the columns and a certain diachronism of many major events, the table illustrates a global pattern of these events and of periods of minor geologic activity. Application of the principles of geochronometric division then enables proposing a subdivision of the Precambrian.

2.3. GEOCHRONOMETRIC UNITS OF THE PRECAMBRIAN

Geochronometric units are designated by the same terms as those of geochronologic hierarchy, i.e. Eon, Era and Period, despite their differences in definition (NORTH AMERICAN STRATIGRAPHIC CODE, 1983). A geochronologic Period corresponds to the duration of deposition of a (chronostratigraphic) System, whereas a geochronometric Period is defined by conventional boundaries expressed in years. To avoid confusion with chronostratigraphic or geochronologic units, and in coherence with chronometric principles, the names of geochronometric units do not have a geographic root.

The Precambrian today is subdivided in two Eons designated by the traditional names of "Archean" and "Proterozoic", created by J.D. Dana in the 19th Century and redefined using geochronometric criteria. The Archean is subdivided in four Eras (LUMBERS & CARD, 1991) whose names are formed by the Greek prefixes Eo, Paleo, Meso and Neo, followed by Archean. The Proterozoic is subdivided in three Eras (Paleo-, Meso- and Neoproterozoic) and ten Periods (PLUMB, 1991) whose names refer to important geologic processes or events for the Period in question. The Cryogenian, for instance, evokes glaciation episodes. One Period name, Neoproterozoic III, is provisional.

2.3.1. Archean

Generally speaking, the lithology of Archean terranes is quite different from that of more recent ones, corresponding mostly to greenstone belts and gneiss complexes.

7. Specific stratigraphies

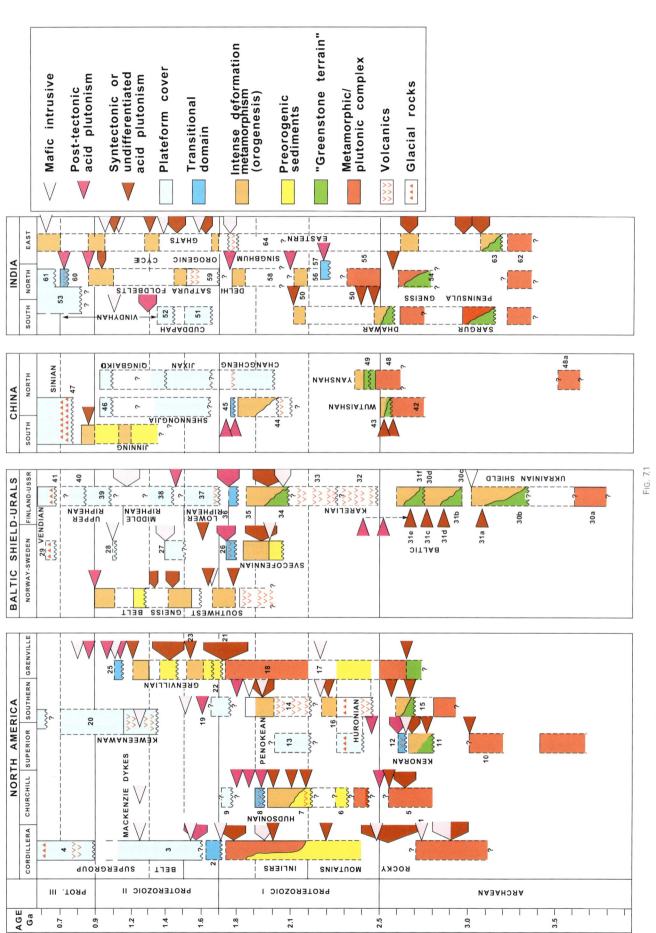

FIG. 7.1 Example of stratigraphic correlations in various precambrian domains (after PLUMB & JAMES, 1986, modified).

Greenstone belts, slightly metamorphic, are enclosed in huge granitoid and orthogneiss units that are depleted in potassium ("TTG" suite of Trondhjemite + Tonalite + Granodiorite). The gneiss complexes (orthogneiss, paragneiss, etc.) show a high degree of metamorphism (WINDLEY, 1986; NISBET, 1987).

Many of such terranes are juvenile, being the direct result of Mantle rock (komatiite, tholeiite, etc.) fusion, or the result of fusion of Mantle-derived products (TTG).

The boundary between Archean and the Proterozoic is set at 2.5 Ga. It corresponds to a period of great geologic calm, preceded by the most important stage of cratonization that Earth has ever known and followed by the start of massive recycling processes of pre-existing crust (WINDLEY, 1984). Even so, the age of stabilization (cratonization) of the Archean crust was diachronous and platform sediments started to be deposited around 3.0 Ga in South Africa (Pongola Supergroup, followed by the Witwatersrand Supergroup) and around 2.8 Ga in Australia (Fortescue Group, overlying the Pilbara craton).

In principle, the Archean started with the beginning of geologic history as recorded in rock. The oldest rocks known today is 3.9 to 4.0 Ga old Canadian orthogneiss (BOWRING & HOUSH, 1995), and rocks predating 3.8 Ga are known from Antarctica, Labrador and Greenland. In addition, 3.9 to 4.2 Ga old detrital zircons of a granitic provenance are known from Australian metaquartzite that was deposited around 3.0 Ga (MAAS et al., 1992). As the "absolute record" of oldest rocks has varied during the past decades, no lower boundary has as yet been fixed for the Archean. CLOUD (1976) suggested creating a third Precambrian Eon, the Hadean, for the time interval between the end of condensation of the Earth (4.49 Ga, ALBARÈDE, 1985) and the arbitrary date of 4.0 Ga, roughly corresponding to the start of the rock record in Earth's history. Finally, NISBET (1991) has proposed to let the Hadean/Archean boundary coincide with the appearance of life.

Analysis of the main sedimentary, magmatic and orogenic cycles of the Archean has led to proposing its subdivision in four Eras: Eoarchean (<3.6 Ga), Paleoarchean (3.6 to 3.2 Ga), Mesoarchean (3.2 to 2.8 Ga), and Neoarchean (2.8 to 2.5 Ga) (LUMBERS & CARD, 1991). At present, there are no plans to define geochronometric units with a lower rank than Era for the Archean.

2.3.2. Proterozoic

The Proterozoic is the longest Eon with a duration of almost 2 Ga. It covers about half of the Earth's history, recording the evolution from Archean-type to modern geodynamic processes. The Proterozoic has been subdivided in three Eras, successively the Paleoproterozoic, the Mesoproterozoic and the Neoproterozoic (PLUMB & GEE, 1987).

The Paleoproterozoic is characterized by the almost simultaneous growth of crustal segments with archaic as well as modern aspects. "Archaic" rocks are associations of greenstones and K-depleted granitoid rocks as well as the highly metamorphic gneiss complexes that are mostly juvenile and in part analogous to Archean terranes, such as the Guiana and West African shields. "Modern" rocks are represented by collision orogens in which Archean basement has been recycled (e.g. the Baltic and North American shields), by sedimentary basins on Archean basement that were little affected since deposition (e.g. the Transvaal Supergroup), by the formation of mega fracture systems corresponding to attempted fragmentation of the Archean cratons and outlined by basic dyke swarms such as the Aphebian dykes of the Canadian Shield (2.5 to 1.8 Ga), or by the first ophiolites (Cape Smith Belt, Québec, 2.0 Ga). The beginning of the Era, 2.5 to 2.05 Ga, corresponds in particular to deposition of sedimentary and volcanic units; its middle part – 2.05 to 1.8 Ga – is characterized by the development of numerous orogens of relatively short duration, and the last part (1.8 to 1.6 Ga) was a period of cratonization and the beginning of platform sedimentation. The upper boundary of the Paleoproterozoic is difficult to determine, but has been conventionally fixed at 1.6 Ga on the basis of available geologic data. Analysis of sedimentary, magmatic and orogenic cycles has further led to a subdivision of the Paleoproterozoic in four Periods: Siderian (2.5 to 2.3 Ga), Rhyacian (2.3 to 2.05 Ga), Orosirian (2.05 to 1.8 Ga) and Statherian (1.8 to 1.6 Ga).

The Mesoproterozoic is characterized by the first significant development in the Earth's history of long intracratonic and pericratonic orogenic belts. Many of these belts are polymetamorphic with major recycling of Archean and/or Paleoproterozoic rocks. The end of the Mesoproterozoic, set at 1.0 Ga, on most continents corresponds to a period of platform sedimentation after a major cratonization event. At this point, the shields had acquired most of their present-day aspect. The Mesoproterozoic is subdivided in three Periods: Calymmian (1.6 to 1.4 Ga), Ectasian (1.4 to 1.2 Ga) and Stenian (1.2 to 1.0 Ga).

The Neoproterozoic generally was a time of relative calm, characterized by major platform deposition. However, a notable orogenic system – the Panafrican/Brazilian chain – was developed in Gondwana with the same characteristics as those of Phanerozoic orogens. This Era also saw the development of several glaciation episodes as well as the appearance of multicellular organisms (the Ediacara fauna). The upper limit of the Neoproterozoic is the base of the Cambrian (542 Ma). The Era is subdivided in three Periods: Tonian (1000 to 850 Ma), Cryogenian (850 to 630 Ma) and Neoproterozoic III, between 630 Ma and the base of the Cambrian.

3. – STRATIGRAPHY OF SURFICIAL FORMATIONS

3.1. DEFINITION

Superficial formations for many geologists are synonymous of the Quaternary, but this is not systematically true. Even though around 90% of superficial formations is effectively of Quaternary age, the genesis of some formations, such as the alterites of Guiana or New Caledonia, goes back to the Cretaceous though in some cases continuing today.

Schematically, such formations are approached from two viewpoints.

At a large scale, such formations are highly variable, but they can be integrated under geomorphologic or climatic denominators that can cover a vast region. Detailed study of a few small but high-quality outcrops – a traverse along a slope may suffice – that show their three-dimensional geometry and lithologic variations, enables understanding the keys to their distribution. Once such keys are recognized and understood at a large scale in their geomorphologic setting, they commonly can be extrapolated to an entire region.

As a second step, the geologist in many cases will seek to understand their genesis and succession through time. Commonly, the study of present-day surface-geodynamics phenomena will provide the keys to understanding the evolution of environments since the end of the Neogene. However, the observed formations are rarely dated with the precision required by a stratigrapher, whereas their commonly very recent age needs very fine dating for a correct assessment of continental evolution. For that reason, all traditional methods must be deployed, such as sedimentology, paleontology, isotope geochronology, etc., as well as archeology and even history. For each case, this approach attempts to define – often indirectly – the chronology of the successive environments, i.e. climate, and erosion and deposition dynamics that have led to the surficial formation being considered.

In addition to several definitions (DEWOLF, 1965; FAURE, 1978; CAMPY & MACAIRE, 1989), the French Geological Survey (BRGM) proposes a formulation that is being validated (LEBRET et al., 1993):

*Surficial formations appear on the surface under the influence of meteoric action. Of continental origin and either loose or secondarily consolidated, they result from the physical disaggregation and/or chemical weathering of pre-existing rock. A surficial formation can either remain in place on the mother rock (*autochthonous *formation), or it may have been displaced over a short distance (*sub-autochthonous *formation), or it can have been remobilized by surface-geodynamics agents (gravity, ice, water, or wind) before secondary deposition (*allochthonous *formation).*

The thickness of such formations generally is up to several meters, but exceptionally they can be over a 100 m thick, the thickness being related to the present-day topography. Paleo-surficial formations, e.g. bauxites or Tertiary alterites, can be disconnected from today's surface and may be exhumed.

Through the effect of pedogenesis, the upper part of surficial formations can evolve into soil that answers to its own rules of internal organization".

3.2. VOCABULARY: SURFICIAL FORMATIONS; STUDY CONCEPTS AND UNITS

The study of surficial formations is similar to that of other sedimentary formations, using the same techniques such as sedimentology, geochemistry, paleontology and stratigraphy. However, their subdivision in lithostratigraphic units presents difficulties that set them apart: surficial formations are mostly of continental origin, they are generally very discontinuous in time and space, and they usually are difficult to date because of an absence of markers. Their common point is that they form a logical part of a landscape organization and morphology (the shape-deposit couple), and that they are intimately related to the structural (recent- or neo-tectonics) and climatic setting that has modelled a region (climate-sedimentation system).

For those reasons, the classification of surficial formations until now has been genetic rather than stratigraphic (BRGM, 1975; LEBRET et al., 1993). They form a continuum that is organized in "topo-sequences", whose organization in a downstream direction can be summarized in nine groups:

– (1) **Interfluve and slope-top deposits** are alterites (isalterite, alloterite), essentially chemical-weathering residues of a substrate subjected to precipitation and temperature fluctuations. Examples: the crusts and arenaceous deposits along the northern edge of the Massif Central, and the Cherty Clay in the Paris Basin.

– (2) **Slope deposits** (scree, periglacial slope deposits, mass wasting) combine a great variety of materials, transiting over a slope before their transport to sedimentary basins. The main transporting agent is gravity, commonly associated with diffuse runoff, or the influence of ice (freeze-and-thaw) in a periglacial climate. The substrate from which such deposits derive still is easily identifiable (shape, texture, chemistry and grain size).

– (3) **Glacial deposits** (till), obviously related to a cold climate, represent the first step of linear transport of erosion products of a region. Source materials show appreciable mixing, but the various provenances are still easily recognized, even though the constituting elements may have been modified through physical disaggregation.

– (4) **Volcano-sedimentary deposits** (see paragraph 5) are classified as surficial formations when they are very thin or where they are disconnected from their original edifice. Such materials are either directly deposited from air falls, or reworked by continental geodynamics, e.g. lahars and ash deposits affected by gelifluction, in which case they can fall under slope deposits.

– (5) **Alluvial deposits** are sediments transported by runoff concentrated in thalwegs. They are fed by lateral inflow from slopes or, in some cases, from glacier fronts (eskers, fluvio-glacial cones). During such transport, the sediments generally undergo mechanical change (fragmentation, wear) and grain-size sorting depending upon the hydraulic regime of the stream, which in itself is influenced by the regional climate. The contents of such deposits reflect input from different sources, forming a representative sample of the rock and soil types found in the catchment area of the stream.

– (6) **Lacustrine sediments**, recent and in most cases Quaternary, are commonly considered as surficial formations. They form a "downstream" stage that answers to the same logic as the sediments in a marine basin, though on a smaller volume and duration scale. Terrigenous sediments are laid down as a delta near stream outlets, and (bio) chemical sediments (limestone, chalk, etc.) are found in the more distal parts.

– (7) **Coastal sediments** include deposits at the continent-sea interface; they may be considered as surficial deposits where they are exhumed and too recent to be easily integrated in the logic of a marine sedimentary basin in a paleogeo-graphic sense. Commonly, these are sediments of fossil estuaries (silt and clay, locally organic), and old gravel bars or back-bar deposits such as lagoons and (mangrove) swamps. Study of such deposits must be based on sedimentological methods applied to recent deposits by integrating a 10^3 – to 10^5 – year framework and the corresponding successive climates.

– (8) **Eolian deposits** group all accumulations resulting from wind transport. Their source areas generally are deserts subject to deflation, which can be hot and dry in (sub) tropical latitudes, cold in a periglacial setting, or coastal strands including areas that fell dry during the Quaternary climatic-eustatic regression. Two groups are generally distinguished:

– a) Eolian sand, generally fine (<300 µm), transported by saltation, forms dunes or great sand plains, such as in Aquitaine or the northern European cover sands;

– b) Loess (periglacial climate) and lithologically comparable units formed in a temperate or warm climate, is composed of silt transported by suspension in the air and has a very homogeneous lithology. Its grain size between 15 and 60 µm is typically unimodal and well-sorted and the chemical composition is quite constant. Its depositional area generally is much larger for a given climatic-sedimentary event than that of eolian sand.

– (9) **Anthropic deposits** (backfill, rubbish, mine dumps, etc.) have become an increasingly important category because of their direct environmental impact.

3.3. STUDY METHODS

The major difficulty in the precise dating of surficial deposits has led to the development of **relative regional lithostratigraphic** frameworks, where the proposed successions are associated to a given geomorphologic framework (in CHALINE, 1980). Hiatuses, i.e. erosion phases, in these lithostratigraphic scales generally are as important as deposition phases for understanding landscape evolution.

Such lithostratigraphic scales are variably integrated in a chronologic framework, where regional sites, such as cave infill and fossil or archeological sites, permit dating by means pf paleontology, palynology, archeology or geochronology. But these scales will always remain a synthesis created from discontinuous and disparate elements (varying types of deposition, different dating methods from site to site, etc.) that hardly answer the recommendations of the International Stratigraphic Guide (SALVADOR, 1994). Today, the regional tables in France are of an uneven precision from one region to the next.

The following paragraph will show that the stratigraphy of surficial formations begins to be well-known for the past 120,000 years as well as around 2.4 Ma (appearance of the ice caps in northern Europe at the end of the Pliocene). However, for the Early Pleistocene and most of the Neogene such scales remain mostly vague for the chronology of surficial formations, despite their importance for understanding the continental evolution of France, i.e. the existence of paleosurfaces before valley incision or the importance of neotectonic movements.

4. – QUATERNARY STRATIGRAPHY

4.1. DEFINITION

Study of the Quaternary seeks to reconstitute the recent, i.e. less than 2 million years, geologic history of the Earth. It is not restricted to an analysis of continental surficial formations, but integrates data from the oceans (oxygen and carbon isotopes, biostratigraphy), and from present-day (Arctic and Antarctic) and fossil (e.g. moraines) ice caps. Its objectives are the creation of a general chronostratigraphic scale, the understanding and quantification of climatic and neotectonic variations, as well as their impact on landscapes and environments. Its short duration and its young age justify a degree of precision that generally is without equivalent in other geologic ages. Its interest, in addition to a better understanding of this period, is to provide geo-predictive elements for managing the impact of Man on his environment.

Since its creation, the Quaternary Era has known numerous definitions: first stratigraphic (the post-Oligocene formations of Touraine, (DENOYERS, 1829), then stratigraphic and paleontologic ("Diluvium" and contemporaneous human remains, SERRE, 1832), and then stratigraphic, paleontologic and climatic (the coincidence between surficial deposits, the development of Man and glaciations).

Since then, much work has shown the difficulties of defining this Era (RAT, 1980), which in fact is only defined by its lower boundary. The IUGS (IGCP 41, 1977), places this boundary by convention at 1.64 Ma; in the Vrica stratotype section (Calabria, Italy), this corresponds to the appearance of the first cold-water bathyal foraminifera in the Mediterranean, just above the Olduvai paleomagnetic inversion (1.67 Ma). However, an international Quaternary reference section in the continental domain remains to be found.

It should be pointed out that many scientists are dissatisfied with this boundary. They consider that, despite the IUGS convention, the base of the Quaternary should be sought around 2.4 Ma, now in the Pliocene, near the Gauss-Matuyama paleomagnetic inversion. They base their position on the fact that major geodynamic and geomorphologic changes, identifiable in the field such as the first glaciations in the northern hemisphere, were recorded for this period. However, we will use the official boundary in this text.

Very early on, the Quaternary was subdivided into Pleistocene and Holocene:

– The Pleistocene (LYELL, 1839) comprises almost all of the Quaternary, corresponding to the deposits overlying Tertiary formations and whose fauna is similar to the present-day one. It is subdivided in Early Pleistocene (1.64 Ma to 750,000 years BP), Middle Pleistocene (750,000 to 120,000 years BP) and Late Pleistocene corresponding to the last climatic cycle (120,000 to 10,000 years BP).

– The Holocene (GERVAIS, 1867-1869) designates fine "postdiluvian" deposits. Its lower boundary is fixed at 10,000 years BP, corresponding to the start of the present-day interglacial period (end of a periglacial climate in temperate Europe and start of the Flandrian transgression).

4.2. STUDY METHODS

4.2.1. *The Quaternary in an oceanic environment*

Isotope curves ($\delta\ ^{18}O/^{16}O$) determined from ocean-drilling cores show since 2.4 Ma about 50 alternating climatic cycles that can partly be correlated with glacial-interglacial climatic alternations (SHACKLETON et al., 1984). Since the start of the Quaternary (i.e. since 1.6 Ma), these curves show the existence of 25 cycles, each of which contains more-rapid and smaller-amplitude oscillations that correspond to small climatic crises within the main cycles. Even though such variations are world-wide, they will manifest themselves differently depending upon latitude (correlations from the equator to the poles remain to be validated) and are diachronous; for instance, the Scandinavian inland ice melted around 9,000 years BP whereas its American counterpart disappeared around 6,000 years BP. Such isotopic climato-stratigraphy curves are associated with the successive inversions of the Earth's magnetic field that have been well dated with radiometric methods (MANKINEN & DALRYMPLE, 1979). This provides precise data that have permitted to draw up a chronostratigraphy in the deep oceanic environment.

4.2.2. *The Quaternary in a continental environment*

Quaternary deposits are especially accessible in the continental domain, an environment that reacts more rapidly to climatic variations and does not present the same inertia as the oceans. For this reason, surficial-formation and Quaternary studies commonly are mixed up. The analysis of many sections shows complex sedimentary (erosion and deposition) or pedologic (weathering) events, the temporal importance of which must be determined before attempting to group them in climato-sedimentary units. The original aspects of each cycle are little by little revealed by the lithostratigraphic succession of such details. This "signature" is obtained by the conjunction of field analyses (lithology and pedology) and laboratory data (sedimentology, palynology, micro-morphology and geochemistry) on the multi-curve diagrams. Since the 1980s, the study of the Late Pleistocene has made great progress, this period now being generally well identified. Comparison with recent work on ice cores from the Greenland ice cap (BOND et al., 1993) has confirmed the validity of the chronostratigraphy for the past 120,000 years.

This approach is gradually being extended to earlier phases, based on complete and well-conserved sedimentary successions found in peat bogs, lake deposits and tectono-sedimentary traps. For the periods from the Pliocene to the Middle Pleistocene, the work carried out in the Netherlands (ZAGWIJN, 1992) and in the Mediterranean (SUC & ZAGWIJN, 1983) proposes a fairly complete palynostratigraphy for the Quaternary. In contrast, only the appearance of frost in Western Europe (around 2.4 Ma) is well identified in the lithostratigraphy of many sites. The other phases of the Quaternary (Early and Middle Pleistocene) remain poorly known in the continental domain. Regional lithostratigraphic compilations show much uncertainty: sections are few and far between, and locally unique for a given region; many discontinuities exist; the relationships are difficult to validate between deposits of different origin ("cold and warm", alluvial and slope deposits, etc.), or for different climatic contexts (northwest France with its low relief and influence from the Atlantic Ocean, against high mountains such as the Alps or the Pyrenees at different latitudes, or the Mediterranean context, etc.); and their exact age is generally imprecise or unknown.

4.3. PRACTICE: STRATIGRAPHIC SCALES FOR THE QUATERNARY

Awaiting the invention of a new method for dating Quaternary deposits, the best way for validating the regional "tables" is a multidisciplinary study of sections. This may then lead to the definition of an integrated stratigraphic scale that is validated for all of France.

Today and for want of anything better, three different chronostratigraphic scales are used in France. Two are applied to the continental domain: the "northern" scale (ZAGWIJN, 1992) and the "alpine" scale (PENCK & BRUCKNER, 1901-1905), the third one being exclusively reserved to the marine circum-Mediterranean domain (KERAUDREN, 1992). The simultaneous existence of these three scales is not only historical, but underscores the difficulty of integrating data of a very different type: glacial advance in the Alps, Dutch palynostratigraphy, or marine transgression-maxima deposits in the Mediterranean Basin. None of these scales is entirely satisfactory, but the fact that they are used shows how difficult it is to draw up a synthetic scale that will apply to all regions.

As the "northern" scale is the closest to the oceanic $\delta\ ^{18}O/^{16}O$ curves, many Quaternary specialists today tend to use this scale despite its imperfections, by associating the corresponding number of the isotopic stage.

5. – STRATIGRAPHY OF VOLCANIC TERRAINS

5.1. PRESENTATION

The specific role played by the volcanic rocks in stratigraphy is due to the bimodality of their origin. Volcanic rocks derive from magmas generated at depth (endogenous process) and therefore they are akin to plutonic rocks. Nevertheless, volcanic rocks derive from magmas erupted at surface (exogenous process) and therefore resemble sedimentary rocks. According to the R. Fisher's formula (in litt.) "rising material is eruptive that descending is sedimentary".

On the continental domains the volcanic material, whose areal distribution is controlled by climatic and topographic parameters, strongly contribute to surface formations (CAMPY & MACAIRE, 1989). On the oceanic domains, primary or secondary volcaniclastic deposits can represent large amounts of sedimentary sequences, including the distal turbidites (CAS & WRIGHT, 1987; CAS & BUSBY-SPERA, 1991). Studying these deposits, terms and vocabulary used by sedimentologists and stratigraphers should be useful reminders (FISHER & SMITH, 1991). Since the last three decade the stratigraphy of the volcanic succession has been feeding an abundant literature in english (cf. FISHER & SCHMINCKE, 1984) rather than in french language. The International Union of Geological Sciences (IUGS) proposed some recommendations for the classification (according to modal mineralogy or chemistry) of igneous rocks (LE MAITRE, 1989). By contrast, the nomenclature of the volcaniclastic products is still matter of debate. This framework should change rapidly since in the last decade petroleum geologists realized that volcanogenic sediments (neglected by exploitation for a long time) are of economic importance, through the discovery of hydrocarbons along the African or north-western European rifts or along the Asia or South America convergent margins.

Note

Volcanism is a discontinuous process, much more than sedimentary process (of detritic or chemical origin, even of continental type) since eruption is instantaneous on the geological scale (from few hours to few months, rarely few years). This involves the following effects:

– a chronologic sequence: the whole volcanic process is a record of istantaneous imput of variable periodicity. The term "stratigraphic record" is here totally justified.

– geometric consequence (spatial): erosion can be the rule, in the space time relations among the products of eruptions. On the basis of topographic context and period of quiescence of the volcanoes, the deposition of volcanic levels can obey to the superposition law or stacking law.

– lithogenic consequence: primary deposits, with special regard to very mobile pyroclastic products can be easily and deeply eroded and reworked. Therefore the study of these products must be joined to that of secondary volcanic deposits (epiclastic deposits). Records of time intervals of such products can be very different on qualitative and quantitative point of view.

5.2. VOCABULARY

5.2.1. *Volcanic and volcanogenic materials*

On the basis of viscosity of the magmas, the geodynamic context and in general of different volcanic succession, the volcanic products give rise to covering or lateral (heretopic) contacts of genetically different categories of products. The description of the latter represents the lithology. Since different volcanic products play different rules in inferring chronology of the events, the identification of genetically different categories of volcanic rocks is of a paramount importance to establish stratigraphic relations.

A distinction between effusive (lavas) and volcanoclastic products will be done as follows.

Lavas (effusive products) are emplaced as more or less viscous silicate melts. The initial environment is a whole liquid (and the final products a whole solid) containing variable amount of solid particles and gaseous vesicles. Their emplacement is controlled by both temperature/mass discharge rate and topography. Cooling of lavas is rapid and therefore these products become solid very soon acquiring the identity of a rock. Reworking of lavas as lava fragments (epiyclasts) needs long period of chemical alteration or erosion.

By contrast the general term **volcanoclastic** products (or vulcaniclastites) is associated to rocks constituted by fragments of volcanic and other origin (not necessary 100% volcanic fragments) independently to their proportion, to the mode of fragmentation and the environment. In the same deposits (i.e. **ignimbrites**) the high temperature can allow welding of fragments during their emplacement. In many other cases, starting from their initial state, the volcanoclastic deposits originate by different diagenetic processes. As all type of sediments and independently to environments of deposition, the initial mobility of the volcanoclastic products facilitate the transport through the exogenous processes. In addition, it is important to distinguish between primary and secondary (reworked) volcanoclastic products being inferred by the abundance of volcanic clasts rather than fragment of other lithologies.

Primary volcanoclastic products are emplaced at the time of the eruption and are the result of eruptive or syneruptive fragmentation processes. They comprise **autoclastic**, **pyroclastic** and **hydroclastic rocks**.

The autoclastic products are the result of mechanic fragmentation of lavas during solidification (e.g. progressive breccias of lava flows, breccias from collapsing domes);

The pyroclastic products (the word means "broken by the fire") or **tephra** (the word means ash) or ejecta are due to fragmentation of the material during the eruption, driven by the magmatic gas. The initial environment is a gas with a load of solid and/or liquid particles. The solid particles can derive from the magma itself (juvenile clasts) or torn up from the country rocks or deeper crustal levels (xenoliths).

Xenoliths can be very abundant among the phreatomagmatic eruptions when a large amount of shallow water is transformed to vapour. They can represent the only erupted products during **phreatic eruptions** consisting of gas released without juvenile magma emission. On the basis of their grain size pyroclast can be classified with direct relations to the size of detritic particles, although with a peculiar nomenclature: bombs or blocks (>64 mm), lapilli (from 64 down to 2 mm), coarse ash (from 2 mm down to 0,063 mm), fine ash (<0,063 mm).

Three fundamental types of pyroclastic products (pyroclastic fall, pyroclastic surge and pyroclastic flow) are distinguished through depositional and eruptive mechanisms. Among the pyroclastic products transitional types of products may exist between fall, surge and flow:

– The pyroclastic fall deposits accumulate according to the general laws of sedimentation. They consist of piling up of ballistic pyroclastic fragments or pyroclastic clasts wich underwent an atmospheric transport. The sedimentologic features of pyroclastic fall deposits are linked to the dynamic state (level of energy) and to the environment (subaerial or water) during sedimentation. Outwards the proximal accumulated products (e.g. volcanic cone), the pyroclastic fall deposits can reach a widespread areal distribution with independent thicknesses from topography.

– The pyroclastic surge deposits are the result of dense aerosols (not highly concentrated in solid) running on the surface topography and characterized by high velocity. They can be compared with deposits from storms or violent currents whose deposition occur in a high to very high energy environment. These deposit can have a moderate areal distribution with thickness wich are partly controlled by the topography (Fig. 7.2): the maximum thickness are on the valleys and depressions.

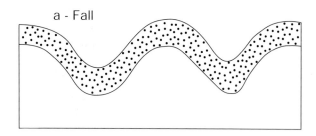
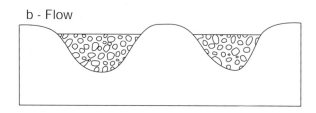
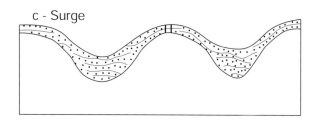

FIG. 7.2
The three principal types of pyroclastic deposits (after WRIGHT).

– The pyroclastic flow deposits or ignimbrites are hyperconcentrated and dense aerosols with behaviour similar to liquids; they are gravity driven during their expression. Strongly controlled by the topography, the pyroclastic flows can be very thick on the valleys and reach high velocity. The mode of transport is mostly "en masse" and the deposits are characterized by a chaotic structure both in proximal (few km) and distal (several km) facies.

The hydroclastic products are the result of quenching fragmentation of hot juvenile particles in a cold subacqueous environment. The process can affect both explosive (hyalotuffs a particular pyroclastic rock) and effusive lava flows (hyaloclastites, resembling autoclastic products, are associated with "pillow-lavas" and constitute widespread and thick accumulus of material in oceanic environment).

The secondary volcanoclastic products (although questionable, epiclastic lato sensu according to CAS & WRIGHT, 1987) accumulated after a discrete time interval of the primary products from which they derive through grain flows or "en masse" mode of transport.

Volcanic material can be reworked shortly after the eruption and therefore they can undergo remobilization before an even weak diagenesis. The grain-grain reworking by the wind, meteoric water or marine currents give rise to deposits similar to primary pyroclastic fall or surge deposits. The latter mainly differ for the total absence of older or non coheval juvenile clasts.

The most frequent "en masse" reworking is represented by **lahars** (mud flows or volcanic debris flows in acqueous phase). In subaqueous (marine or lacustrine) environment the reworking take place shortly after the emplacement of the primary products. In subaerial environment the lahars frequently occur during the eruption itself (para-volcanic

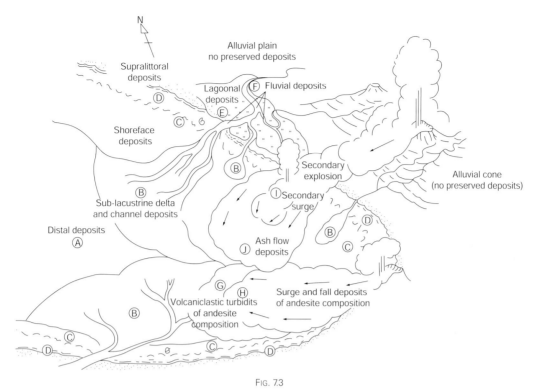

FIG. 7.3
Facies organization and reconstruction of the palaeo-environments related to the Koura Formation, Mihonoseki district, Japan (from KANO in CAS & BUSBY-SPERA, 1991).

events). Due to reworking lahars can be also remobilized several years later the main accumulation. Along the coastal area they are substituted by the turbidity currents.

On a geological time scale the materials reworked shortly after the eruption can be always considered penecontemporaneous with the primary deposits in a sequence of facies these reworked products represent the distal (external) equivalent.

The detritic volcanic material (**epiclastic rocks** stricto sensu) according to the general sedimentologic view of FISHER & SCHMINCKE, 1984 and FISHER & SMITH, 1991) consists of older volcanic rocks formed by the classic alteration and erosion processes. Thus, stratigraphic study of these epiclastic volcanic deposits does not differ at all from that applied to the whole detritic sedimentary sequences. Chronologic significance of detritic volcanic material is therefore totally separated from the occurrence of the eruptive processes (epiclastic deposits stricto sensu, are only successive to the eruptive activity).

5.2.2. Facies associations and lithostratigraphic units

The main unit of event is the **eruption**, during which primary product and the reworked equivalent (shortly after) are joined within a sequence of facies (Fig. 7.3). The coherent distribution of such a sequence (facies model) is defined in relation to a "dynamic type" of event occurred during a present-day or recent eruptive activity (e.g. strombolian type, plinian type, peleean type). All the reference eruption types is not "a priori" established and the list is growing, on the basis of new observations of eruptions on volcanoes (e.g. surtseyan type, 1963; S.Helen type 1980); because of new interpretations of eruption dynamics some eruption type had to be changed (e.g. volcanian type). It's therefore unavoidable some misunderstandings in nomenclature of eruption styles of volcanoes; some volcanoes can change its eruptive dynamics during the same eruption on a time interval span ranging from weeks to months.

The volcanic products has source vents or eruptive fissures from which they are outpoured. Thus, the stratigraphic study of these deposits should take into account their areal distribution starting from the eruptive source (proximal, medium distance and distal products). In a similar way of studying sediments, on the basis of grain size and isopach map of the volcanic deposit, source vent can be located. The more eruption is old, the more difficult will be to define the source vents of a volcanic product.

Proximal products build the main volcanic edifice. Proximal volcanics are studied and stratigraphically described with similar approaches adopted for sedimentary terrains. The main difference consists of lacking of lateral continuity for volcanic products. For example, only the whole conical shape of a volcanic scoria cone makes it different from a cone-shape landslide body. Moreover, a non-stratified distinct volume of a volcanic dome can be compared to a bioherma. A lava flow units, pumice units or debris flow units can be compared to strata. Lavas, pyroclastic products or generally volcanic products reworked shortly after the eruptions are three fundamental types of materials that can be genetically treated in terms of facies. Specific features may however derive from the mode of emplacement of volcanic products from the eruptive sources (facies of event). Distinguishing the frequence of lava-type emplacements as subvolcanic shallow intrusions (dikes, sills, necks) or mode of emplacement of pyroclastic bodies (e.g. diatremes) are very important. Errors on the chronologic sequence of volcanic products may derive for example from misunderstandings between a sill and a lava flow.

Volcanic products at intermediate distance from the eruptive centres (kilometers to tens of kilometers) can be represented by lava flows, pyroclastic flows, pyroclastic fall and their reworked counterpart. These latter are growing with distance from the volcanic source. If distance from the eruptive vents is greater, the non-volcanic clastic material will mix with volcanic debris or alternatively may form individual, reworked volcaniclastic levels. These intermediate volcanic

UNITS OF VOLCANIC ACTIVITY	PRODUCTS AND MORPHOLOGY	GENERAL GENETIC CLASSIFICATION	LITHOSTRATIGRAPHIC UNITS
Eruptive pulse (seconds, minutes) / (Time interval between negligible phases) 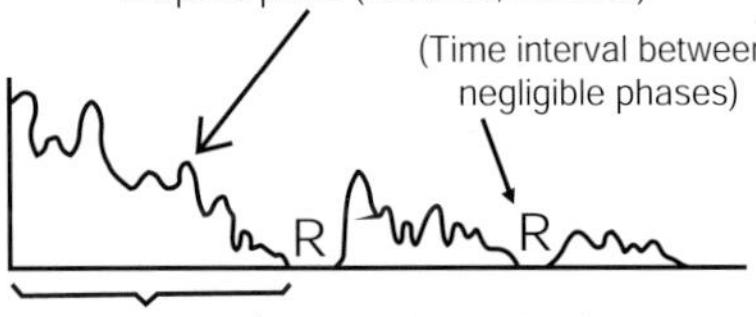 / Eruptive phase (minutes, hours, days) / ERUPTION (days, months, years)	Pyroclastic fall levels, pyroclastic flows, lahars, lava flows. Monogenetic volcanoes (scoria cones, domes, maars)	Eruption units (may consist of one or more levels group of levels, flows)	Lavas and volcaniclastic products can be grouped according to lithostratigraphic units. These are characterized by physical and mappable features generally corresponding to units of volcanic activity of upper rank:
Time interval long enough within the eruptions to provide a soil or erosive morphology 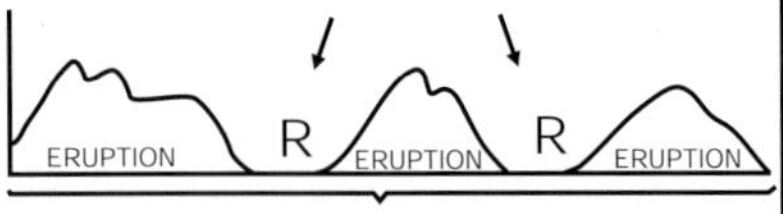 / ERUPTIVE STAGE (ten, hundred or million years)	Polygenetic volcanoes (stratovolcanoes, shield volcanoes, calderas); plateau formation (ignimbrites or basalts)	Multiple eruptive units or group of units. They can be grouped into cycles.	Group / Formation (fundamental unit on the field) / Member / Level, group of levels, lenses, flows
Time interval long enough to allow tectonic events 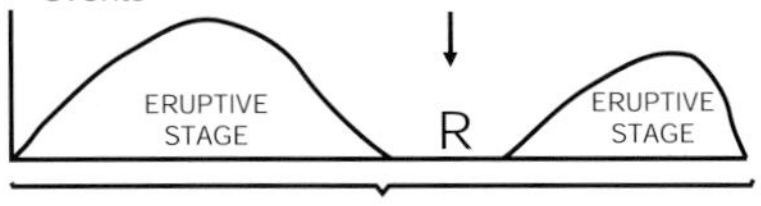 / ERUPTIVE PERIOD (thousand or million years)	Volcanic fields and regions characterized by numerous volcanoes or other volcanic edifices	Genetic classification on the basis of chemistry, petrochemistry, petrography or tectonics	

R : "rest"; the rest periods vary from few minutes to million years.

FIG. 7.4

Terminology of the volcanic products (from FISHER & SMITH, 1991).

facies constitute the main bodies of the volcano-sedimentary complexes (Fig. 7.4).

The distal and ultradistal volcanic products are mainly represented by fine-grained pyroclastic fall deposits being transported by the wind to tens up to thousands of kilometers. They consist of well-sorted, fine and/or ligth pumices and ash. These fine volcanic products are found in very thin levels or laminae interstratified with non volcanogenic sediments. The distal volcanic products can have a wide areal distribution and they can be transformed (e.g. bentonites). In additions, they can represent very useful horizons in the reconstruction of sedimentary piles of material (e.g. "tonsteins" of carboniferous basins, cinerites within lacustrian sediments). Due to their atmospheric mode of transport and their wide areal distribution, the ultradistal pyroclastic fall deposits are very useful tools for interbasins correlations. The term "tephra" (ash, JUVIGNÉ, 1990), is roughly considered synonym of pyroclastics.

Tephrochronology and **tephrostratigraphy** (THORARINSSON, 1944) mean therefore the dating of the sedimentary successions on the basis of studying fine grained volcanic tephra (stratigraphic markers, chrono-horizons). The tephrochronology is very useful in quaternary stratigraphy (SELF & SPARKS, 1981).

During structural descriptions or mapping complex volcanic stratigraphy (in particular stratovolcanoes), it is necessary to group these genetically linked sequence of facies into a hierarchic schemes. The volcanic products rarely obey to the severe rules that are used for the sedimentary terrains. The use of the formal lithostratigraphic units (group, formation, member, bed) is still common among the volcanologists. On the basis of the different case of study the stratigraphic approaches can be addressed to lithological, chemical, mineralogical and structural differences, according to the scientist's speciality and local convenience. The terms "formation", "series", "set" do not generally corresponds to a precise hierarchic definition and can be named with a type locality, a dominant lithology, a chronostratigraphic level, a relative position (lower, upper) or simply a serial number (series I, II, III). Up to now, volcanology does not seem to have a feeling with an international stratigraphic nomenclature.

When possible, interruptions of the volcanic activity are established on the basis of a major formation or for the lacking of volcanic levels which are able to mark eruptions. Particular volcanic events are marked by distinct chemical compositions, mineralogy, lithology and or a wide areal distribution (e.g. the "great level" of rhyolitic fibrous pumices of Mont-Dore, Auvergne). A similar volcanic activity which is often taken as the beginning or closing of an "eruptive cycle" may consist of a catastrophic eruption and therefore it may be represented by a "major discordance". When a paroxysm eruption occurs, a volcanic cone can be truncated for hundred meters on a few hours (e.g. 500 meters in the Mt S.Helens, Washington, USA during the 18 may 1980 eruption) and the great corresponding volume running downward. In this way, in the marginal sectors of the stratovolcanoes, huge **debris avalanches**, generated by gravitational instability produce flank collapses (sector collapses). The resulting debris flow (and lahar) deposits can represent olistostromes between proximal and intermediate volcanic facies.

5.2.3. Questionable vocabulary

How and when can we use the terms "volcano-sedimentary"? In french language volcano-sedimentary has been often used to define various objects, with different genetic significance:

– a sequence of volcanic and non-volcanic levels;

– litified and stratified volcanic deposits, even is isolated laminae;

– a "melange" of material having magmatic and non magmatic origin.

As stated above, all the volcanoclastic products are sediments, i.e. the result of a physical process of sedimentation, including the pyroclastic material. Although the deposits deriving from eruptive particles obey to the general laws of sedimentation, the interpretation of depositional mechanisms of volcanoclastic products can be the result of a mixing of different kind of explosive activity. In other words the volcaniclastic products can not be interpreted as simple **reworking** or sedimentary processes occurring in marine, lacustrine or fluvial environment. Due to this ambiguity in interpreting depositional features, reconstruction of environment of deposition of volcaniclastic product can lead to misunderstanding and confusion.

The term volcano-sedimentary deposit should be given to stratified **complexes** where primary or reworked pyroclastic units (even lavas), epiclastic units and non volcanogenic (limestones, clays, radiolarites, diatomites) coexist at a various scale. A volcano-sedimentary deposit is therefore a lithostratigraphic term with a significance similar to formation or member. By contrast, the term volcano-sedimentary deposit (or succession) can not be used to describe lithological features such as:

– volcanic debris (epiclastic products);

– litified pyroclastic fall or surge deposits;

– volcaniclastic rocks where the "melange" between magmatic and sedimentary material is the result of a fragmentation due to explosions within a sedimentary substratum (mixed pyroclastic rocks as the case of "peperites" from Limagne);

– ash levels or ash laminae, interstratified within the non volcanic sedimentary series, transformed (or not) to clayey deposits (by diagenesis) after their deposition (e.g. bentonites).

5.3. PRACTICE

5.3.1. Periodicity

Eruptions are short events separated by long period of **quiescence**, if we exclude few exceptions like Stromboli characterized by a persistent activity since 2500 years ago. If we consider, for example, the volcanoes of Puys in Auvergne the cumulative lasting of the eruptions have been lasting as cumulative time a maximum of twenty years althoug they started the activity 95.000 years ago. Moreover the last eruption of these volcanoes is dated at about 7.000 years ago, i.e. half time the two longest periods of rest along the eruptive history of the volcano. Since discontinuity is the rule in the record of volcanic eruptions, we consider the stratigraphy of the volcanic series as factual. **Volcanic periodicity** can be measured at different scale as FISHER & SCHMINCKE (1984) proposed a normalized terminology of periodic units along the volcanic activity (Fig. 7.4).

The fundamental single event is the eruption that can last from few days to several mounths and exceptionally few years. The average member of eruptions per years throughout the subaerial volcanoes from a catalogue since 1950 is about 50 eruptions is rarely a continuous phenomenon: it can be divided into different eruptive phases, from few minutes to few days. These eruption phases can be separated by period of inactivity lasting few hours to few weeks. At the end of each eruptive pahses the eruptive style of the eruption can be dramatically changed. Each phase consists of eruptive pulses whose frequency is on the order of minutes to seconds.

The chronologic unit given by the eruption allows to distinguish the following volcanic edifices:

– **monogenic volcanoes** which are built during a single eruption (a scoria cone, a maar, a dome constititing as single edifices the volcanoes of the Puys in Auvergne);

– **polygenic volcanoes** which are built through a series of eruptions often characterized by different eruptive style (e.g. the **stratovolcanoes** of Mt. Dore or Cantal).

The polygenic volcanoes develop through eruption separated each other by a sufficient time interval (from tens to thousand years) to allow erosion, pedogenetic processes or partial hardening of the volcanic products to take place. However eruptive epoch can be also established, grouping a volcanic activity on the order of hundred to thousand of years. These epochs can be considered eruptive cycles (a concept similar to sedimentary cycle) since they suggest an eruptive scenario which can repeat periodically along the eruptive history of the volcano.

Finally, at the scale of a volcanic region or a large polygenic volcano the eruptive epochs can be grouped into eruptive period consisting of a time interval of million of years during which large scale geologic phenomena other than eruptions (e.g. tectonic process) are also comprised.

5.3.2. Dating

The evaluation of the age of the volcanic formation is carried out on direct or indirect methods.

The direct method is the absolute geochronological dating which is possible through the mass spectrometer and it follow radiogenic isotopic laws (the most used is the K/Ar method). The choice on the specific geochronological method to be used strongly depend on the type and age of material to date. It can be possible to date the volcanic product itself (K/Ar, U/Th, TL), an accidental organic material (^{14}C), or a mineral heated during the eruption (TL, Fission Tracks). On continental domains, magnetostratigraphy is frequently used to check the validity or precision of the radiochronological datings.

If the absolute age is not needed, the indirect method (the oldest one) provides an age of the volcanic products on the basis of the age of the associated (interstratified) sedimentary formations. In other words the principles of the biostratigraphy are used. In order to make correlations, it is possible to match radiochronological datings on the proximal products and biostratigraphic data on sedimentary formations interstratified with distal volcanic products. If possible, the two methods can be used, both on the intermediate facies (Fig. 7.5).

For many years, in regions where volcanic products and fossiliferous formation are interstratified, only the biostratigraphy (although with some uncertainties in continental domains) allowed the relative datings of the volcanic events.

By contrast, recently, radiochronological datings of specific volcanic levels gave a strong contribution to better define the biostratigraphic scale. Nowadays, precise radiochronological datings (e.g. $^{39}Ar/^{40}Ar$ on single crystals) can be used to ultradistal volcanic products (interstratified with sedimentary formations) whose eruption source vents are often unknown.

Translated by
A. RENZULLI (paragraph 1),
R. STEAD (paragraphs 2, 3, 4)
and P. SANTI (paragraph 5).

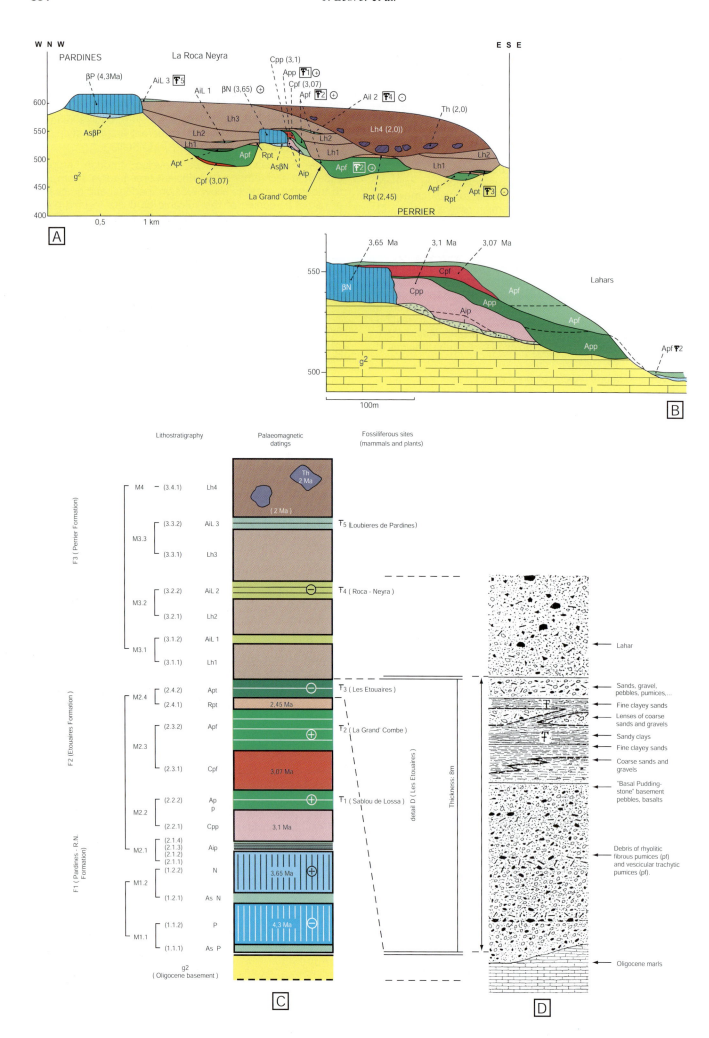

7. Specific stratigraphies

Fig. 7.5

The plateau of Pardines – Perrier, close to Issoire (Puy-de-Dôme, France): a synthetic example of "integrated stratigraphy" in a volcanic environment (from Ly, 1982, modified by POIDEVIN et al., 1984).

Perrier is a locality famous for its upper Pliocene ("Villafranchian") mammalian fauna and for its large lahars (volcanic debris flows). This area represents a series of erosive channeling, and channel deposition in a time interval of 2 Ma along the ancient course of the Allier river (flowing today some kilometers eastward of its ancient location). The deposits testify both the alluvial succession of the Allier river, and the mixed alluvial/volcanic products belonging to the left side tributaries (the "Couzes") coming from the hills of Nord-Cézalier and Monts Dore which were growing.

We are in an "intermediate complex" with respect to the two eruptive edifices which are 15 and 35 km far, respectively. The volcanigenic products are represented by lava flows, pyroclastic flows and falls, deposits reworked shortly after the eruption (ash, pumices, lahars) or by catastrophic debris avalanche and volcano-detrital materials (which are present in all the alluvial deposits).

The interpretation of the stratigrphy of this complex needs an integrated approach: geomorphology, sedimentology, volcanology, petrography, palaeontology, palynology, magnetostratigraphy, geochronology and correlation with other geological processes occurring in the regions close to the volcanic area.

– **7.5-A** – Sketch of the southern sector of Pardines – Perrier (the fossiliferous sites from F1 to F5 are artificially plotted in the section, although some of these are really located in the western and northern flanks). Vertical exaggeration = 5,5.

(3.7): Ages in millions years (Ma) – K–Ar method on the lavas and $^{39}Ar-^{40}Ar$ on the pyroclastic pumices.

(+): Direction of the palaeomagnetism – normal (+) or inverse (–) of the basalts or sediments.

F: fossiliferous site (mammals and plants).

– **7.5-B** – Detail of the stratigraphic relations between the Roca Neyra and the Grand'Combe units.

– **7.5-C**- Synthetic Stratigraphic column of the Pardines – Perrier group. Located on the basement constituted by Oligocene terrains of the Limagne d'Issoire (marls and lacustrine limestone with gastropods), the Pardines – Perrier group consists of the following lithostratigraphic succession:

-1: Basaltic formation of Pardines – Roca Neyra, generated by the eruptions of the volcanoes of Leiranoux, in the Nord-Cezalier, 15 km far.

-2: Volcano-sedimentary formation of Etouaires, recording the Monts Dore volcanism in the neighbourhood of La Barboule and Mont-Dore, 35 km far.

-3: Lahar formation of Perrier, which testify the construction and destruction of the Aiguiller massif (Monts Dore), 30 km far (this formation outcrops along the Allier river up to 20 km of distance).

Formation 1 – Basaltic formation of Pardines – Roca Neyra
Member 1.1 (Pardines)
Levels 1.1.1 – AsbP: Basaltic fluvial deposits of Pardines
Level 1.1.2 – bP: Basaltic flow of Pardines (4,3 Ma mag–)
Member 1.2 (Roca-Neyra)
Levels 1.2.1 – AsbN: Basaltic fluvial deposits of Roca-Neyra
Level 1.2.2 – bN: Basaltic flow of Roca-Neyra (3,65 Ma mag+)

Formation 2 – Volcano – sedimentary formation of Etouaires
Member 2.1
Levels from 2.1.1 to 2.1.4 – Aip: Fluvial deposits with pumices
Member 2.2
Level 2.2.1 – Cpp: Pyroclastic flow consisting of porphyritic rhyolitic pumices with quartz (3,1 Ma) (1)
Levels 2.2.2 – App: Fluvial deposits represented by reworked porphyritic pumices (mag+) and the fossiliferous horizon F1 (Sablou de Lossa)
Member 2.3
Level 2.3.1 – Cpf: Pyroclastic flow consisting of rhyolitic fibrous pumices (3,07 Ma) (2)
Levels 2.3.2 – Apf: Fluvial deposits represented by reworked fibrous pumices (mag+) and the fossiliferous horizon F2 (Grand'Combe)
Member 2.4
Level 2.4.1 – Rpt: Pyroclastic fall made of trachytic pumices (2,45 Ma) (3)
Levels 2.4.2 – Apt: Fluvial deposits represented by reworked trachytic pumices (mag–) and the major fossiliferous horizon F3 (Etouaires) (4)

Formation 3 – Lahar formation of Perrier
Member 3.1
Level 3.1.1 – Lh1: The first lahar (debris flow) (5)
Level 3.1.2 – AiL1: Alluvial deposit within the lahar 1 (discontinuity)
Member 3.2
Level 3.2.1 – Lh2: The second lahar (debris flow)
Levels 3.2.2 – AiL2: Inter-lahar deposits 2 (fluvial deposits and solifluction flows) and the fossiliferous horizon F4 (Roca Neyra) (mag–) (6)
Member 3.3
Level 3.3.1 – Lh3: The third lahar (debris flow)
Levels 3.3.2 – AiL3: Inter-laharic alluvial deposits 3 (sands, gravels and ashes) and the fossiliferous horizon F5 (Loubiere de Pardines)
Member 3.4
Level 3.4.1 – Lh4: The fourth "lahar" (debris avalanche constituted by mega blocks made of ordanchite) (2 Ma) (7)

Notes

(1) The rhyolitic quartz (bipyramidal crystals) of the Cpp and Cpf eruptions, widely dispersed by the atmospheric plume (distal and ultra-distal falls) represent a chronological marker between the pliocene sediments of the Massif Central and also faraway in the Velay and Bourbonnais regions.

(2) The fibrous pumice flow Cpf represents the largest ignimbrite deposit of the massif of Monts Dore. This ignimbrite constitute the first general reference for the stratigraphy of this massif, and its emplacement was followed by the collapse of the caldera of the Haute-Dordogne.

(3) Different fall deposits of trachytic pumices are present. The ages mesured in the Rpt give a time interval from 2,4 and 2,6 Ma. The reference age is 2,45 Ma because of these falls correspond to the palaeomagnetic inversion Gauss-Matuyama, considered by some Quaternary scientists as the Pliocene-Pleistocene boundary.

(4) In the gorge of Etouaires (*locus tipicus*) the group of alluvial deposits Apt, with 8 m of thickness, directly cover the oligocene basement. This deposition seems to be accompanied or followed by trachytic pumices Rpt, allowing to fix fossils ages. The mammalian-rich fauna of the Etouaires makes this site being the reference for the Villafranchian of the western Europe. A dating of the tooth enamel of mammals through the "electric spin" method gives an age of 2,5 Ma, a little bit older than the age represented by the inverse palaeomagnetism of the sediments.

(5) An uncertainty comes from the ages of the three first lahars Lh1, Lh2, Lh3 and from the geological processes generating these formations. They seem to precede the Lh4 lahar, belonging to the Aiguiller massif which was probably built, 2 Ma ago, in a short time interval (of the order of 0,1 Ma).

(6) With respect that of Etouaires (F3), the fossiliferous horizon of the Roca Neyra (F4) represents a climatic change: from a forest environment dominated by stags to a prairie landscape with antelopes.

(7) The term lahar is rather improper for the Lh4 level. In the field this formation is characterized by a cliff with troglodyte houses. It is constituted by a debris avalanche with mega-blocks (several tens of meters), running out 50 km. This avalanche, considered the second stratigraphic reference level for the north-eastern sector of the Monts Dore, represents a flank collapse of the volcano edifice, causing the istantaneous destruction of the Aiguiller massif. Its age is inferred by the frequence of the basaltic mega-blocks, mostly represented by two lava types whose datings in the Aiguiller massif counterparts are $2 \pm 0,1$ Ma.

– **7.5-D**-Detail of the Apt level (fossiliferous horizon F3) in the gorge of Etouaires.

This alluvial group, 8 m thick, direcly cover the oligocene basement. Its deposition seem to be preceded by the thachytic pumice fall of the Monts Dore and the palaeomagnetic inversion of Gauss-Matuyama, both corresponding to 2,45 Ma. The mammalian-rich fauna of the Estouaires makes this site being the Villafranchian reference for western Europe.

CHAPTER 8
CHRONOSTRATIGRAPHIC UNITS AND CORRELATIONS
from chronostratigraphic classification to Earth history

J. Rey (Coord.), L. Courel, J. Thierry, J.-F. Raynaud & S. Galeotti

CONTENTS

1. – DEFINITION 117
2. – TERMINOLOGY 117
   2.1. The chronostratigraphic units 117
      2.1.1. Chronostratigraphic, geochronologic and geochronometric units 118
      2.1.2. The Stage 118
      2.1.3. The System 118
      2.1.4. The Era 118
      2.1.5. The chronozone 118
   2.2. The chronostratigraphic scale 118
      2.2.1. The standard chronostratigraphic scale 118
      2.2.2. The regional chronostratigraphic scales 119
      2.2.3. The global chronostratigraphic scale and other stratigraphic scales 119
   2.3. The stratotype 119
      2.3.1. Definition 119
      2.3.2. The unit stratotype 119
      2.3.3. The boundary stratotype 119
   2.4. The chrono-correlations 120
3. – PRACTICE 121
   3.1. Duration of chronostratigraphic units 121
   3.2. Denomination of chronostratigraphic units 121
   3.3. A semi-quantitative approach: the composite standard reference section 121
   3.4. The procedure of integrated stratigraphy 122
   3.5. Geochronometry of the Precambrian 122
   3.6. Chronostratigraphy of the Plio-Quaternary interval 123
4. – CONCLUSIONS: THE STRATIGRAPHIC TOOL 123
5. – CONVENTIONS 123
   5.1. Main rules in the definition of chronostratigraphic units . 123
   5.2. Procedures for the ratification of chronostratigraphic units and their stratotypes 123

1. – DEFINITION

Chronostratigraphy is a branch of stratigraphy concerned with the age of rocks and time relations. It is aims at:

– (1) subdividing the sequence of rocks which accumulated successively in the history of Earth in units (chronostratigraphic units) on the base of their age;

– (2) hierarchically and chronologically ordering these units, by using a firm ***chronostratigraphic classification***. This process requires framing local or regional scale observations into a global context with the goal of establishing a continuous succession without gaps and overlaps. Such a succession can eventually serve as a global standard chronostratigraphic scale on which any observed geological event or object can be placed.

The obtained Time Scale represents a framework rigorously traced and conventionally adopted by international authorities. However, in a broader sense it also expresses a series of units derived from the various approaches of stratigraphy (lithostratigraphy, genetic stratigraphy, chemostratigraphy, biostratigraphy, magnetostratigraphy, isotope geochronology…, etc.). Taking into account all these data must allow, through an integrated and unifying approach, to reconstruct the evolution of the superficial strata of Earth which, in turn, allows refinement of correlations and datings. In this sense, chronostratigraphy opens vast perspectives and represents the logical result of the stratigraphic method.

2. – TERMINOLOGY

2.1. THE CHRONOSTRATIGRAPHIC UNITS

2.1.1. *Chronostratigraphic, geochronologic and geochronometric units*

In contrast to the continuity of time, the largely discontinuous character of the time record in rocks (resulting in the presence of sedimentary hiatuses), leads to distinguish ***chronostratigraphic units***. These corresponds to the bodies of rocks, processes and geological phenomena recorded during a certain interval of time (the amount of sand that passes through a hourglass for a given time interval) and represent the physical and tangible counterparts of the ***geochronologic units***. The latter, in fact, are abstract and intangible units, which express time directly ("the time interval during which a certain quantity of sand passes through the hourglass").

This is how, following various international propositions since the meeting of 1885, the International Subcommission of Stratigraphic Classification (HEDBERG, 1976, SALVADOR, 1994) proposed the use of distinct terms for chronostratigraphic and geochronologic units:

Chronostratigraphy:	Geochronology:
Eonothem	Eon
Erathem	Era
System	Period
Series	Epoch
Stage	Age
Substage	Subage

This distinction is useful because it expresses the contrast between the data resulting from natural processes and the physical laws. However, it has turned out to be not practical in the current usage because certain terms are rarely employed (e.g., Erathem); others are not always used properly (e.g., Era); finally, others are frequently met in the texts, but with altered significances (age, period, time, series). To simplify the terminology and avoid the improper use of these terms, ZALASIEWICZ *et al.* (2004) have recently proposed ending the distinction between the dual stratigraphic terminology of time-rock (chronostratigraphic) units and geologic time (geochronologic) units. As pointed out by the latter authors, besides the misunderstood usage, the widespread adoption of the global stratotype sections and points (GSSP, see paragraph 2.2.3.) principle in defining intervals of geologic time within rock strata has rendered unnecessary the traditional distinction between these two essentially parallel time scales.

Precambrian stratigraphy requires the definition of a further kind of units, the **geochronometric units**. Similarly to the geochronologic units, geochronometric units are abstract entities, corresponding to a subdivision based on absolute ages expressed in millions of years. However, differently from the geochronologic units, they do not have any physical counterpart. The terms **Eon**, **Era** and **Period** of the geochronologic classification can be used to designate the geochronometric units.

2.1.2. *The Stage*

The **stage** is a chronostratigraphic unit of relatively minor rank in the chronostratigraphic hierarchy and represents a relatively short time interval.

The stage, which is the most often used unit, is the basic working unit of chronostratigraphy because it is suited in scope and rank to the practical needs and purposes of intraregional chronostratigraphic classification. The stage is, indeed, the smallest chronostratigraphic unit that can be recognized on a global scale.

A stage can be subdivided, partially or entirely, into **substages,** which usually have a more local value.

The stage is a concrete unit and includes all rocks formed during an age. It is expressed by various lithologies (possibly allowing it to be mapped), by paleontological contents and events and by physical and chemical characteristics recognizable by using the various stratigraphic tools. Stage boundaries were historically based on natural changes, recognized at least in certain points, often associated to stratigraphic gaps. As a consequence, classical stage boundaries often correspond to apparent changes and fall within time spans which are not represented in the stratigraphic succession of the type locality. The need for a complete representation of geological times by the chronostratigraphic units favours the definition of the stage by bracketing it within boundaries defined in continuous sequences. Such boundaries must correspond to changes in the biological, sedimentological, physical or chemical record. To allow long distance correlations, these changes must be significant, of global extension and easily recognizable by paleontological, mineralogical, lithological, geochemical or magnetostratigraphic markers. Some boundary stages are also used to separate chronostratigraphic units of higher rank.

2.1.3. *The System*

The **system** is a reference unit of higher rank grouping several stages.

Systems were historically recognized in key areas where they correspond to physical entities. They were defined by their biofacies or their lithofacies, their boundaries therefore corresponding to important changes in the lithological (e.g. Carboniferous, Permian) or paleontological (e.g. Jurassic, Cretaceous) record.

System boundaries have been for a long time poorly defined, uncertain or discussed, because of the rather vague original definitions and later recognition of gaps or overlaps between adjacent units, or due to the lack of universally accepted concepts in their definition. The definition of these boundaries – which must coincide with stage boundaries – is committed since more than 20 years by the International Commission of Stratigraphy (ICS) of the International Union of Geological Sciences (IUGS).

The system can be subdivided into **series**. The latter comprise two or more stages and their upper and lower boundaries are defined by the lower boundary of the oldest stage and by the upper boundary of the youngest stage, respectively.

2.1.4. *The Era*

Among those commonly used, the **era** is the largest geochronologic unit (and, in the current use, of chronostratigraphy). Its boundaries are associated to major breaks in the paleontological record and, therefore, in the history of life. For this reason, this subdivision is possible only for the Phanerozoic where three eras are identified: Paleozoic, Mesozoic and Cenozoic.

2.1.5. *The chronozone*

Two, partially contradictory, definitions of **chronozone** have been proposed. According to the first one, the chronozone is the lowest rank unit (the elementary subdivision) in the hierarchy of the chronostratigraphy (HEDBERG, 1976; BATES & JACKSON, 1980; FOUCAULT & RAOULT, 1995). The second defines this unit as a formal chronostratigraphic unit of unspecified rank, not part of the hierarchy of conventional chronostratigraphic units. According to this definition, the chronozone is the body of rocks formed anywhere during the time span of some designated stratigraphic unit or geologic feature (HEDBERG, 1976; SALVADOR, 1994).

In any case, the time span of a chronozone is the time span of a previously designated stratigraphic unit or interval, such as a lithostratigraphic, biostratigraphic, or magnetostratigraphic polarity unit. While the stratigraphic unit on which the chronozone is based extends geographically only as far as its diagnostic properties can be recognized, the corresponding chronozone includes all rocks formed everywhere during a given time span, without reference to any particular stratigraphic section. Therefore, the chronozone has a universal chronostratigraphic value, independent of the geographical distribution of the marker(s) defining the corresponding unit. However, though the geographic extent of a chronozone is, in theory, worldwide, its applicability is limited to the area over which its time span can be identified, which is usually less.

The name of a chronozone takes is derived from the stratigraphic unit on which it is based, e.g., Exus albus Chronozone, based on the Exus albus Range Zone (Fig. 5.1).

2.2. THE CHRONOSTRATIGRAPHIC SCALE

2.2.1. *The standard chronostratigraphic scale*

The succession of the various units of different hierarchical levels through time, constitutes the **global standard chronostratigraphic scale**. This scale must cover the entire time and surface of Earth allowing to date any rock and any event or phenomenon in any given place.

Theoretically, every unit may have a global extension. However, only higher rank units have a global character and the extension of the geographic area of application of the units decreases with their hierarchical rank: while the systems and subsystems are universally recognized, the stages are not always identifiable on a global scale.

The validity of the units and the relative boundaries are discussed and voted in the light of the most recent advances in the field of stratigraphy by the sub-commissions of the IUGS.

2.2.2. *The regional chronostratigraphic scales*

Paleontology remains are essential tool in defining stages and their boundaries. However, the identification on a global scale of the stages on the base of their paleontological content is obviously hampered by the geographically uneven distribution of organisms. As an example, continental formations cannot be correlated with the stratotypes of marine stages on the base of their fossil content. The same problem arises when comparing marine sequences from different domain with low paleontological affinities (e.g. Boreal vs. Tethys domain). Regional stratigraphic classifications require the definition of **regional stages**, which can be organized in a **regional chronostratigraphic scale** covering part of the geological time. The definition of these stages follows the same principles regulating the construction of standard scale. However, they are approved by national commissions.

The regional chronostratigraphic units are useful or even essential when they are not entirely equivalent to the units of the standard scale. It is indeed preferable to correlate intervals with accuracy to regional units, rather than incorrectly integrate them into a reference scale.

However, it must be considered that regional stages represent only a transitory phase and that they should be abandoned once a sound correlation with the standard chronostratigraphic scale is established. On the contrary, should they reveal to be better suited for global correlation than the corresponding stage in the standard scale they should be used to replace the former standard unit (e.g.: Olenekian, Spathian).

2.2.3. *The global chronostratigraphic scale and other stratigraphic scales*

The units of the global chronostratigraphic scale include all the rocks formed during a given time interval, therefore the body of rocks of a given age whose boundaries are coeval in any place. These rocks embody several traits having a chronological value, such as, for instance, their paleontological contents or certain physical characteristics which allow the definition of thematic scales based on different records:

– biochronological scales, composed of successive biozones which are likely to be applicable to different environments when different fossil groups are used;

– magnetostratigraphic scales composed of successive magnetozones of alternating normal and reverse polarity;

– numerical scales based on measured values.

These thematic scales can be related to and complement the chronostratigraphic scale aiming at the construction of a **geological time scale** which is a unifying scale since it is based on units which do not have the same nature.

This integrated scale represents a large working site to stratigraphy. In fact, every scale is subject to improvements, refinements or emendations. Moreover, many rigorous tie points between different scales must be identified.

2.3. THE STRATOTYPE

2.3.1. *Definition*

For the boundaries of a stratigraphic unit of any kind (litho-; bio-, chrono-, etc.) to be recognized, reference sections, serving as standards, have been identified since the beginnings of stratigraphy. These are called **stratotypes**.

Stratotypes are defined in a given site which represents the standard locality. Stratotypes can define, either a stratigraphic unit (unit stratotype), or unit boundaries (**boundary stratotype**).

This chapter solely discusses the stratotypes of chronostratigraphic units and chronostratigraphic boundaries, particularly those of the stages that constitute the basic units of stratigraphic classification.

The general progresses in stratigraphy related to the development of new techniques and study of previously unexplored areas have shown that a stratotype very often gives only a partial representation of the stage. Nevertheless, after approval by the international committee, the standard section represents the reference for correlations at a global scale.

2.3.2. *The unit stratotype*

The stratotype of a chronostratigraphic unit (**unit stratotype**) such as the stage corresponds to the standard section of the entire unit. This section should be characterized by continuous deposition and a good exposure of the lower and upper boundaries of the unit, thus allowing to define its duration. In order to allow the correlation of the defined unit out of the standard locality, the stratotype must contain various stratigraphic markers: biological (fauna with a sufficiently wide geographical distribution), geochemical, mineralogical (magnetic minerals, volcanic or cosmic markers, geochronometers…).

These principles have been often tackled by the practical application due to the development of stratigraphy. Indeed, many reference section constituting the **historical stratotypes** have been defined by their original author in the 19th century, when the stratigraphical concepts and methods where largely different from those currently adopted. Moreover, historical stratotypes are often defined in macrofossil rich sequences deposited in neritic setting where the lithological and paleontological changes are particularly evident. Unfortunately, neritic environments are generally characterized by a rather discontinuous sedimentation and fossil assemblages markedly affected by provincialism. These conditions hardly correspond to the prerequisites necessary for a global standard chronostratigraphy.

Thus, these historical stratotypes have often been supplemented, modified or replaced, giving place to a nomenclature inspired to the taxonomic nomenclature. For instance, **composite stratotype** indicates a historical stratotype composed of several outcrops that represent the original stage once assembled. A **parastratotype** is a reference section cropping out in the same area and coeval to the stratotype, and supplements the information given by the latter. The use of this term is strongly discouraged by the International Commission of Stratigraphy (COWIE *et al*, 1986). The **hypostratotype** is a reference section described in a basin different that the initial stratotype. The **neostratotype** is a reference section replacing a historical stratotype which is no longer accessible or became unusable.

The multiplication of these reference sections, though enriching the description of the stage, defies a precise definition and a clear perception of its vertical extension.

2.3.3. *The boundary stratotype*

The time interval corresponding to a chronostratigraphic unit is comprised between the higher and lower boundaries of

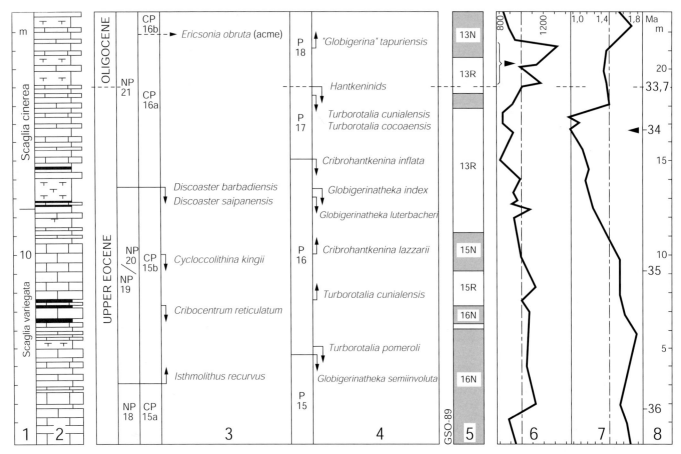

Fig. 8.1

Integrated stratigraphy across the Eocene/Oligocene bàundary in the global stratotype section of Massignano (after ODIN and MONTANARI, 1989). The boundary is denoted by the disappearance of Hantkeninids and Cribrohantkeninids (planktonic foraminifereal, column 4). It coincides with the onset of a positive shift in the Sr content (colum 6), it is placed 1.3 metres below a positive shift in the carbon isotope record (column 7), that is 1,5 m above the top of Chronozone 13R (column 5), within calcareous nannofossil Zone NP 21. Absolute datings indicate that this chronohorizon has an age of 33,7 Ma.

the unit. It is therefore the stratotype of these boundaries that better expresses the duration of a stage. These boundaries cannot be defined on the same section, or represented in the same area. However, a boundary is defined based on the elements that it separates. The boundary stratotype is therefore not only a point, but must be included in a continuous set of beds, below and above it, that should have sufficient thickness. This concept is uttered by the expression "*Global Boundary Stratotype Section and Point*".

The boundary stratotype has to be based on the identification of different markers, which do not necessarily occur at the same horizon and are not therefore necessarily synchronous. The most favourable for long distance correlation, is conventionally chosen as a point defining the unit boundary in the type locality.

This concept of stratotype offers several undeniable advantages:

– (1) to propose precise, concrete and fixed reference markers, materialized on the section by the so called "golden spike", the global stratotype point;

– (2) to maintain the possibility of filling the possible gaps in the geological record. Theoretically, a single boundary stratotype may be used to define the top of one stage and the base of the following. Ideally, this ensures the two boundaries to be not confused. However, the presence next to the defined boundary, of a stratigraphic gap not detectable by the available stratigraphic methods cannot be ruled out. For this reason, the boundary stratotypes defines only the base of the overlying chronostratigraphic unit.

In some cases, stage boundary stratotypes serve also as stratotype of boundaries between higher rank units (subsystems, systems…).

The establishment of boundary stratotypes requires detailed analyses and an accurate discussion, which often implies several years for an official ratification to be completed.

At present over 50 boundary stratotypes have been ratified by the International Commission of Stratigraphy (e.g., the Eocene/Oligocene boundary stratotype at Massignano, Fig. 8.1; The Frasnian/Famennian boundary stratotype in the Coumiac quarry, near Cesson in the "Montagne Noire" Fig. 8.2; see also Chapter 9). Since it has been first proposed the concept of boundary stratotypes has contributed important developments in the chronostratigraphy by making it possible to a clear definition to successive time constrained "bars" in the chronostratigraphic scale. In an ideally complete and continuous chronostratigraphic succession the boundary stratotype concept supplements – but not replaces – the unit stratotype. The latter, however, still maintains the role of defining, even partially and imperfectly, the content of a stage.

2.4. THE CHRONO-CORRELATIONS

On the basis of boundary stratotype definitions (with the various hierarchical ranks), unit boundaries can be extended geographically beyond the type section, thus allowing chronostratigraphic correlations or **chrono-correlations**.

FIG. 8.2
The Frasnian/Famennian boundary in the Coumiac quarry (Montagne Noire). A « golden spike » marks the boundary between bed 31g ("upper Kellwasser", Frasnian) and bed 32a of Famennien age. The base of the latter stage coincides with the first appearance of the conodont *Palmatolepis triangularis* (picture by R. FEIST).

The boundaries of a chronostratigraphic unit are, by definition, delimited by synchronous surfaces. This does not mean that the rocks embraced by the unit boundaries have exactly the same age in various geographical localities, but only that they deposited during the same time interval.

The synchronism of the correlation lines ("time lines") is relative, depending on the resolution of the stratigraphic method used. Generally, the precision of the chrono-correlation decreases by increasing the geographical distances, and the diversity of facies and the depositional settings. To improve the validity of chrono-correlations, it is essential to employ all the available stratigraphic markers: distribution of different fossil groups, lateral continuity and superposition of beds, tephra layers, extraterrestrial microtektites or magnetite, seismic data and well logs, geochemical and paleomagnetic signatures, isotope dating, etc. The composite standard reference section (see paragraph 3.4) and the integrated stratigraphy (see paragraph 3.5.) are methods or procedures fulfilling this requirement.

3. – PRACTICE

3.1. DURATION OF THE CHRONOSTRATIGRAPHIC UNITS

The subdivision of chronostratigraphic units reflects significant natural changes, whose duration is variable and certainly does not respect a strict hierarchy: based on isotope age determination, the duration of classical stages (not including the Quaternary), generally ranges between 2 to 10 Ma, with an average value of about 5 Ma (that is about 1% of the duration of the entire Phanerozoic). The currently adopted Series span time intervals ranging between 10 to 35 Ma. The Phanerozoic systems have a duration ranging between 25 to 70 Ma, with an average value of about 50 Ma. With its duration of 1.7 Ma, the Quaternary represents and exception.

3.2. DENOMINATION OF CHRONOSTRATIGRAPHIC UNITS

The names of the stages in the Standard Time Scale are generally derived from the geographic name of the area where they were originally defined. This can be a region (e.g. Aquitanian) or a locality next to the type section (e.g. Aptian). The adjective form with a "ian" termination of the geographical name is used. However, there exist some exceptions like the Tithonian stage whose name is derived from the Greek mythology. Moreover, the name of regional stages may be derived from their characteristic fossil content or lithofacies (Virgulian, Corallian).

System names may have a different origin: lithologic (Carboniferous, Cretaceous), ethnic (Ordovician, Silurian) or geographical (Devonian, Jurassic), or they may be derived from the position in the geological time (the informal term of Quaternary). Different designations such as "ian", "ic", "ferous" are therefore used for system names.

Series are named according to their position in the system (lower, middle, upper), or the region of the type series (Wenlock, Ludlow) sometimes followed by the suffix "ian" (Pennsylvanian, Mississippian), or they can have Greek roots (Eocene, Miocene).

3.3. A SEMI-QUANTITATIVE APPROACH: THE COMPOSITE STANDARD REFERENCE SECTION

The definition of a **composite standard reference section** (SHAW, 1964; MILLER, 1977; EDWARDS, 1989) by the **graphic correlation** is aimed to reconstruct the most likely content of a given lithostratigraphic unit, particularly on a basin scale. The method is based on compiling all the events (biological, mineralogical, geochemical, etc.) observed in several sections or wells on a single profile. A stratotype can be also defined. The transfer of this profile over one linear time scale (arithmetic) provides a "composite reference section" (CRS).

The semi-empirical method of graphic correlation to define a CRS can be briefly described as follows:

A section selected on the base of the thickness and the quality of data (in particular paleontological) is placed on a 2D diagram as the X axis, whereas another is used as the Y axis. The stratigraphic position of selected events (e.g. base or top occurrences of fossil taxa, volcaniclastic layers, etc.) from the two sections is plotted on the graph. The distance from the base of each section gives the X-Y coordinates for each of the selected events. The succession of points obtained allows to trace a **correlation line** that provides a comparison of the geological histories of the two sections, by highlighting gaps or changes in the sedimentation rate.

The composite section resulting by this procedure can thereafter be enriched by events observed in only one of the selected section, therefore adding further information and allowing to predict the range of rare species in the section where they are not observed.

In the same way, the composite profile can be enriched by adding successive sets of data from other sections or wells.

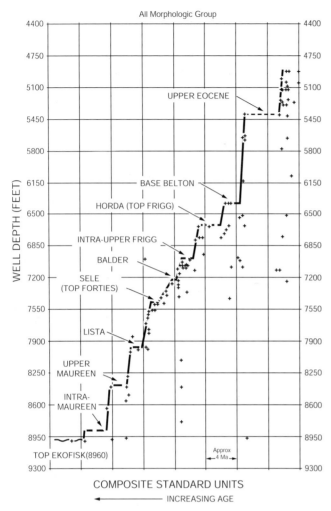

FIG. 8.3

Graphic correlation between a well from the North Sea and a composite reference section of the Paleogene. Biostratigraphic data are represented by a series of points corresponding to the successive occurrences of various taxa* in the well (X-axis) and in the reference profile (Y-axis). The position of the different points allows to identify eight plateau corresponding to sequence boundaries (document by Amoco Production Company. From STEIN, J.A., GAMBER, J.H., KREBS, W.N. & LA COE, M.K. (1992). – A composite standard approach to biostratigrapic evaluation of the North Sea Paleogene. – Petroleum Geology of North Western Europe, Aberdeen 1992, Abstr.).

* Benthic and planktonic foraminifera, diatoms, radiolarians, calcareous nannofossils, dinoflagellate cysts, spores and pollens.

Measured against a unit stratotype, the composite reference section allows to evaluate its representativeness compared to the corresponding stage.

While the stratotype symbolizes the specific record of a time interval in a limited area, the composite profile section represents a tool to evaluate the succession of events observed in several sections and to check the validity of their relative position in time over a wider region. For the stratigrapher, the CRS represents a valuable tool to portray his results for the use of the sedimentologists and geophysicists, and particularly adapted to highlight key levels in the sequential interpretation.

3.4. THE PROCEDURE OF INTEGRATED STRATIGRAPHY

Integrated stratigraphy consists in associating all the data made available by the different methods of stratigraphy. Goal of the integrated stratigraphy is to order and correlate, in space and time, the different kinds of stratigraphic units of the various scales.

Starting from an outcrop or a subsurface section, the method is based on the survey, in 2D or 3D, of any available information and signals. Among those currently used, some have a primarily geometric significance (lithologic and seismic data), others environmental (sedimentological, geochemical, paleoecological data). Finally, some have a temporal significance (biochronological, paleomagnetic or geochronological data).

The comparison of data from different sites makes it possible to carry out methodical correlations. Certain markers allow to define chronologic time lines.

Indirect correlations, with the aid of the chronostratigraphic reference scale, allow to gradually obtain a representation of a geological object into four dimensions: the three dimensions of space and the dimension of time. This procedure, primarily based on the identification and the interpretation of descriptive data, makes it possible to obtain datings and chrono-correlations of increasing reliability and precision, to improve our knowledge of the differences between the different stratigraphic scales and to define with a good precision chronostratigraphic unit boundaries.

3.5. GEOCHRONOMETRY OF THE PRECAMBRIAN

The rocks of the Precambrian, which represents more than 85% of the entire geological time scale, have not been subdivided into chronostratigraphic units of global value. Indeed, Precambrian chronostratigraphy deserves a separate discussion, because the data on which the subdivision of the Phanerozoic is based, are missing for this time interval: the original lithological characters are often obliterated by metamorphism; fossils are rare or absent; the magnetostratigraphic signal is rarely preserved. The only available tool remains isotope geochronology.

In practice, Precambrian stratigraphy primarily rests on major features of geometrical nature (discordances between different orogenic cycles) and on lithological breaks delimiting units of minor importance. It is difficult, if not impossible, to place the units identified regionally in a global frame. To create a common language, the Subcommission of Precambrian Stratigraphy proposed to establish conventional boundaries of exclusively temporal significance. There are no boundary stratotypes for this interval and the resulting units (Systems,

It is helpful to adjust the correlation line on the base of carefully located physical data, which can be considered to be synchronous (magnetic inversion, isotopic shifts, radioactivity peaks…).

The last step in the procedure consists in calibrating the composite section by using numerical dating if available. This step would allow to correlate the CRS with a numerical time scale.

By the method of graphic correlation, any section placed on the Y axis can be compared to the a CRS used as the X axis. This procedure reveals the sedimentary history of each section relative to the CRS. Changes in relative rate of accumulation between sections, in fact, will cause the slope of the correlation line to change, whereas a gap in depositional history will result in a plateau. The latter are identified on the section by their position in the Y axis, and their duration can be measured on the CRS, which represents the X axis (8.3).

Eras) are neither chronostratigraphic nor geochronologic but geochronometric units (see chapters VII and IX).

3.6. CHRONOSTRATIGRAPHY OF THE PLIO-QUATERNARY INTERVAL

The chronostratigraphy of the Pliocene and Quaternary intervals is based on the same fundamental principles as those of older units.

It is however, largely different in terms of stratigraphic resolution, nature and/or quality of the information available. On one hand, the overall duration of both the Pliocene and the Quaternary is the same of the average stage duration in older intervals; yet, further subdivisions are possible, implying the erection of formal units of much shorter duration than the average in the geological time scale. On the other hand, the quality of data from more recent intervals is generally higher. Moreover, some data are available only for the uppermost part of the geological column, particularly for the Quaternary: organic remains, human manufactures and repeated climatic changes of known duration. Continental deposits and in ice-core records (well accessible for the Quaternary) play a key role in this context. They offer the possibility to obtain a high-resolution record of these changes, which is useful for regional or global stratigraphic correlations depending on the extension of the climatic events. However, this richness of data stimulated the development of numerous detailed scales based on data of different nature from different areas and paleoenvironments, which results in the lack of a firm reference scale. Nevertheless, many chronostratigraphic units can be correlated by means of numerical ages. They can therefore still be used with profit resulting in a very detailed and multiform chronostratigraphy (see paragraph 4 in chapter VII).

4. – CONCLUSIONS: THE STRATIGRAPHIC TOOL

By establishing a temporal framework (the geological time scale) which describes the space-time evolution of the various facies, stratigraphy provides the elements to trace the succession of geographical settings on the surface of our planet and to analyse the geodynamic conditions which induced them, therefore to reconstruct the evolution of the continents and the oceans.

The major tendency of stratigraphy (that the procedures of the genetic and integrated stratigraphy illustrate particularly well) is to try building unitary models in the organization of the geological objects which are likely to show and explain the existence of logical relations between different stratigraphic signals.

This approach can improve the quality of the chrono-correlations and the precision of the dating. Anticipated by d'ORBIGNY and admirably expressed by GIGNOUX, the search for the causes controlling the evolution in space and time of the systems that are part of our Planet best expresses the fundamental role and significance of stratigraphy.

5. – CONVENTIONS

5.1. MAIN RULES IN THE DEFINITION OF CHRONOSTRATIGRAPHIC UNITS

The creation of chronostratigraphic units (stages, series, systems), submitted to international authorities, is guided by some fundamental rules:

– (1) stage names must be derived from the name, modern or ancient, of the standard locality, followed by the suffix "ian". Two names of different stages cannot be derived from the same standard locality;

– (2) chronostratigraphic units must necessarily be defined starting from stratotypes, which makes it possible to illustrate their contents and to precisely identify their boundaries. These stratotypes, selected following the principles described in paragraph 3.3., are subject of multidisciplinary studies including all the available stratigraphic methods;

– (3) publication priority must be respected. For the stages, however, it is advisable to consider the literature published only after an established year. The French Committee of Stratigraphy (SIGAL & TINTANT, 1962), for example, proposed that, for the Jurassic, only the publication after the 1850, date of the publication of the stages by d'ORBIGNY, should be considered. The principle guiding this decision is that priority alone does not justify to replace a widely used name with a poorly used name. Conversely, a name which is not appropriate should not be preserved only due to priority;

– (4) besides a ratification by the competent authorities, the creation of a new chronostratigraphic unit requires a publication in a recognized scientific journal;

– (5) the revision or the redefinition of a unit correctly defined without changing its name requires the same justification, the same types of information and the same steps as the proposal of a new unit.

5.2. PROCEDURES FOR THE RATIFICATION OF CHRONOSTRATIGRAPHIC UNITS AND THEIR STRATOTYPES

The International Commission of Stratigraphy of the IUGS is the organization responsible for the coordination of the choice and ratification of the stratotypes for the units of the standard chronostratigraphic scale. The International Commission of Stratigraphy is organized into Subcommissions specifically focused on each system of the Phanerozoic for the standardization of stratigraphic units. With the purpose of defining unit or boundary stratotype, Working Groups are organized, generally under individual Subcommissions. Commonly, a Working Group is created for the selection and definition of the lower boundaries of geochronologic/chronostratigraphic units. Task Groups may also be created for the purpose of replacing and/or selecting new boundary definitions, stage units or other stratigraphic units. Working Groups submit formal proposals for the definition or redefinition of stratigraphic units to the appropriate Subcommission which votes it in turn, then submitted it to the International Commission of Stratigraphy. If the commission accepts the proposal, the latter is submitted to the ratification of the IUGS during of geological international conferences.

CHAPTER 9

THE GEOLOGICAL TIME SCALE

F.M. Gradstein, James Ogg & Gabi Ogg

---

1. – INTRODUCTION 125
2. – BOUNDARY STRATOTYPES 125
    2.1. Definition 125
    2.2. Klonk 126
    2.3. Progress with GSSP's 126
3. – RECONCILE PROTEROZOIC ROCK RECORD WITH ABSTRACT TIME 128
4. – UNITS OF TIME 128
5. – BUILDING THE GEOLOGICAL TIME SCALE 130
6. – SEDIMENTARY CYCLES 130
7. – DECAY OF ATOMS 130
8. – INTERPOLATION AND STATISTICS 132
9. – THE GEOLOGICAL TIME SCALE 132
    9.1. GTS 2004 132
    9.2. GTS 2010 133
10. – TIME SCALE CREATOR 134

1. – INTRODUCTION

A good understanding of geological time is vital to every scientist that strives to understand processes and events that shaped and changed our earth. This understanding takes place in the framework called Earth Geological History, the super calendar of local and global events since Earth was born. The challenge to this understanding is reading, organizing and sorting the calendar pages in stone, and, last but not least, reconstructing its missing pages. Correlation is a vital part of the reconstruction process.

One of the earliest constructors was Nicolas STENO (1631-1687) who made careful and original observations on regional rock sequences, and concluded that the strata of the earth crust contain the superimposed records of a chronological sequence of events that can be correlated worldwide. Since that time, this science of stratigraphy has developed a sophisticated earth history framework at increasingly detailed levels of resolution.

As clearly outlined in this book, geological correlation formally is expressed in terms of five consecutive operations and units:

– Rock units, like formations or well log intervals = lithostratigraphic correlation

    e.g. Kimmeridge Clay Formation of England

– Fossil units, like zones = biostratigraphic correlation

    e.g. *Globuligerina oxfordiana* planktonic foraminiferal zone

– Relative time units = geochronologic ("Earth time") correlations

    e.g. Jurassic Period, Tithonian Age, magnetic polarity chron C29r

– Rocks deposited during these time units = chronostratigraphic (time-rock) correlation

    e.g. Jurassic System, Tithonian Stage, magnetic polarity zone C29r

– Linear time units or ages = geochronometric or geochronologic correlation

    e.g. 150 Ma, 10 ka

Without correlation to a global reference scale, successions of strata or events in time derived in one area, are unique and contribute nothing to understand Earth history elsewhere. The rules of hierarchy in geological correlation, from rocks and fossils to relative and linear time are carefully laid down in the International Stratigraphic Guide; an abbreviated copy of this "rule book" with further references can be found on the web site of the International Commission on Stratigraphy (ICS; www.stratigraphy.org).

Before we deal with linear geological time, a few words about the common geological calendar built from relative age units. This calendar called chronostratigraphic calendar is not unlike a historical calendar in which civilization periods, such as the Minoan Period, Reign of Louis XIV or American Civil War, are used as building blocks, devoid of a linear scale. Archeological relicts deposited during these intervals, such as the Palace of Minos on Crete, Versailles or spent cannon balls at Gettysburg, comprise the associated physical chronostratigraphic record. The chronostratigraphic scale is assembled from rock sequences stacked and segmented in relative units based on their unique fossil and physical content. Through correlation of the unique fossil and physical record to other sections across the globe, this scale becomes meaningful and useful.

The standard chronostratigraphic scheme, in downloadable graphics format available from ICS at their standard website, is made up of up of successive stages in the rock record, like Cenomanian, Turonian, then Coniacian, etc., within the Cretaceous system.

2. – BOUNDARY STRATOTYPES

2.1. DEFINITION

Originally, each stage unit was typified by a well-defined body of rocks at a specific location of an assigned and agreed upon relative age span, younger than typical rocks of the underlying stage and older than the typical rocks of the next higher stage. This is the concept of defining stage units with stratotype sections. The principles and building blocks of this chronostratigraphy were slowly established during centuries of study in many discontinuous and incomplete outcrop sections. Inevitably, lateral changes in lithology between regions and lack of agreement on criteria, particularly which fossils were characteristic of which relative rock unit, have always resulted in a considerable amount of confusion and disagreement on stage nomenclature and stage use. Hence, a suite of global subdivisions with precise correlation horizons were required.

One of the major champions of practical and rational thinking an stratigraphy, Hollis Hedberg en 1967 set the stage for a veritable upgrade if this venerable geoscience, when he wrote:

"in my opinion, the first and most urgent task in connection with our present international geochronology scale is to achieve a better definition of its units and horizons so that each will have a standard fixed-time significance, and the same tme significance for all geologists everywhere. Most of the named international stratigraphic (geochronology) units still lack precise globally accepted definitions and consequently their limits are controversial and varably interpreted by different workers. This is a serious and wholly unnecessary impediment to progress in global stratigraphy. What we nedd is simply a single permanently fixed and globally accepted standard definition for each named unit or horizon, and this is where the concept of stratotype standards (particularly boundary stratotypes and other horizon stratotypes) provides a satisfactory answer".

Now, relatively rapid progress is being made with definition of GSSP's (Global Boundary Stratotype Sections and Points) to fix the lower boundary of all geologic stages, using discrete fossil and/or physical events that correlate well in the rock record. For the ladder of chronostratigraphy, this GSSP concept switches the emphasis from marking the spaces between steps (stage stratotypes) to fixing the rungs (boundaries of stages).

ICS has set clear guidelines for the establishment of GSSP's. The information required is outlined in detail in stratigraphic guides, and also available from the website of ICS. These include the ICS guidelines listed in table 9.1.

TABLE 9.1

Information required for the establishment of a Global Boundary Stratotype Section and Point (GSSP).

1	Name of the boundary
2	GSSP definition
3	Stratigraphic rank and status of the boundary
4	Stratigraphic position of the defined unit
5	Type locality of the GSSP
6	Geologic setting and geographic location, incl. coordinates
7	Lithology/sedimentology/paleobathymetry
8	Map and GPS (World Geodetic System 84 datum) coordinates
9	Accessibility, incl. logistics, national politics and property rights
10	Conservation
11	Identification in the field
12	Stratigraphic completeness of the section
13	Global correlation, using where applicable biostratigraphy, magnetostratigraphy, stable isotope stratigraphy, and other stratigraphic tools and methods
14	Reference to historical background studies
15	Scientific publication of the new GSSP

2.2. KLONK

It is now 25 years since the first stratigraphic boundary was defined by a boundary stratotype or "golden spike", inaugurating the concept of the Global Boundary Stratotype Section and Point (GSSP). This event of historic proportions for chronostratigraphy and geochronology involved the boundary between the Silurian and Devonian Systems, or rather the lower limit of the Devonian at a locality called Klonk in the Czech Republic (CHLUPÁČ, I. 1993).

The problem of the Silurian-Devonian boundary and its consensus settlement in the Klonk section, hinged on a century old debate, known as the "Hercynian Question" that touched many outstanding geoscientists of the previous century. The issue came to the foreground after 1877, when Kaiser stated that the youngest stages (étages) of Barrande's 'Silurian System' in Bohemia, correspond to the Devonian System in the Harz Mountains of Germany and other regions. Kaiser's findings contrasted with the conventional 19th century wisdom that graptolites became extinct at the end of the Silurian. Eventually, it became clear that so-called Silurian graptolites in some sections occur together with so-called Devonian fossils in other sections, leading to the modern consensus that graptolites are not limited to Silurian strata.

A bronze plaque in the Klonk outcrop shows the exact position of the modern Silurian-Devonian Boundary that also represents the base of the Lochkovian Stage, which is the lowest stage in the Devonian. The base of the Lochkovian Stage is defined by the first occurrence of the Devonian graptolite *Monograptus uniformis* in bed # 20 of the Klonk Section, NE of the village of Suchomasty. The lower Lochkovian index trilobites with representatives of the *Warburgella rugulosa* group occur in the next younger limestone bed # 21 of that section (CHLUPÁČ, 1993).

2.3. PROGRESS WITH GSSP'S

At present a majority of phanerozoic stages have formal boundary definitions (Fig. 9.1), details of which may be found in the website of ICS, the international geoscience body that guides, formally votes and aproves the names of stages and boundary definitions. In 2004, 45 stages had ratified lower boundaries and 31 in 2000. Auxiliary boundary stratotype sections are of use for regional correlations, and both Russia and New Zealand, to name just two countries, are erecting a set of key outcrops that define the lower boundaries of their local stages utilized on regional geological maps.

The reasons that not all stages already have boundary stratotypes are varied. Several factors pay a role, of which the three main ones are:

– (a) For each stage boundary many sections worldwide may have to be studied in detail.

– (b) Many stratigraphers hope to find the criteria in the rock record for the perfect GSSP definition, rather an adequate and practical one.

– (c) Prejudices exist against selection of global physical events over (often more regional) fossil events, and

– (d) Selection of a GSSP is to overcome regional or historical controversies.

Dogged discussions tend to forget or ignore the fact that (fortunately) there is no rule of historical priority in stratigraphy, and that all stratigraphy ultimately is guided by subjective consensus. There is an abundance of occasions where a set of zones traditionally associated with one stage eventually was re-assigned to the next underlying or overlying stage. Hence, beds assigned to such zones were re-assigned a younger or older age. Preference for stratigraphic priority is laudable when selecting GSSP's, but subsidiary to scientific and practical merit. The search for perfection also ignores the fact that emendations of a GSSP definition are feasible under ICS rules, using consensus voting on emended proposals.

It can be easily understood that it is important that a GSSP is selected in a continuous stratigraphic section, without obvious hiatuses. After a completed proposal has been

9. The geological time scale

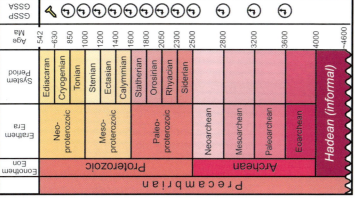

FIG. 9.1

The International Stratigraphic Chart, with "golden spikes" denoting which stage and period boundaries are defined by a Global Boundary Stratotype Section and Point (GSSP). The time scale is that of Geologic Time Scale 2004 (GTS 2004), as of December, 2007. The status of the information is as of December, 2007. The status of a Phanerozoic stage boundary has led to a (minor) shift in linear age; those updated Phanerozoic levels lack error bars on the chart, and await proper re-calculation in GTS 2010.

* The status of the Quaternary is not yet decided. Its base may be assigned as the base of the Gelasian and extend the base of the Pleistocene to 2.6 Ma.

** Middle Cretaceous is informal.

formulated it is submitted to ICS for approval; the final step is formal publication of the proposal.

An interesting exception with a regional historical precedent for a basal disconformity in a GSSP is the lower limit of the Zanclean Stage and Pliocene Series, ratified in 2000. Because the onset of the Pliocene in the Mediterranean traditionally is the marine transgression following the Messinian drying out of that region, the GSSP section for the Miocene/Pliocene boundary reflects this.

It may be desirable to add some documentation to a GSSP proposal about the historical significance of a new stage GSSP in the hierarchy of stratigraphic units of higher rank. Careful consideration needs to be given to the fact that, for example, the GSSP for the base of the Pleistocene Series in the Vrica Section in southern Italy automatically defines the boundary of the Pliocene/Pleistocene, and the base of the Calabrian Stage, the lowest Pleistocene stage. In the same vein, since the Induan is the lowest stage in the Triassic System, and the Triassic the lowest System in the Mesozoic Erathem, the base of the Induan Stage as defined in the Meishan Section, China also defines base Triassic and the Paleozoic/Mesozoic boundary. It is along these lines of logical stratigraphic reasoning that definition of a GSSP should entail some historical background research, to ensure consensus stratigraphic hierarchy, built on scientific and practical merit.

A good example of the benefit of the GSSP concept to time scale building is the current set of ages for the base Paleozoic at 542 Ma, base Mesozoic at 251 Ma and base Cenozoic at 65.5 Ma. The detailed stratigraphic definitions have guided geochronologists to sections and cores where volcanic tuffs dated with the U-Pb or Ar-Ar methods alternate with fossiliferous intervals containing the index criteria for recognition of these major boundaries. Details are in the "Geological Time Scale 2004" (GRADSTEIN et al., 2004). Another benefit is that studies on tempo and modes of evolution or the dynamics of a catastrophic process – i.e., the properties that help differentiate stratigraphic units – cannot be undertaken without having the definition of the key stage boundaries in place.

3. – RECONCILE PROTEROZOIC ROCK RECORD WITH ABSTRACT TIME

The Precambrian is subdivided in the Archean and Proterozoic Eras (Fig. 9.1). It encompasses 88% of the geological record on Earth, from about 4000 Ma to exactly 542 Ma, and has a deficient stratigraphic subdivision. Due to the fact that most of the Proterozoic record lacks adequate fossils, and physical event correlation in rock units across the globe is a vastly complex task, a different type of boundary stratotype was developed for that enormously long interval of time on Earth. The new boundary stratotype is called Global Standard Stratigraphic Age (GSSA), an abstract term for an abstract, non-geological concept. The definition of a boundary by its linear age is the consequence of the fact that the Proterozoic now includes units of global stratigraphic subdivision, where the boundaries are defined in terms of the age in millions of years. Summaries of the ten ratified GSSA's at 2500, 2300, 2050, 1800, 1600, 1400, 1200, 1000, 850, and 650 Ma may be found in the website of IGC. However, although there appears to be consensus that the subdivision of the Proterozoic in three Eras – Paleoproterozoic, Mesoproterozoic and Neoproterozoic is excellent, the finer Period subdivisions at these abstract age levels often contain no datable rocks, which makes their use haphazard. Even more damaging, uncertainties in radiometric dates makes that in Precambrian these abstract geochronologic time lines levels may be uncertain over ten or more million years.

A consensus is developing in ICS that abstract Period definitions that cannot be sustained by the Precambrian rock record on Earth should ultimately be reconciled with the latter. GSSA's must become GSSP's using observable and correlative geological events. Fig. 9.2 (after BLEEKER in GRADSTEIN et al., 2004) shows a "radical" proposal for a "natural" Precambrian time scale. Earth history is divided into six eons, with boundaries defined by what can be considered first-order "watersheds" in the evolution of our planet. The six eons can be briefly characterized as follows:

– (1) "Accretion & Differentiation", planet formation, growth and differentiation up to the Moon-forming giant impact event.

– (2) Hadean, intense bombardment and its consequences, but no preserved supracrustals.

– (3) Archean, increasing crustal record from the oldest supracrustals of Isua to the onset of giant iron formation deposition in the Hamersley basin, likely related to increasing oxygenation of the atmosphere.

– (4) "Transition", starting with deposition of giant iron formations up to the first bona fide continental red beds.

– (5) Proterozoic, a nearly modern plate-tectonic Earth but without metazoan life.

– (6) Phanerozoic, characterized by metazoan life forms of increasing complexity.

The latter definition of the Phanerozoic would involve moving its lower boundary to encompass the Ediacaran Period, which was ratified as a formal period in 2004. The GSSP of the Ediacaran is the base of the Marinoan cap carbonate (Nuccaleena Formation), immediately above the Elatina diamictite in the Enorama Creek section, Flinders Ranges, South Australia.

4. – UNITS OF TIME

Ages are given in years before "Present" (BP). To avoid a constantly changing datum, "Present" was fixed as AD 1950 (as in C14 determinations), the date of the beginning of modern isotope dating research in laboratories around the world. For most geologists, this offset of official "Present" from "today" is not important. However, for archeologists and researchers into events during the Holocene (the past 11,500 years), the current offset (50 years) between the "BP" convention from radiometric laboratories and actual total elapsed calendar years becomes significant.

For clarity, the linear age in years is abbreviated as "a" (for annum), and ages are generally in ka for thousands of years or Ma for millions of years before present. The elapsed time or duration is abbreviated as "yr" (for year), and durations are generally in kyr or myr. Therefore, the Cenozoic began at 65.5 Ma, and spans 65.5 myr (to the present day).

Opinions are raised that one and the same unit should be used for absolute and relative measurements. Hence, elapsed time or duration should also be abbreviated as in *ka* or *Ma*. Therefore the Cenozoic began at 65.5 Ma, and spans 65.5 Ma. This is similar to the use of m (meter) for both absolute depth/distance and a depth/distance difference, and the use of ^0C for both temperature and temperature difference. However, such usage for geologic time is far from standardized between scientific journals and organizations; hence we refrain from it.

The uncertainties on computed ages or durations are expressed as standard deviation (1-sigma or 63% confidence) or 2-sigma (95% confidence). The uncertainty is indicated by

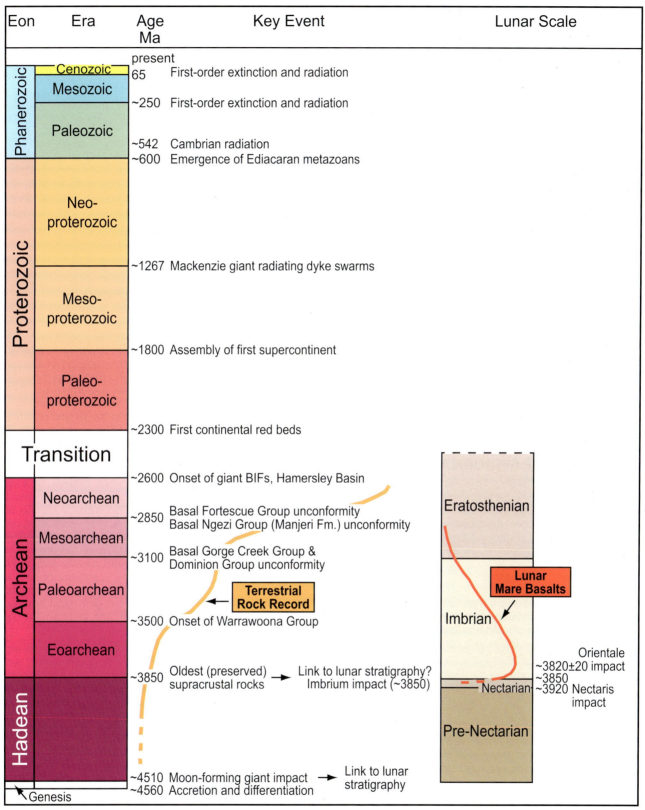

Fig. 9.2
A "radical" proposal for a "natural" Precambrian time scale. Earth history is divided in six eons, with boundaries defined by what may be considered first-order "water-sheds" in the evolution of planet Earth (after W. Bleeker, in GTS 2004).

"±" and will have implied units of kyr or myr as appropriate to the magnitude of the age. Therefore, an age cited as "124.6 ±0.3 Ma" implies a 0.3 myr uncertainty (1-sigma, unless specified as 2-sigma) on the 124.6 Ma date. We present the uncertainties (±) on summary graphics of the geologic time scale as 2-sigma (95% confidence) values.

Geologic time is measured in years, but the standard unit for time is the second (s). Because the Earth's rotation is not uniform, this "second" is not defined as a fraction (1/86400) of a solar day, but as the atomic second. The basic principle of the atomic clock is that electromagnetic waves of a particular frequency are emitted when an atomic transition occurs. In 1967 the Thirteenth General Conference on Weights and Measures defined the atomic second as the duration of 9,192,631,770 periods of the radiation corresponding to the transition between two hyperfine levels of the ground state of the Cesium atom 133. This value was established to agree as closely as possible with the solar-day second. The frequency of 9 192 631 770 Hertz (Hz), which the definition assigns to the Cesium radiation was carefully chosen to make it impossible, by any existing experimental evidence, to distinguish the atomic second from the ephemeris second based on the Earth's motion. The advantage of having the atomic second as the unit of time in the International System of Units is the relative ease, in theory, for anyone to build and calibrate an atomic clock with a precision of 1 part per 1011 (or better). In practice, clocks are calibrated against broadcast time signals, with the frequency oscillations in Hz being the "pendulum" of the atomic time keeping device. 1 year is approximately 31.56 megaseconds (1 a = ~31.56 Ms).

5. – BUILDING THE GEOLOGICAL TIME SCALE

The ideal geologic time scale makes use of precise and standardized age estimates at successive stage boundaries, but we are still some distance from that goal. Although significant progress has been made during the last decade in defining Phanerozoic stage boundaries in a stratigraphic sense, many stage limits, particularly in the middle part of Phanerozoic, leave to be desired, or are being re-defined to ensure better biostratigraphic and geochronologic calibration. Apart from the fundamental chronostratigraphic issues, there is the problem of sufficient and stratigraphically meaningful age dates, and their calibration to a common standard (see Section on Decay of Atoms).

The modern geological time scale not only utilizes stratigraphically meaningful radiometric dates in millions of years, but also increasingly successfully earth-orbit tuned sedimentary cycles in thousands of years. Both types of measurements are achieved in rocks that are stratigraphically discontinuous, and need a certain amount of luck to find core or outcrop sections that are sufficiently continuous, sufficiently rich in geological events like index fossils or stable or unstable isotope measurements, and correlative to other sets of strata with a known stratigraphic (relative) age.

The steps involved in modern time scale construction may be summarized as follows:

– Step 1. Construct up-to-date chronostratigraphic scale from the global earth rock record;

– Step 2. Calibrate chronostratigraphic scale in linear age units, using radiometric age dates, and/or astronomically tuned cyclic sedimentary sequences or stable isotope sequences;

– Step 3. Interpolate the combined chronostratigraphic and chronometric scale where direct information isinsufficient;

– Step 4. Calculate or estimate error bars on the combined chronostratigraphic and chronometric scale to obtain a geologic time scale with estimates of uncertainty on boundary ages and on unit durations;

– Step 5. Peer review the Geologic Time Scale through the International Commission on Stratigraphy (ICS).

Fig. 9.3 provides an overview of methods through the Phanerozoic that created GTS 2004, as discussed below. Prior to the Cambrian, the first Phanerozoic Period, the geologic time scale is less sophisticated, and based on sparse radiometric ages dates only. The time and effort involved in constructing a new geological time scale and assembling all relevant information is quite considerable, with many geoscience specialists involved. Because of this, and because continuous updating in small measure with new information is not advantageous to the stability of any standard, new Phanerozoic time scales tend to come out sparsely.

6. – SEDIMENTARY CYCLES

The principle of sedimentary cycles approach to time scale building rests on the gravitational interactions of the Earth with the Sun, Moon and other planets that cause systematic changes in the orbital and rotational system of our planet. These interactions give rise to cyclic oscillations in the eccentricity of the Earth's orbit, and in the tilt and precession of the Earth's axis, the so-called Milankowich cycles. The cycle of eccentricity relates to the changes in shape of the Earth'orbit from circular to elliptical with main periods of 100,000 and 400,000 years. The tilt and the clockwise rotation of the Earth's axis have main periods of 41,000 and 26,000 years respectively. Because precession has the opposite motion to the orbit of the Earth around the Sun, the actual cycle is 21,000 yaers long. The associated cyclic variations in annual and seasonal solar radiation onto different latitudes alter long-term climate in colder versus warmer and, wetter versus drier periods that lead to easily recognizable sedimentary cycles, such as regular interbeds of limy and shaly facies. Massive outcrops of hundreds or thousands of such cycles are observed in numerous geological basins, for example around the Mediterranean, and in sediment cores from ocean drilling wells.

Counting of these centimeter- to meter- thick cycles in great detail over land outcrops and in ocean drilling wells, combined with the additional correlation aids provided by magnetostratigraphy, oxygen isotope stratigraphy and biostratigraphy, is producing a detailed cycle scaling for the Neogene (youngest period). The critical step is the direct linkage of each cycle to the theoretical computed astronomical scale of the 21,000, 41,000 and 100,000 year paleoclimatic cycles. This astronomical tuning of the geological cycle record from the Mediterranean and Atlantic has led to unprecedented accuracy and resolution for the last 11 million years. By applying a combination of cycle stratigraphy, astronomical projections, oxygen isotope stratigraphy and magnetostratigraphy to the deep sea record, the scale is being extended back to about 23 Ma ago (LOURENS et al. in GRADSTEIN et al., 2004), with excellent promise for extension back through most of Paleogene (see RÖLH et al., 2001; PÄLIKE et al., 2006). The stability of the eccentricity cycles allaows orbital tuning to be extended back into Mosezoic, to the benefit of the geologic time scale.

In New Zealand, T. NAISH and colleagues have calibrated the upper Neogene record to the standard Neogene time scale. Using the high-resolution land-based cycle, isotope

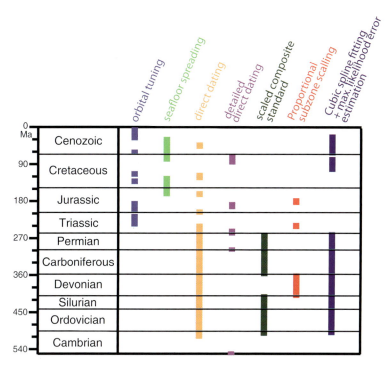

FIG. 9.3
Combination of methods used to construct the Geological Time Scale 2004 (GTS 2004).

and magnetic record in the Wanganui Basin, these authors thereby transferred precise absolute ages to local shallow marine sediments and demonstrated the link between sequence and cycle stratigraphy.

A special application of orbitally tuned cyclic sediment sequences is to "rubber-band" stratigraphically floating units, like parts of Paleocene, Albian, and parts of Lower Jurassic. A quantitative estimation of the duration of all cycles within a stratigraphic unit allows estimation of its duration.

7. – DECAY OF ATOMS

For rocks older than Neogene, the derivation of a numerical time scale depends on the availability of suitable radiometric ages. Radiometric dating generally involves measuring the ratio of the original element in a mineral, like sanidine feldspar or zircon, to its isotopic daughter products. The age of a mineral may then be calculated by means of the isotopic decay constant. Depending on the half-life of the element, several radiometric clocks are available; $^{87}Rb/^{86}Sr$, $^{40}K/^{40}Ar$ and U/Pb are the most commonly used.

In 1976, the Subcommision on Geochronology recommended an inter-calibrated set of decay constants and isotopic abundances for the U-Th-Pb, Rb-Sr, and K-Ar systems with the uranium-decay constants by JAFFEY et al (1971) as the mainstay for the standard set (STEIGER & JÄGER, 1977). This new set of decay constants necessitated systematic upward or downward revisions of previous radiometric ages by 1% to 2%.

For time scale building radiometric dating of volcanic flows and tuffs in a stratified sedimentary succession is most meaningful. Radiometric dating techniques with less than 1% analytical error are providing suites of high-precision U/Pb for the Paleozoic and early Mesozoic. The integration of this level of chronometric precision with high-resolution biostratigraphy, magnetostratigraphy or Milankovich cyclicity is a major challenge to time scale studies and interpolation mathematics. Even the most detailed biostratigraphic scheme probably has no biozonal units of less than 0.5 – 1.0 my duration, not to speak of the actual precision in dating a particular "stratigraphic piercing" point, for which an U/Pb age estimate would be available with an analytical uncertainty of 0.5 my. Similarly, combination of analytically less precise Ar/Ar dates with much more precise U/Pb dates in statistical interpolations creates a strong bias towards the latter, despite the fact that both may have equal litho-, bio-, and chronostratigraphic precision.

On the other hand, the combination of precise stratigraphic definitions through GSSP's and accurate radiometric dates near these levels is paving the way for a substantial increase in the accuracy of the Geological Time Scale. Base Phanerozoic (= base Cambrian), base Ordovician, base Carboniferous, base Mesozoic (= base Triassic), base Cenomanian, base Maastrichtian, base Cenozoic, base Eocene, and several others can now be pinned down in the Geological Time Scale with much greater precision than before.

The U-Pb zircon ages using the Thermal Ionization Mass Spectrometric (TIMS) methods are highly advantageous because of two independent radioactive decay schemes: ^{238}U to ^{206}Pb and ^{235}U to ^{207}Pb. This allows three dates to be calculated for concordant analytical points $^{206}Pb/^{238}U$, $^{207}Pb/^{235}U$ and $^{207}Pb/^{206}Pb$. Analysis can be done on zircons and on monazite, the latter being rare in volcanic rocks. Using zircons, contamination with older xenocrysts can become a problem. Populations of dates on many single crystal per rock sample are favoured.

The U-Pb High Resolution- Secondary Ion Mass Spectrometry (HR-SIMS) method tunnels in zircons to detect growth ages. The method is important for geosciences, but has a larger analysis error than the isotope dilution method due to the fact that the minute amount of Pb measured strongly limits the precision of the $^{235}U - ^{207}Pb$ ratio. To overcome this problem multiple measurements on single grains are necessary. There is also the limitation that the instrument requires calibration to a monitor standard.

Intra-laboratory variability of the measurements using ^{40}Ar-^{39}Ar analyses and dating is highly precise and about 0.2%. On the other hand, the inter-laboratory uncertainty may not be better than 2%. This is due to the fact that age dates are relative to a standard that monitors the transformation of ^{39}K into ^{39}Ar during neutron irradiation of the sample. The standard must be chosen such that it gives concordance between the U-Pb and Ar-Ar dating systems. Unfortunately, there is no consensus in the scientific community on which standard to use, and which numerical value to assign to it.

In a detailed study RENNE et al. (1994) address the issue of inter-calibration of standards in ^{40}Ar-^{39}Ar analysis, their absolute ages and uncertainties, and make the following recommendations: 523.1 ± 4.6 Ma for MMhb-1 (Montana hornblende), 28.34 ± 0.28 Ma for TCs (Taylor Creek sanidine) and 28.02 ± 0.28 Ma for FCs (Fish Canyon sanidine). The errors (1 sigma) are full external errors, including uncertainties in decay constants. The Mmhb-1 standard age was found to be inhomogeneous at the single grain, sub – 15 mg level, making it now unsuitable as an inter-laboratory standard. Individual age dates are reliable if large aliquots of the standard are applied. GTS2004 accepts the above standard values for ^{40}Ar-^{39}Ar analysis; appropriate corrections were carried out on ^{40}Ar-^{39}Ar age dates using different standards.

An unresolved issue with ^{40}Ar-^{39}Ar dating is that the nuclear physics community since 1973 uses a decay constant of 5.428 ± 0.034 x 10^{-10}/ year, but a value of 5.534 x 10^{-10}/ year is used in geochronology (STEIGER & JÄGER, 1977). In a recent study, the decay constant and the age of a standard were evaluated simultaneously, using statistical methods on a series of isotopic dates (KWON et al., 2002). The authors conclude a value of 5.4755 ± 0.0170 x 10^{-10} (1 sigma) year for the decay constant and an age of the FCT of 28.269 ± 0.0661 (1 sigma) Ma. Acceptance of these values that are up to 2% different from the values used for GTS2004 requires further study and ultimately should lead to a consensus decision by the geochronological and nuclear physics science communities. Acceptance of an astronomically dated neutron irradiation monitor will potentially decrease the overall uncertainty in $^{40}Ar - 39Ar$ analysis (F. HILGEN, pers. comm. 2007).

Analytical advancements during the past 15 years now allow wide application of the ^{187}Re (rhenium)–^{187}Os (osmium) radioactive isotope system in Earth Science, both as a geochronometer and process tracer. Through improved analytical methodologies for digestion (organic-selective dissolution techniques) and optimal sampling strategies high-precision (< ± 1% 2s) Re-Os ages are being attained. In addition, it is now established that Re-Os geochronometer in black shale remains undisturbed through hydrocarbon maturation and in some cases chlorite-grade metamorphism allowing precise depositional ages to be determined from a wider range of shales than previously thought possible (e.g., units with TOC contents ~ 0.5%).

Accurate Re-Os black shale dates are best illustrated by comparison to units or boundaries for which precise U-Pb zircon age determinations exist. A black shale at the Devonian-Carboniferous (DC) boundary in Western Canada, has yielded a Re-Os age of 361.3 ± 2.4 Ma (2 s, including λ uncertainty), in accord with the most recent U-Pb zircon age interpolations. Rhenium-Osmium (Re-Os) dating of black shale of the proposed Global Stratotype Section and Point (GSSP) for the Oxfordian-Kimmeridgian boundary (Staffin shale at the Isle of Skye, northern UK) yields an age of 154.1 ± 2.2 Ma (SELBY, 2007). These results demonstrate that the Re-Os shale geochronometer has a significant role to play for stratigraphic intervals and its rock sections with limited potential for ashbed U-Pb zircon dating.

K-Ar and Rb-Sr methods for age dating are prone to a variety of errors and in general only provide minimum values for the rock ages. Dating of authigenic sedimentary minerals, mainly involving glauconite, found widespread in many marine sediments, is being phased out. Mild heating or overburden pressure after burial may lead to loss of argon, the daughter product measured in the $^{40}K/^{40}Ar$ clock in glauconite. Another problem is that glauconite also contains an abundance of tiny flakes that allow diffusion of Ar at low temperatures. The result is that glauconite dates may be too young. Because of such problems, which may be difficult to detect, modern geologic time scales avoid age dates based on glauconite.

8. – INTERPOLATION AND STATISTICS

Despite the progress in standardization and dating, parts of the Mesozoic and Paleozoic geological record have sparse radiometric coverage. Ideally, each of the 100 or so stage boundaries that comprise the Paleozoic, Mesozoic and Cenozoic Eras of the Phanerozoic Erathem would coincide with an accurate radiometric date from volcanic ashes deposited at each of the boundaries. However, this coincidence is rare in the geological record. Only base Phanerozoic (= base Cambrian), base Ordovician, base Carboniferous, base Mesozoic (base Triassic), base Cenomanian, base Turonian, base Maastrichtian, base Cenozoic and base Eocène can now precisely pinned down in the Geological Time Scale. The combined number of fossil events and magnetic reversals far exceeds the total number of radiometrically datable horizons. Therefore, a framework of bio-, magneto- and chronostratigraphy provides the principal fabric for constrained stretching of the relative geological time scale between dated tie points on the loom of linear time.

For such stretching, interpolation methods are employed that are either geological or statistical in nature. Among the geological scaling methods, an assumption of relative constancy of seafloor spreading over limited periods of time is a common tool for parts of the Late Jurassic, parts of Cretaceous and for the Paleogene. Magnetic polarity chrons, the units of magnetochronology, can be recognized both on the ocean floor as magnetic anomalies measured in kilometers from the spreading center, and in marine sediments as polarity zones that contain biostratigraphic events and assemblages. Knowing the age of a few ocean crust magnetic anomalies (earth magnetic reversals or magnetochrons), allows interpolation of the ages of the intervening magnetic pattern, which in turn can be correlated to the fossil record and geological stage boundaries. The method may be phased out when superior scaling with orbitally-tuned land and deep sea sequences is complete.

The subduction of pre-Late Jurassic oceanic crust precludes such an interpolation approach for older Mesozoic and the Paleozoic strata. Here statistical interpolation methods come to play an important role.

Three such methods are:

– (1) Maximum likelihood estimation of the age boundaries from large suites of dates in successive stages.

– (2) Cubic spline fitting of sets of successive series of age dates versus stages, where the stages are scaled with constraints like number of biozones or composite units.

– (3) Optimum smoothing factors are calculated with cross-validation.

Using some of this geomathematical methodology adapted for time scale construction by AGTERBERG (1994), GRADSTEIN et al., (1995) showed error bars on boundaries for all 31 Mesozoic stages. The error bars reflect both radiometric and stratigraphic uncertainty. The authoritative biochronologic compilation for the tethyan realm of HARDENBOL et al. (1998) was calibrated to this time scale also; its has now been recalibrated to GTS2004, using TSCreator© (see below).

9. – THE GEOLOGICAL TIME SCALE

9.1. GTS 2004

Using the full potential of the three geomathematical procedures listed in the previous section under 1-3, and more detailed radiometric and stratigraphic information, the Geological Time Scale (GTS2004) provides error bars on boundaries for a majority of Phanerozoic stages. In addition, error bars are calculated on stage duration. Uncertainty in the duration of the age units is less than the error in age of their boundaries.

Fig. 9.3 shows the elaborate combination of methods required to arrive at GTS2004, produced under the auspices of ICS. The Paleogene time scale is primarily calibrated from biostratigraphic correlations to magnetic polarity chrons, which in turn are scaled according to marine magnetic anomaly profiles from the South Atlantic seafloor, pinned to an extensive set of Ar/Ar radiometric ages. Interpolation, using constancy of seafloor spreading between radiometrically constrained profile segments, assigns ages to magnetic polarity chrons, which, in turn calibrates zonal events and stage boundaries. The powerful combination of magnetostratigraphy and orbital cycle chronology presently refines both the Paleocene and particularly the Neogene time scale in a continuous and linear manner (LOURENS et al, in GRADSTEIN et al., 2004).

In contrast, the Mesozoic time scale lacks a unifying interpolation concept, because marine magnetic anomaly profiles only extend back to the Callovian stage, and the middle Cretaceous Period lacks a magnetic anomaly signature. In addition, the radiometric data set has inadequate precision to constrain the age assignments of many pre-Aptian stage boundaries. Therefore, whereas portions of the Mesozoic time scale can now be exactly determined by a combination of precise radiometric ages, many published in the last five years on biostratigraphically constrained sections, the majority of the stage boundaries have been assigned ages through geological and mathematical interpolation methods. Hence, uncertainty can be considerable, particularly in the sparse dated Upper Jurassic through Lower Cretaceous interval. Here, Re-Os black shale dates are making a difference (SELBY, 2007) and reduce uncertainty of the age of the Oxfordian / Kimmeridgian boundary by an order of magnitude.

Since the chronostratigraphic and geochronologic compilation for the Paleozoic in HARLAND et al. (1990), several studies have been published that use relatively precise isotope dates to improve Paleozoic chronology below the Carboniferous (e.g. TUCKER & MCKERROW, 1995; Roberts et al., 1995). Nevertheless, there remains a general paucity of relatively precise and stratigraphically meaningful age dates in parts of the Paleozoic, particularly since U-Pb TIMS and HR-SIMS dates at present cannot be easily used together, an issue addressed in detail in GTS2004. The new Paleozoic time scale in GTS2004 was contributed with chapters by J. SHERGOLD et al. on the Cambrian, R. COOPER et al. on the Ordovician, M. MELCHIN et al. on the Silurian, M. HOUSE et al. on the Devonian, V. DAVYDOV et al. on the Carboniferous, and B. WARDLAW et al., on the Permian.

The Mesozoic was contributed by J. OGG et al., F. GRADSTEIN and J. OGG calculated the Paleogene (in LUTERBACHER et al., 2004) and LOURENS et al. the Neogene scale.

All Cenozoic, Mesozoic and Paleozoic age assignments are assembled in the International Stratigraphic Chart of figure 9.1, which displays the current Geologic Time Scale. The "golden spikes" denote which stage and period boundaries are defined by a Global Boundary Stratotype Section and Point (GSSP). The status of this information is as of December, 2007. The actual ages are of GTS2004, except where a new and improved definition of a Phanerozoic stage boundary has led to a (minor) shift in linear age; those chronostratigraphic levels lack an uncertainty estimate for the age. Intra-Cambrian stages with * are informal.

9.2. GTS 2010

At the time of this writing a major update of the Geologic Time Scale is underway, targeted for publication in 2010 in collaboration with Cambridge University Press. It will be a full colour version. The book will be online with a sophisticated search and browsing front menu. Newly planned chapters include the Planetary scale, the Cryogenian-Ediacaran Periods scale, a Prehistoric scale, a sequence stratigraphic scale, and extensive emphasis on stable isotope chemostratigrapy. Use will be made of the most modern insights in error analysis of radiogenic isotope analysis for age dating. Orbital tuning will be extended from Neogene downward into Paleogene and Cretaceous. Significant linear scale update itself may come both from new age dates in Ediacaran, Cambrian, Devonian, Carboniferous, Triassic, Jurassic and Cretaceous, and from substantial progress in ICS with the formal definition of the lower boundaries of stages and periods in the Phanerozoic.

10. – TIME SCALE CREATOR

There is now an electronic version of the Geologic Time Scale with the international standard bio-magneto-sequence time scale charts. There are charts for Lower Paleozoic, Upper Paleozoic, Triassic, Jurassic, Cretaceous and Cenozoic. In addition, users can interactively generate custom stratigraphic charts. This JAVA package, called TS-Creator©, can be freely downloaded from the ICS website (www.stratigraphy.org).

In 1998, a team of specialists led by JAN HARDENBOL (Exxon) published a detailed set of large-format charts summarizing the Mesozoic-Cenozoic correlations and ages of biostratigraphic (dozens of fossil types), sequence stratigraphic, geomagnetic and other events through the Mesozoic and Cenozoic Eras of the past 250 million years (GRACIANSKY et al., 1998). These charts were scaled to the numerical age scale available in 1995. In 2005, the ICS undertook conversion and enhancement of the extensive HARDENBOL et al. charts, plus inclusion of Paleozoic stratigraphic data. The project was sponsored by the U.S. National Science Foundation, ExxonMobil, Chevron, Shell and BP. Now these charts have being updated and re-calibrated to Geologic Time Scale 2004 (GTS2004), and form part also of TSCreator©. Figure 4 shows the updated bio-magneto-chrono-sequence stratigraphy for the Cenozoic, with extra columns for North Sea basin biochronology. All data are part of the program.

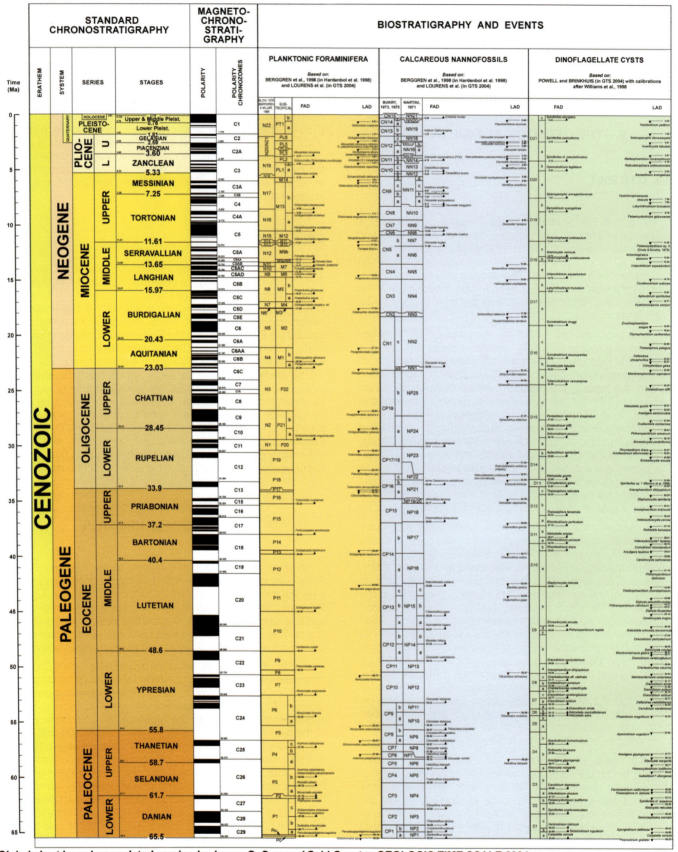

FIG.
Bio-magneto-chrono-sequence stratigraphy for the Cenozoic (updated from GRACIANSKY et al., 1998),

GLOBAL and NORTH SEA

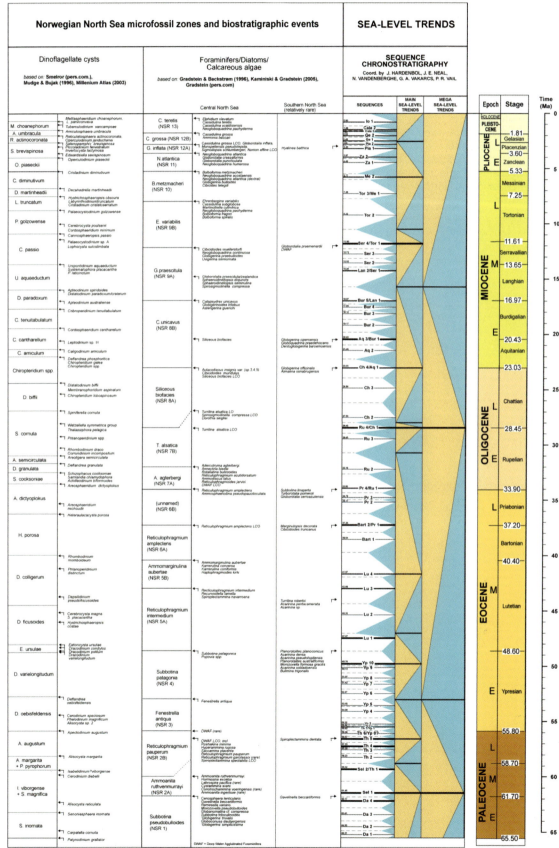

9.4 with extra columns for North Sea basin biochronology. Chart prepared with TSCreator©, using the data stored in the program.

In the program, tables of Cambrian through Holocene stratigraphic events calibrate to GTS2004 ages. There are over 11000 (!) biostratigraphic, sea-level, and magnetic zones and datums, plus a suite of geochemical curves. Documentation of zonal definitions, relative age assignments, and how these events were recalibrated to the 2004 time scale was also compiled. This includes updating cross-correlations and enhancing detail for selected stratigraphic methods (e.g., trilobites, conodonts, graptolites, ammonoids, fusulinids, chitinozoans, megaspores, nannofossils, foraminifera, dinoflagellates, radiolarians, diatoms, strontium-isotope and C-org curves, etc.). Paleozoic sea-level sequences included the global compilation prepared for the Cambrian-Triassic Arabian Platform charts of SHARLAND et al. (2004), and HAQ & AL-QAHTANI (2005). Numerical ages are calculated within the database using the calibrations; therefore, all ages can be automatically recomputed when control ages are improved in future time scales. Regional scales of selected areas (e.g., Russia, China, Australia, New Zealand and North America) are also included.

TS-Creator© automatically loads the reference database, gets instructions from the user on the stratigraphic interval and stratigraphic information to be displayed, and then generates both on-screen and scalable-vector graphic (SVG and PDF) renditions that directly input into Adobe Illustrator and other drafting programs. As explained above, when future enhancements are made to the database of stratigraphic scales or ages, then new graphics can be instantly produced. Next, the user can click on a value, zone or boundary in the charts on the computer screen, and a window opens with an explanation of the calibration, definition and interpolated age. This "hot-linked" chart suite is currently a back-looking reference to information in the source tables, but in the future will also provide links to other tables and text from the Geologic Time Scale book, images of stage-boundary sections and fossil taxa, and the additional enhancements anticipated during Geologic Time Scale updates.

(*) Constructive criticism on parts of this text was provided by F.P. AGTERBERG and M. VILLENEUVE.

GLOSSARY

A

***Absolute dating* – Datations numériques – (Page 2):**
The action of situating terrains or events on a time-scale divided into thousands, millions or billions of years.

***Abundance biozone* – Biozone d'abondance – (Page 69):**
The body of strata within which the abundance of one or more taxa is significantly higher than recorded in the adjacent beds.

***AF demagnetization* – Démagnétisation AF – (Page 59):**
Demagnetization obtained by placing the specimen in a slowly decreasing alternating magnetic field.

***Accomodation space* – Espace disponible – (Page 21):**
Space avalaible for sediment accumulation, i.e. volume between the sea surfaces and a reference datum (such as substrate at the onset of sedimentation or sequence boundary).

***Aggradation* – Agradation – (Page 24):**
A stratal architecture defined by the vertical stacking of the facies belts, without migration of the offlap break landward or basinward. It occurs when the sediment supply and rates of the new space created by the accommodation are roughly balanced.

***Allocycle* – Allocycle – (Page 34):**
It refers to cycles controlled by factors acting externally to the basin, such as eustasy and climate. They are typical of oceanic basins where the influence of local factors and the superficial dynamics is negligible.

***Anagenesis* – Anagenèse – (Page 72):**
Progressive evolution within a lineage; originally considered as an evolutionary mode (synthetic theory of the evolution), it rather represents a consequence of the evolution.

***Anomaly* – Anomalie – (Page 44):**
Abnormal concentration of a chemical element or isotope ratio value which is far from the average. The most known example is the anomaly in the iridium content at the Cretaceous/Paleocene boundary.

***Apparent age* – Âge apparent – (Page 91):**
Raw numeric result obtained by one isotope dating method before geological interpretation; it is normally provided with an analytic uncertainty.

***Assemblage biozone* – Biozone d'association – (Page 69):**
The body of strata characterized by the coexistence of at least 3 fossil taxa, which together distinguish it from adjacent layers without considering their total distribution.

***Assise* – Assise – (Page 10):**
Informal term, mainly used in hydrocarbon exploration studies, indicating a lithologic level or horizon (e.g.: Assise de Bruay, Westphalian C).

***Astrochronology* – Astrochronologie – (Page 36):**
Method using the occurrence of cycles of known periodicity (*i.e.* Milankovitch cycles) to establish the time span represented by a given stratigraphic interval. This method, which provides a high resolution chronology, is independent from the radiometric dating and has been used to establish the astronomical polarity timescale (APTS).

***Astronomical tuning* – Calage astronomique – (Page 36):**
The calibration of sedimentary cycles or cyclic variations in climate proxy records to target curves derived from astronomical solutions for the solar-planetary and Earth-Moon systems.

Astronomical solutions:
Models based on the mechanics of the to compute past variations in Earth'orbital parameters (precession, obliquity and eccentricity).

***Autoclastic rocks* – Autoclastites – (Page 110):**
The product of mechanical fragmentation during the solidification of a lava flow. In this phase, stressed and deformed by the movement, the cooling crust may fracture producing angular, smooth-faced clasts.

***Autocycle* – Autocycle – (Page 34):**
It refers to cycles controlled by factors acting internally to the basin or within the depositional system (migration of a meandering fluvial system, tectonic subsidence, compaction, sediment supply, biotic production, etc.). Autocycles mainly characterize continental and shallow-water environments.

B

***Backstepping* – Rétrogradation – (Page 24):**
The process in which sediments and the related depositional environments migrate landward. This configuration occurs during a rapid rise of the relative sea level.

***Bank* – Banc – (Page 8):**
Informal lithostratigraphic term indicating a prominent level intercalated in an exposed succession (e.g.: calcareous bank); in a quarry indicates an exploited level (e.g.: Banc royal, Lutetian).

***Base level* – Niveau de base – (Page 22):**
The equilibrium surface below which sediment settles and above which erosion occurs.

***Bed* – Lit – (Page 8):**
The smallest formal unit in the hierarchy of sedimentary lithostratigraphic units, e.g. a single stratum lithologically distinguishable from other layers above and below. Usually only marker beds particularly useful for local or regional correlation are given proper names and considered as formal lithostratigraphic units.

***Biochron* – Biochrone – (Page 69):**
The relative duration of a biozone. Its absolute duration can be indirectly derived with a certain precision from radiochronological data.

***Biochronological correlation* – Corrélation biochronologique – (Page 66):**
Comparison of different stratigraphic units (strata, biozones, etc.) based on their paleontological content aiming to position them in a reference chronological system. The correlation is direct if the biostratigraphic markers are found in the same interval and locality. In this case they obviously have the same age. The correlation is indirect when the markers are found in different localities. In this case it implies the construction of a general (regional) stratigraphic scheme based on biostratigraphy, lithostratigraphy, chemostratigraphy, sequence analysis, etc.

In a strictly spatio-temporal sense, two objects are bio-chronologically correlated if, based on their paleontological characteristics, they can be assumed to be contemporaneous or they cannot be chronologically separated (simple correlation). The correlation can result from the comparison of two objects. If the position of one of them into the reference scheme is known, this allows to indirectly assume the position of the other into the same reference scheme (relative age). In practice, the correlation is often obtained by this method of relative dating, the contemporaneousness of two distinct and not directly correlatable objects is inferred from their comparable positions on the reference scale.

Biochronozone – **Biochronozone** – (Page 74):
Time interval defined with the means of the unitary association method.

Biohorizon – **Biohorizon** – (Page 77):
This term has two possible meanings:
– (1) boundary, surface or interface corresponding to a remarkable change in the biostratigraphic record;
– (2) thin stratigraphic unit characterized by a distinctive fossil association within which no finer subdivision is possible.
Although only the first one is accepted by the International Commission on Stratigraphy, the second definition is commonly used by macropaleontologists.

Biomarker – **Biorepère** – (Page 77):
The term biomarker is preferred to "*datum*" whose etymological meaning designates the phyletic, non fortuitous, first/last occurrence of one or more taxa. Except for the micropaleontologists, the term "*datum*" has gradually lost its evolutionary implications to become synonym of local first/last occurrence of a species. This term is thus commonly used by biostratigraphers when elaborating zonations on the basis of fossil groups where the evolutionary relationships between the different species (***morphospecies***) are poorly known; however, the first/last local occurrence of a species, even if fortuitous (i.e. related to the various environmental parameters or to preservational factors) is objective because based on the observation, and in this sense very useful for correlations.

Biostratigraphy – **Biostratigraphie** – (Page 2):
Study of the spatial and temporal organization of sedimentary stratigraphic successions based on paleontological data.

Biozone – **Biozone** – (Page 65):
The basic unity of biostratigraphy, based on the vertical (temporal) and horizontal (geographical) distribution of one or more characteristic taxa. Biozones can vary greatly in thickness, geographic extent, and represented time span.

Boundary stratotype – **Stratotype de limite chronostratigraphique** – (Page 119):
The specified reference stratigraphic section that contains the specific point that defines a boundary between two stratigraphic units. The latter is marked by a "golden spike" placed on the outcrop.

Bundle – **Faisceau** – (Page 10):
This term is used, particularly in hydrocarbon exploration, to indicate a package of strata.

C

Calibration – **Calibrage** – (Page 4):
The (stratigraphic) calibration consists in the correlation of a local or regional scale with the standard geologic time scale. This is obtained by using a combination of different stratigraphic data from a single locality (e.g. magnetostratigraphy, biostratigraphy, radiometric dating, sequence stratigraphy).

Carbonate cycle – **Cycles de carbonates** – (Page 35):
Quasi periodic flucuations of the $CaCO_3$ content in pelagic sediments which can be be related to the Milankovitch cycles.

CDT – **CDT** – (Page 42):
Reference standard for sulfur isotope ratio analyses ($^{34}S/^{32}S$); it is a troleite (FeS) from the Canyon Diablo meteor.

Characteristic Remanent Magnetization (ChRM) – **Magnétisation rémanente caractéristique** – (Page 69):
The direction of the magnetization of a rock specimen after magnetic cleaning. ChRM is supposed to represent the direction of the Earth magnetic field at the time of the formation of the rock.

Chemozone – **Chimiozone** – (Page 42):
Sediment or time interval characterized by the variation of the concentration or of the isotope ratio of a given element (Renard, 1985). Isotopic stages are an example of chemiozones.

Chemostratigraphy – **Chimiostratigraphie** – (Page 41):
Stratigraphy based on variations of isotope ratios or concentrations of chemical elements, measured on bulk sediment or fossils. Previously used as an informal definition, this term was first used by Berger and Vincent (1981) and then Renard (1985). It is an extension of isotope stratigraphy.

Chrono-correlation – **Chronocorrélation** – (Page 120):
Demonstration or identification of an age equivalence between stratigraphic units defined in different places.

Chronological stratigraphy – **Stratigraphie chronologique** – (Page 2):
Stratigraphic procedure which consists of organizing terrains or geological events in chronological order and constructing a geological time-scale, without numerical references.

Chronometrical stratigraphy – **Stratigraphie chronométrique** – (Page 2):
Stratigraphie chronométrique – Stratigraphic process which consists in measuring time on geological layers, phenomena or processes and in constructing a numerical geological timescale.

Chronospecies – **Chrono-espèce** – (Page 72):
Successive evolutionary stages caused by progressive modifications of one or more morphological features in a gradual (anagenetic) evolutionary lineage and having a relative chronological value. The "confines" delimiting the various morphologies allow to recognize "stages" or "chronological subspecies"; a succession of these biological entities constitutes a "chronospecies".

Chronostratigraphic classification – **Classification chronostratigraphique** – (Page 117):
Chronological and hierarchical organization of chronostratigraphic units according to their duration.

Chronostratigraphic unit – **Unité chronostratigraphique** – (Pages 103, 117):
A body of rocks, layered or unlayered, formed during a specified interval of geologic time. The units of geologic time during which chronostratigraphic units were formed are called geochronologic units.

Chronostratigraphy – **Chronostratigraphie** – (Page 117):
Branch of stratigraphy which deals with the organization of geological units and events in time.

Chronozone – **Chronozone** – (Pages 66, 118):
Chronostratigraphic unit corresponding either to the body of rocks formed in an any place during a given time interval (whatever is the hierarchical rank), or to the elementary subdivision in the hierarchy of chronostratigraphic units. Because of these different definitions and its relatively theoretical character it is difficult to use this term properly.

Cladogenesis – **Cladogenèse** – (Page 72):
Divergent evolution bringing to the splitting of a lineage into two descending branches; originally considered an evolutionary mode (synthetic theory of the evolution), it rather represents a consequence of speciation.

Complex – **Complexe** – (Page 103):
A unit composed of various kinds of any class or classes or rocks (sedimentary, igneous, metamorphic) and characterized by irregularly mixed lithology or by highly complicated structural relations.

***Composite standard reference section* – Profil composite de référence** – (Page 121):
Method of artificial superposition of paleontological or other events placed on a linear time scale, established starting from an existing standard reference section, enriched by successive comparison with all the available profiles following the procedure of graphic correlation.

***Composite stratotype* – Stratotype composite** – (Page 119):
A stratotype formed by the combination of several specified intervals of strata combined to make a composite standard of reference.

***Concurrent-range zone* – Zone de distribution concomitante** – (Page 71):
Body of strata corresponding to the overlapping parts of the range of at least two taxa.

***Conformity* – Concordance** – (Page 125):
Continuous stratigraphic sequence without hiatus.

***Continuous biochronological scale* – Échelle biochronologique continue** – (Page 69):
A scale defined by the combination of various kinds of zones. The subdivisions are: either defined by interval zones separating bioevents (First Appearance Datum = FAD and Last Appearance Datum = LAD), or by the relative abundance of taxa (abundance zones). The continuous biochronologic scales are, therefore, formed by relative time units.

***Coprecipitation* – Coprécipitation** – (Page 47):
Process where atoms of a chemical compound are replaced by atoms of another species (e.g., Ca replaced by Sr in $CaCO_3$). This process can take place only if the atoms have similar electric charges, and, most important, dimensions. In natural conditions, coprecipitation is really difficult to distinguish from adsorption and occlusion.

***Correlation line* – Ligne de corrélation** – (Page 121):
In the method of graphic correlation, the line that describes the relations between two sedimentary sequences in a X-Y graph. It is produced by joining the various events observed considered homologous and synchronous in the two sequences.

***Cosmogenic isotope* – Isotope cosmogénique** – (Page 91):
Isotope produced by cosmic radiations by a mechanism of spallation.

***Curie temperature* – Point de Curie** – (Page 59):
Temperature above which the ferromagnetic properties of a mineral disappear and the mineral become paramagnetic.

***Cyclostratigraphy* – Cyclostratigraphie** – (Page 2):
A stratigraphic discipline that uses lithologic cycles to establish long-distance correlation and to define the time duration of sedimentary successions. The first step may consist in defining cycles controlled by local factors. In this case, cycles have a short-distance correlation value. The cycles triggered by regional or global factors, such as the Milankovitch cycles, need to have a good age calibration and well defined period and number, within a sedimentary interval. The most used cycles are those associated to the precession, with duration of 21 Kyr.

D

$\delta^{13}C$ – $\delta^{13}C$ – (Page 45):
Reporting notation (in ‰) of the carbon isotope ratio of a compound relative to a reference standard (PDB or SMOW), defined as $\delta^{13}C = [^{13}C/^{12}C]$ sample – $[^{13}C/^{12}C]$ standard) / ($[^{13}C/^{12}C]$ standard) x 1000.

$\delta^{18}O$ – $\delta^{18}O$ – (Page 42):
Reporting notation (in ‰) of the oxygen isotope ratio of a compound relative to a reference standard (PDB or SMOW), defined as $\delta^{18}O = ([^{18}O/^{16}O]$ sample – $[^{18}O/^{16}O]$ standard) / ($[^{18}O/^{16}O]$ standard) x 1000.

***Debris avalanche* – Avalanche de débris** – (Page 112):
A sudden and catastrophic collapse of volcanic material from an unstable side of a volcano. Volcanologists coined this term after the event of Mount St Helen (Washington, USA) (May 18, 1980) to indicate the destabilization of the side of a volcano and its instantaneous slide on distances of several tens of kilometers. Volumes involved in the landslide go from several km^3 to several tens (or even hundreds) of km^3. Initially discovered on andesitic stratovolcanoes of subduction zones, debris avalanches were thereafter found to be a common feature of big insular volcanoes.

***Depositional sequence* – Séquence de dépôt** – (Page 24):
Stratigraphic unit of a relatively conformable succession of genetically related strata bounded at its top and base by unconformities or their correlative conformities. It forms during a 3rd order cycle of relative sea level change, and it is composed by the systems tracts.

***Depositional system* – Système de dépôt** – (Page 19):
3D framework of genetically related facies sequences, whose formation occurs under the control of common environmental conditions (eustasy, climate, tectonics,...) or through the same depositional mechanisms.

***Diachronism* – Diachronisme** – (Page 67):
The difference in age between different places of a geological event or object. The term diachronism is to be preferred to heterochronism, which is used in evolutionary biology and in paleontology to indicate the shift in the development of ontogenetic sequences or the displacement between ontogenetic events.

***Diastem* – Diastème** – (Page 10):
A very small break in the depositional process represented by a surface separating two beds (A. LOMBARD, 1956).

***Disconformity* – Discontinuité séquentielle** – (Page 17):
A discontinuity associated either to a remarkable change in one or various aspects of the sedimentary record or to a significant chronological jump.

***Discontinuous biochronological scale* – Échelle biochronologique discrète (discontinue)** – (Page 69):
A successions of non contiguous, discrete, biostratigraphy units. In this scale the subdivisions (unitary associations) are separated by "intervals of separation" and characterized by exclusive groupings of species which are mutually excluded.

***Domain* – Domaine** – (Page 68):
General paleogeographic term, indicating either a resticted area of any nature (a natural region, a morpho-structural geological unit, etc.), or the area of geographical distribution (observed or inferred) of a fossil group or taxon. In the latter case it is used with the same meaning of (paleo)biogeographical province.

E

***Earth magnetic field* – Champ magnétique terrestre** – (Page 54):
The magnetic field generated inside the Earth. About 99% of the magnetic field measured at the surface of the Earth generates in the Earth's liquid core. The most important component of the Earth's magnetic field can be described as a magnetic dipole located at the center of the Earth and inclined at about 10.6° to the Earth's rotation axis.

***Ecostratigraphy* – Écostratigraphie** – (Page 76):
The procedure of subdiving a sequence by using ecozones whose characteristic fossil composition depends on the facies and, therefore, reflect the environmental evolution through time.

***Ecozone* – Écozone** – (Pages 69, 76):
The stratum (or the body of strata) characterized by a peculiar fossil association called ecotype which implies a nonfortuitous association (mixtures, preservational effects, etc.) and reflects certain environmental parameters. It has a limited chronostratigraphic. The concept of ecozone comprises a lithological component and contributes to the description of the sedimentary succession and to the environmental reconstruction. The use of ecozones must be carried out in a context of time units, with the aid of zonal biostratigraphy. In this sense ecostratigraphy (the use of ecozones) and zonal stratigraphy (the use of zonal subdivision) are two distinct, but not opposite, branches of biostratigraphy. Although of limited use for biostratigraphic correlations due to their limited lateral extent, ecozones are fruitfully used for local subdivisions especially of subsurface successions.

***Elementary sequence* – Séquence élémentaire**
– (Page 17):
Metric or plurimetric sequence made up of a well defined composing element (petrographic, mineralogic or electric sequence of objects), or a group of elements (facies sequence).

***Eon* – Éon** – (Page 118):
The geochronological unit of highest rank. The geological time is subdivided into three Eons: Archaean, Proterozoic and Phanerozoic.

***Epiclastic rocks* – Épiclastites** – (Page 111):
Detritic sedimentary rocks constituted by fragments of older rocks. A part only of the constituent (epiclasts) can formed by chemical or mechanical weathering of volcanic rock (volcaniclasts), mixed in any proportion with elements of other nature (alloclasts).

***Era* – Ere** – (Page 118):
The highest rank geochronological unit of the Phanerozoic. In the current usage, this term defines also the corresponding chronostratigraphic unit (Erathem).

***Erosional surface* – Surface d'érosion** – (Page 25):
Surface testifying the removal of previously deposited sediments. It can occur either at the base of the lowstand system tract (sequence boundary) or at the base of the transgressive system tract (submarine erosion).

***Eruption* – Éruption** – (Page 111):
The process by which solid, liquid and gaseous material are released from the Earth's interior onto the Earth's surface and into the atmosphere. An eruption event may have a largely variable duration from a few days to a few months or even years. When the proportion of gaseous material is minimal lava is released quietly. In the presence of high proportions of gas the eruption event is violent.

***Event stratigraphy* – Stratigraphie événementielle**
– (Page 2):
A term first proposed by D.V. Ager in 1973 for the identification, characterization and correlation of the effects significant physical, chemical or biological events on the regional and global stratigraphic records. The correlation of these various effects recognized in the sedimentary record would lead to understand the time relationships between them and help in establishing truly synchronous horizons, thus improving the resolution and accuracy of the chronostratigraphic scale.

F

***Facies model* – Modèle de faciès** – (Page 19):
Three-dimensional theoretical model of genetically related facies. It may be applied to one or more associated lithostratigraphic units.

***Facies sequence* – Séquence de faciès** – (Page 19):
Sequence of vertical and horizontal facies succession of different age, whose organization reflects the depositional processes (e.g.: turbiditic sequence), the characteristics of the depositional environment (e.g.: evaporitic sequence) or the sedimentary evolution within a basin (e.g.: klupfélian sequence). Facies sequence can be defined at different scales: elementary sequence, mesosequence, and major sequence.

***First Appearance Datum (FAD)* – Biorepère d'apparition (BA) ou Biorepère de première apparition (BPA)**
– (Page 77):
Lowest stratigraphic occurrence of a taxon, corresponding to the lower boundary of the biozone defined on the base of its presence. This term is very theoretical and often confused with the concept of "first observation" of an index taxa within a given stratigraphic succession. The latter expression, is realistic because directly derived from data. However its chronostratigraphic significance should always be carefully evaluated because of its possible only local meaning. The term first appearance datum should not be used before detailed study verifying its chronostratigraphic value.

***First occurrence (FO)* – Première observation (PO)**
– (Page 68):
The lowest stratigraphic level at which one or more taxa are observed in a sedimentary succession.

***Formation* – Formation** – (Page 8):
It is the primary unit of intermediate rank in the hierarchy of lithostratigraphic classification, which formally indicates a body of rocks defined by uniformly distributed facies associations. It is the basic lithostratigraphic unit used for the geological mapping.

***Frost effect* – Effet glaciaire** – (Page 45):
A variation of the oxygen isotope values of sea water due to accumulation of ice sheets which preferentially store the lighter isotope.

G

GAD hypothesis – (Page 54):
The Geocentric Axial Dipole hypothesis assumes that the Earth magnetic field, averaged over a period of a few thousand years, corresponds to that of a dipole whose axis coincides to the Earth rotation axis.

***Genetic sequence* – Séquence génétique** – (Page 19):
Sedimentary unit that reflects the evolution of the factors controlling the depositional process. More in detail, the genetic sequence represents the evolution of a cycle controlled by the change in the accommodation space. The boundaries of this sequence are represented by the inversion points in the cyclic evolution of the accommodation space and correspond to unconformities according to sequence stratigraphy, or to the maximum flooding surface according to the high resolution facies stratigraphy.

***Genetic stratigraphy* – Stratigraphie génétique** – (Page 2):
It analyses, defines and describes the driving factors that control the sedimentary processes and the geometric and chronostratigraphic relationships of deposits.

***Geochemical cycles* – Cycles géochimiques** – (Page 42):
Quasi-periodic recurrent variations of a geochemical signal with a uniform or variable duration. This term is not really appropriate because the initial state is never achieved again.

***Geochemical event* – Événement géochimique**
– (Page 42):
Positive or negative geochemical variation, generally unique and spanning a time interval from thousand to million of years, clearly different from the average signal. (see *shift*)

***Geochemical excursion* – Excursion géochimique**
– (Page 43):
A more or less rapid geochemical variation which is followed by the re-establishment of conditions relatively close to the

initial ones. Some authors consider this term as a synonm of event, while others use them for distinctly different durations of time.

***Geochemical fossils or Geochemical biomarkers – Fossiles géochimiques ou marqueurs géochimiques* – (Page 52):**
Organic compounds preserved in the sediments and clearly derived from biological molecules.

***Geochemical shift – Accident géochimique* – (Page 44):**
Rapid and abrupt (200 kyr) change (either increase or decrease) of a geochemical proxy, which might not be reversible.

***Geochronologic unit – Unité géochronologique* – (Page 103):**
Abstract unit of subdivision of the geological time (thousands, millions, or billions of years) materialized by the corresponding chronostratigraphic unit.

***Geochronometers – Géochronomètre* – (Page 91):**
Mineral or rock characterized by physico-chemical properties which make it suitable to be dated by geochronological methods.

***Geochronometric unit – Unité géochronométrique* – (Page 103):**
Abstract unit of subdivision of the geological time (thousands, millions, or billions of years) without an corresponding physical unit.

***Geographic coordinates – Coordonées géographiques* – (Page 61):**
Direction of the magnetization of a specimen after correction for its orientation recorded during the sampling.

***Geological time scale – Échelle des temps géologiques* – (Page 119):**
A sequence of divisions of geological time comprising in order from oldest to youngest units. It is presented in a table where chronostratigraphic units are placed next to the Global Standard Scale, the numerical ages, the magnetostratigraphic and biochronologic units, and possibly the markers in the geochemical, eustatic, climatic or other records.

***Geomagnetic polarity time scale (GPTS) – Échelle de polarité magnétique* – (Page 55):**
Scale of dated Earth magnetic field reversal, mostly obtained from magnetic oceanic anomalies.

***Geomagnetic pole – Pole géomagnétique* – (Page 53):**
Point formed by the dipole field axis and Earth surface.

***Geometric stratigraphy – Stratigraphie géométrique* – (Page 2):**
A stratigraphic procedure which consists in the study of the geometrical relationships between geological units or groups.

***Gilbert – Gilbert* – (Page 37):**
Cyclostratigraphic unit introduced by DE BOER & WONDERS (1984). It corresponds to the duration, variable through the geological time, of a precession cycle

***Global standard chronostratigraphic scale – Échelle chronostratigraphique mondiale de référence* – (Page 118):**
Continous succession of contiguous chronostratigraphic units of different hierarchical rank. It is the reference scale for the stratigraphy of the whole Planet.

***Graphic correlation – Corrélation graphique* – (Page 121):**
Geometrical method of comparison of two sedimentary successions. Following this method an X-Y diagram, whose axes are represented by the two series compared, is constructed. The various events observed are placed in the diagram according to their stratigraphic position in the two series.

***Groupe – group* – (Page 8):**
Formal lithostratigraphic unit next in rank above a formation. It groups two or more contiguous and associated formations with significant and diagnostic lithologic features in common.

H

***High resolution stratigraphy – Stratigraphie haute-résolution* – (Page 4):**
Process which consists in defining stratigraphic units which are of as high a quality as possible according to the tool used and/or the geological period taken into consideration. The use of this expression is not advisable.

***High-resolution variations – Variations de haute fréquence* – (Page 4):**
Short-term (10^4 to 10^5 years) climate fluctuations controlled by Milankovitch cycles.

***Historical stratotype – Stratotype historique* – (Page 119):**
Section that was indicated and used for the historical definition of a stratigraphic unit.

***Horizon – Horizon* – (Page 8):**
A particular level without thickness in a stratigraphic sequence; in lithostratigraphy, however, it may indicate a distinctive and very thin bed (lithohorizon) clearly distinct from adjacent layers and having an significant lateral extention.

***Hydroclastic rocks – Hydroclastites* – (Page 110):**
The product of fragmentation of a magma due to its sudden chilling when it is in contact with water (quenching).

***Hypostratotype – Hypostratotype* – (Page 119):**
Reference section described in a region different from that of the original stratotype.

I

***Ideal reference system – Référentiel idéal* – (Page 74):**
Ideal sequence of unitary associations or biochronozones where the real duration of existence of a taxa assemblage is known. Its exact delimitation can only be approximate because of the usually discontinuous character of the paleontological record.

***Idealized spatial profile – Séquence paysage* – (Page 17):**
It indicates a facies model illustrating, in plan or section views, the distribution, in space or time, of the sedimentary environments within a depositional system. The idealized spatial profile corresponds to the horizontal representation of a virtual sequence.

***Idealized vertical profile – Séquence virtuelle* – (Page 17):**
It corresponds to a theoretical facies model illustrating, on a lithostratigraphic log, the most complete succession as possible of stacked elementary sequences that characterize a region or a stratigraphic interval.

***Ignimbrite – Ignimbrite* – (Page 110):**
A generic term indicating a volcanic deposit formed by a pyroclastic flow, essentially constituted of ash and lapilli, often (but no necessarily) welded in its lower part, and disposed in sheets of large extension (up to 25 000 km² large and 500 of thick).

***Indice – Indice* – (Page 44):**
Geochemical theoretical parameters including the sum, product or ratio of different geochemical signals.

***Integrated biochronology – Biochronologie intégrée* – (Page 74):**
Concurrent researches of the stratigraphic distribution of various fossil groups (subclasses, classes, etc.) and their relationships to establish a **composite biochronological scale**.

***Integrated stratigraphy* – Stratigraphie intégrée**
– (Page 122):
Process which consists of simultaneously using the various stratigraphic methods and which is based on the relations between the various geodynamic processes which occur in the sedimentary recording.

***International Geomagnetic Reference Field* (IGRF) – Champ géomagnétique international de référence**
– (Page 53):
Representation of the Earth magnetic field that includes Gauss coefficient up n = 10. This global model includes most of the characteristic of the Earth internal field.

***Interval biozone* – Biozone d'intervalle** – (Page 69):
The body of strata delimited by the first or last appearance of taxa. The lower boundary is marked by the first or last appearance of one characteristic taxon, and the upper boundary by the first or last appearance of another characteristic taxon. This kind of biozone may be, thus, defined and recognized even if it does not contain any significant paleontological association.

***interval of co-occurrence* – Intervalle de coexistence**
– (Page 71):
The interval corresponding to the maximum overlaps of the stratigraphic range of two or more taxa. The intervals of co-occurrence are separated by the so-called "intervals of separation". The identification of both and its reproducibility forms the basis of the "unitary association" method.

***Interval of separation* – Intervalle de séparation**
– (Page 72):
In logical biostratigraphy, the shortest interval separating the "co-occurrence intervals" of taxa.

***Inverse polarity* – Polarité inverse** – (Page 54):
Earth magnetic field with an orientation opposite to the present day orientation (North magnetic pole close to North Geographic pole).

***Iridium (anomaly in)* – Iridium (anomalie en)** – (Page 49):
High iridium concentrations recorded at the Cretaceous/Paleocene boundary. It is generally related to the impact of a meteorite whose crater is believed to be the Chicxulub (Central America).

Isochron *(isotope geochronology)* **– Isochrone** (géochronologie isotopique) – (Page 93):
In isotope geochronology, the aligment of data points reported on a X-Y diagram whose X-axis represents the parental to daughter isotope ratio and the Y-axis ratio between the daughter isotope and the non-radiogenic isotope of the same element as the daughter. The slope of the alignment gives information of the absolute age.

***Isochronous* – Isochrone** – (Page 31):
Term commonly used to indicate stratigraphic units or surfaces (discontinuities) having the same geochronologic, biochronologic or chronostratigraphic age. Its meaning would be "of equal duration" and its use to describe units or surface having the same is not correct. However, because its very common use and the non informative character of the possible substitutes (homochrone, opposite to heterochron), it is proposed to keep this term and its derivations (isochrony, isochronism).

***Isotope geo(radio) chronology* – Géo(radio-)chronologie isotopique** – (Page 2):
Isotopic dating of rocks minerals used for the establishment of the geological time scale.

***Isotopic age* – Âge isotopique** – (Page 92):
Numerical value obtained by isotope dating methods and considered as a geological age. The uncertainty is mainly linked to analytical aspects.

***Isotopic homogenization* – Homogénéisation isotopique**
– (Page 96):
Geochemical mechanism, within a mineral, tends to distribute the same isotopic composition to the various elements crystallizing at the same time.

***Isotopic ratio* – Rapport isotopique** – (Page 42):
Ratio of the heavy to the light isotopes of an element in one compound, e.g., the oxygen isotope ratio ($^{18}O/^{16}O$).

***Isotopic stage* – Stade isotopique** – (Page 42):
Stratigraphic unit representing a fluctuation of the oxygen isotope ratio. The stage boundaries are located at the inflexion points between an isotopic maximum and minimum. The definition of isotopic stages, based on the pioneering work of Emiliani (1955), has been historically considered as the beginning of chemiostratigraphy.

***Isotopic stratigraphy* – Stratigraphie isotopique**
– (Page 45):
Stratigraphic subdivision based on the variations of the oxygen isotope ratio.

L

***Lahar* – Lahar** – (Page 110):
Lahar is a term of Indonesian origin applied to rapidly flowing mixture of rock debris and water originated on the slopes of a volcano. This kind of mudflow can accompany a volcanic eruption (primary lahar) or follow it (secondary lahar). They form in a variety of ways, chiefly by the rapid melting of snow and ice by pyroclastic flows, intense rainfall on unconsolidated volcanic rock deposits, breakout of a lake dammed by volcanic deposits, and as a consequence of debris avalanches. Lahars are the cause of most volcanic fatalities.

***Lamina* – Lamine** – (Page 8):
Lithologic unit of millimetric thickness.

Last Appearance Datum index (LAD) – (Page 77):
Biorepère de disparition (BD) ou **Biorepère de dernière apparition (BDA)** –
Highest stratigraphic occurrence of a taxon ("*Last Appearance Datum*", "LAD") corrresponding to the upper boundary of the biozone defined on the base of its presence.

***Last occurrence (LO)* – Dernière observation (DO)**
– (Page 68):
The highest stratigraphic level at which one or more taxa are observed in a sedimentary succession.

***Lava* – Lave** – (Page 109):
Lava is the word for magma (molten rock) which has reached the surface through a volcanic eruption. The term is commonly applied to streams of liquid rock that flow from a crater. Geologists also use the word to indicate the solidified deposits of lava flows. The magma can cool below Earth's surface (intrusion). Onto Earth's surface, according to its viscosity, it can accumulate over the site of eruption (extrusion), or pour out himself on variable distances (effusion).

Life span *(of a taxon)* **– Durée d'existence** ou **Durée de vie** (d'un taxon) – (Page 67):
The time span between the origination of a taxon and its extinction.

***Lineage biozone* – Biozone de lignage** – (Page 72):
The body of strata defined by the total range of one taxon representing a fragment of a phyletic lineage.

***Lithodeme* – Lithodème** – (Page 102):
It is the fundamental unit of the lithodemic nomenclature. A body of rocks that underwent a process of high degree of methamorphism obliterating its primary structures or a body of plutonic rocks of mappable dimension. It is constituted of rocks belonging to one single genetic class (e.g. metasediments, metavolcanites, intrusive association, etc.) and defined on the basis of its mineralogical, textural

and structural characters. Its boundaries coincide with sharp or gradual contacts (extrusive, intrusive, tectonic or metamorphic contact or sedimentary cover).

***Lithological cycle* – Cycle lithologique** – (Page 34):
Association of two or more facies or textures, which periodically repeats, returning to the facies or texture that define the base of the cycle. One cycle composed by two terms is also indicated as rhythm.

***Lithostratigraphic stratotype* – Stratotype lithostratigraphique** – (Page 37):
A lithostratigraphic unit designed as a standard of reference for a region or a wider geographic domain. A type section has to be defined through both surface or subsurface data (*stratotype, type section* or *type locality*).

***Lithostratigraphic unit* – Unité lithostratigraphique** – (Page 8):
A body of rocks defined and recognized on the basis of its observable and disctintive lithologic properties or combination of them and its stratigraphic relations.

***Lithostratigraphy* – Lithostratigraphie** – (Page 2):
The study of sedimentary sequences, of their hierarchy and the temporal record based on their lithological characteristics.

***Long term trend* – Tendance à long terme** – (Page 42):
Constant change of a geochemical signal that lasts several millions of years (see high frequency or high resolution variations).

M

***M-sequence* – Séquence M** – (Page 56):
Sequence of oceanic magnetic anomalies (first studied in the Pacific Ocean) where the magnetic polarity sequence from upper Jurassic to lower Cretaceous has been established (M = Mesozoic).

***Magnetic declination* – Déclinaison magnétique** – (Page 58):
The angle between horizontal component of the magnetic field (the magnetic meridian) and the geographic North (the geographic meridian).

***Magnetic inclination* – Inclinaison magnétique** – (Page 58):
The angle between the Earth magnetic field vector and the horizontal.

***Magnetic polarity-transition zone, polarity-transition zone* – Zone de transition de polarité magnétique, magnétozone de transition de polarité** – (Page 53):
Interval of stratified rock that has recorded the transition between two magnetozones with opposite polarity.

***Magnetostratigraphy, magnetic stratigraphy* – Magnétostratigraphie** – (Page 2):
Analysis of the evolution of the natural magnetic characteristics, such as magnetic susceptibility, intensity and direction of the Natural Remanent Magnetization (NRM), etc. in a rock sequence. Magnetostratigraphy organizes the rock strata according to their magnetic properties that were acquired at the time of deposition. Since the polarity of the Earth's magnetic field has reversed repeatedly in the geological past, the most useful magnetic property in stratigraphic work is the change in the direction of the NRM of the rocks due to reversals in the polarity of the Earth's magnetic field.

***Magnetozone* – Magnétozone** – (Page 65):
Main magnetostratigraphic unit formed by a the body of rocks with the same magnetic polarity, which is different from that of adjacent strata.

***Major sequence* – Séquence majeure** – (Page 19):
Ten or hundred meters thick sequence of objects or facies made up of several elementary sequences, reflecting the evolution of depositional environments at the scale of the whole basin or of a part of it (equivalent to the megasequence).

***Maximum flooding surface, mfs* – Surface d'inondation maximale** – (Page 25):
It corresponds to the surface at which the rate of rise of relative sea level is maximum and to the turning point from increasing and decreasing volumes of the accommodation. Bathymetry is maximal, and non-depositional surfaces or condensed intervals may occur concomitantly with this surface. Above this surface progradation begins again. According to sequence stratigraphy, this surface corresponds to the boundary between the transgressive system tract and the highstand system tract; whereas it represents the sequence boundary of the genetic depositional sequences.

***Member* – Membre** – (Page 8):
Formal lithostratigraphic unit next in rank below a formation, and used to subdivide formations.

***Mesosequence* – Mésoséquence** – (Page 19):
Sequence of objects or facies sequence of a rank between the elementary sequence and the major sequence.

***Milankovitch cycle* – Cycles de Milankovitch** – (Page 32):
Astronomical cycles of the Earth, such as the eccentricity (the divergence of the orbit of the Earth around the Sun), the obliquity of the plane of the ecliptic, and the precession of equinoxes. The eccentricity cycle is constant through time, while precession and obliquity have time durations varying with geologic time, because of the interaction between Earth and the Moon. These cycles induce fluctuations of the solar energy radiated at different latitudes and reflected in a cyclical record of climatic changes (e.g.: Quaternary glacial cycles).

***Mixing coefficient or dividing coefficient* – Coefficient d'incorporation ou coefficient de partage** – (Page 47):
Replacement ratio of a major element by a trace element in a chemical compound. This coefficient is generally a function of the precipitation temperature, of the mineralogical nature of the compound and of some vital effect, if the compound is biologically produced.

***Monogenic volcano* – Volcan monogénique** – (Page 113):
A volcano built by a single eruption.

N

NASC (North American Shales Composite) – (Page 42):
Reference standard used to normalize content in Rare Earth Elements (REE). The REE concentrations of a sample are generally represented by their atomic number (REE spectrum) after being normalized to a standard which represents the average composition of Earth crust.

***Natural remanent magnetization (NRM)* – Aimantation rémanente naturelle (ARN)** – (Page 53):
Magnetization of a rock naturally acquired during its formation or at later time by the effect of the Earth magnetic field.

***Neostratotype* – Néostratotype** – (Page 119):
A newly described stratotype replacing the original one when the latter has became inaccessible or no longer utilizable.

***Normal polarity* – Polarité normale** – (Page 54):
Earth magnetic field with an orientation similar to the present day orientation (South magnetic pole close to North Geographic pole).

***NSB987* – NSB987** – (Page 42):
Reference standard for strontium isotope analyses ($^{87}Sr/^{86}Sr$). NSB987 is an extremely pure strontium carbonate.

P

***Parasequence* – Paraséquence** – (Page 31):
Package of strata genetically related and bounded at the base and at the top by flooding surfaces, and formed during a 4th and 5th order cycle of relative sea level change.

***Parastratotype* – Parastratotype –** (Page 119):
An additional section, cropping out in the same area of the original reference stratotype which, in turn, takes the name of holostratotype. The use of this term is here discouraged.

***PDB1* – PDB1 –** (Page 42):
International reference standards used to report variations in oxygen and carbon isotope ratios. The PDB standard originates from the calcium carbonate of the rostrum of a Cretaceous belemnite *[Belemnitella americana]* collected in the Peedee formation of South Carolina (USA).

***Peak* – Pic –** (Page 44):
See anomaly.

***Period* – Période –** (Page 118):
Geochronologic unit used for the primary subdivision of the Era. The corresponding chronostratigraphic unit is the system.

***Phreatic eruption* – Phréatique (éruption) –** (Page 110):
An explosive volcanic eruption that occurs when water and heated volcanic rocks interact to produce a violent expulsion of steam, water, ash, blocks and bombs. In this case fresh magma is not involved. When significant amounts of magmatic material are ejected, the eruption is called phreatomagmatic.

***Polarity horizon* – Horizon d'inversion de polarité magnétique –** (Page 55):
The stratigraphic level separating two bodies of strata characterized by opposite magnetic polarity.

***Plutonic bed* – Couche plutonique –** (Page 103):
A body of rocks, classified as a discrete element because of its peculiar petrographic character. Its upper and lower boundaries correspond to sharp petrographic changes.

***Plutonic cyclic unit* – Unité cyclique plutonique –** (Page 103):
Rhythmic unit characterized by the repetition of layers showing discrete modal mineralogies reflecting, within each cycle, the sequence of fractional cristallisation of the parental magma.

***Plutonic divisions* – Divisions plutoniques –** (Page 103):
Grouping of stratified intrusive rocks defined on the basis of structural and petrographic criteria (e.g. appearance and disappearance of key cumulus minerals in the stratigraphic column). Divisions can be subdivided into series, zones and subzones.

***Plutonic group* – Groupe plutonique –** (Page 103):
A primary grouping of plutonic units sharing some characters and stratigraphically next to each other, though not necessarily adjacent.

***Plutonic member* – Membre plutonique –** (Page 103):
A unit constituted by a series of plutonic beds which form a natural association.

***Plutonic rythmic unit* – Unité rythmique plutonique –** (Page 103):
Regular succession of layers of comparable nature or succession of discrete layers presenting a pattern which is repeated stratigraphically in a regular way.

***Plutonic series* – Série plutonique –** (Page 103):
A first order structural unit grouping a major succession of intrusive rocks. The sequence can be subdivided into zones.

***Plutonic units* – Unités plutoniques –** (Page 103):
Parts or sets of parts of a stratified intrusion defined on the basis of petrographic criteria. The main units are the group, member, layer and the rhythmic unit.

***Plutonic zone* – Zone plutonique –** (Page 103):
First order stratigraphic division of a plutonic series. The lower and upper boundaries of the zones are identified by the successive appearances and disappearances of the various key cumulus minerals in the sequence of crystallization. The lowermost zone in a tholeiitic series would therefore be delimited by the appearance and the disappearance of olivine which is the first mineral to be crystallised from the parental magma. To be used, a zone must have mappable dimensions. A zone can, in turn, be subdivided into subzones.

***Polarity chron* – Chron de polarité –** (Page 56):
Main subdivisions of the GPTS.

***Polarity subchron* – sous-chron de polarité –** (Page 56):
Very short (~ 0.1 Ma) polarity intervals occurring within a chron.

***Polarity superchron* – super-chron de polarité –** (Page 56):
Very long periods of single polarity such as the Cretaceous Quite Zone.

***Polygentic volcano* – Volcan polygénique –** (Page 113):
A volcano built by successive eruptions.

***Ppb* – Ppb –** (Page 42):
Concentration unit (part per billion or ng/kg)

***Ppm* – Ppm –** (Page 42):
Concentration unit (part per million or mg/kg)

***Precision* – Précision –** (Page 66):
A termed derived from physical sciences (i.e. the quality of an instrument to measure the true value with a low approximation), the highest degree of precision (high resolution or high resolution power) could be obtained starting from a composite succession of biomarkers established by correlation of several biozonations based on different fossil groups.

***Progradation* – Progradation –** (Page 23):
The process in which sediments and the related depositional environments migrate seaward. Downlap terminations may characterize the base of prograding strata at the contact with horizontal strata, while clinoforms can represent the geometric configuration of the progradation.

***Protoreferential* – Protoréférentiel –** (Page 74):
A term used in logical biostratigraphy to describe the synthetic range chart constructed on the base of real data and which takes into account all the stratigraphic relationships (coexistences/superpositions) observed between all the taxa at all the studied localities; it is an ordered sequence (succession) of unitary associations which represents an obligatory methodological step leading to the real reference system, and to the zonal reference system. The respective chronological value for relative dating of the unitary associations of such a sequence depends on the control of superposition and the reproducibility of each of them.

***Punctuated equilibria* – Équilibres ponctués –** (Page 72):
The theory, proposed in 1972 by Niles Eldredge and Stephen J. Gould, that evolution is characterized by long periods of equilibrium during which little speciation occurs, "punctuated" by periods of rapid evolutionary change with the origination of a new species.

***Pyroclastic rocks* – Pyroclastites –** (Page 110):
Synon.: **Tephra**, **Ejecta**.
Material constituted by an assemblage of pyroclasts. It can be deposited on land or in a subacqueous environment. According to transportation, dispersal and depositional modes, which are function of the eruption dynamic, they can be classified as follows:
– *pyroclastic fall deposits*: resulting from the accumulation on Earth's surface of fragments balistically ejected into the atmosphere or transported by winds.
– *pyroclastic flow deposit*: the accumulation of volcanic material caused by a the lateral flowage of a turbulent mixture of hot gases and unsorted pyroclastic material.
– *pyroclastic surge deposits:* produced by the accumulation of large volumes of pyroclastic material acting as heavy fluids and controlled in their movement by gravity and the topography of

underlying land surface. Characterized by a high density the material flows relatively slow. Pyroclastic surges travel faster than pyroclastic flows, cover less distance and the resulting deposits are thinner layers of volcanic material than those accumulated with pyroclastic flows.

***Pyroclast* – Pyroclaste** – (Page 110):
Fragment of lava or any other rock or crystal ejected by an eruptive, generally explosive, process.
The granulometric classification of pyroclasts reflects that of detrital particles:
– > 64 mm: block (a bomb is a juvenile block modeled in the plastic state);
– between 64 mm-2 mms: lapilli;
– between 2 mm-62,5 µm: coarse ash;
– < 62,5 µms: fine ash = dust.

Q

***Quiescence (repose)* – Repos** – (Page 113):
The interval of time between volcanic eruptions.

R

***Radioactive isotopes* – Isotopes radioactif** – (Page 91):
Unstable isotope which spontaneously decays into a stable isotope known as radiogenic through one or more intermediate stages.

***Radiogenic isotopes* – Isotopes radiogéniques** – (Page 91):
Stable isotope derived from the spontaneous decay of a radioactive isotope.

***Range biozone* – Biozone de distribution** – (Page 69):
The body of strata corresponding, either to the total known distribution of one or more taxa (total "range" Zone), or to the overlapping parts of the range of at least two taxa ("partial range Zone").

***Real reference system* – Référentiel réel** – (Page 74):
A sequence of subdivisions (unitary associations or biochronozones) whose reproducibility is proven; in practice, it is a protoreferential accompanied by its reproducibility matrix.

***REE, rare earth elements* – Terres rares** – (Page 50):
Series of 14 elements between La and Lu. The REE concentrations are normalized to the average concentration of the Earth crust (see NASC) and are generally represented with their atomic number in ascending order.

***Regional chronostratigraphic scale* – Échelle chronostratigraphique régionale** – (Page 119):
Chronostratigraphic scale, covering a more or less extended time interval, composed of regional stages, which is used as time-framework for areas of limited extension (basin, continent, ocean).

***Regional stage* – Étage régional** – (Page 119):
Chronostratigraphic unit having the same rank of the stage but defined on a regional sedimentary succession that cannot be definitely and precisely placed into the Global Standard Chronostratigraphic Scale. The continous succession of regional stages allows to establish a regional chronostratigraphic scale.

***Relative biochronological scale* – Échelle biochronologique relative** – (Page 66):
A succession of intervals (biozones), each of them characterized by a peculiar paleontological content. The time span of each interval is not known. In this case the concept of duration is replaced by those of anteriority, (sub)contemporaneousness and posteriority.

***Relative dating* – Datations relatives** – (Page 2):
The determination of the position of a rock or a geological event within the geological time scale, without numerical values.

***Residence time* – Temps de résidence** – (Page 41):
Time necessary for an atom of a given element that entered the ocean to be eliminated. This parameter, based on the principle of a steady state of the ocean, is calculated as the total oceanic inventory divided by the input/export balance per year. Many existing data are biased because based only on the balance between the input from continental weathering and export by sedimentation. Recently taken into account, the impact of hydrothermal processes has been considerable on the estimation of the residence time of certain elements.

***Resolution power* – Pouvoir de résolution** – (Page 66):
The capacity of a stratigraphic scale to separate the stratigraphic record in the shortest possible intervals.

***Reworking* – Remaniement** – (Page 113):
See Epiclastites.

S

***Sample coordinates* – Coordonées d'échantillon** – (Page 58):
Direction of the magnetization in the specimen coordinates.

***Seismic stratigraphy* – Stratigraphie sismique (ou sismostratigraphie)** – (Page 2):
It is the study of the geologic bodies based on the geometric organization and relationship of seismic reflectors.

***Sequence* – Séquence** – (Page 8):
A succession of at least two lithologic units, forming a natural succession without important stratigraphic interruptions.

***Sequential analysis* – Analyse séquentielle** – (Page 17):
Stratigraphic technique used to define the sequence of objects and facies.

***Sequence Boundary* – Surface de baisse des eaux** – (Page 25):
Surface triggered by a fall in the relative sea level, characterized by a basinward shift in facies, a downward shift in costal onlap, and onlap of the overlying strata. For large-scale falls (below the base level) of the relative sea level, the sequence boundary is characterized by subaerial exposure, concurrent subaerial erosion associated to the stream rejuvenation and submarine fan deposition.

***Sequence of objects* – Séquence d'objets** – (Page 17):
Sedimentary unit characterized by the gradual evolution of the constituent elements (or objects) in a depositional succession. A synonym is stratigraphic sequence.

***Sequence stratigraphy* – Stratigraphie séquentielle** – (Page 2):
Stratigraphic tool whose purpose is to identify the genetic sequences on the basis of the geometric configuration of the deposits. A cycle in the relative sea level change is considered to be the primary mechanism triggering the formation of these genetic sequences (depositional sequences); these are bounded by unconformities or their correlative conformities, which are associated to a fall of the relative sea level.

***Series* – Sous-Système** – (Page 118):
A chronostratigraphic unit of intermediate rank between the Stage and the System.

***Short term variation* – Tendance à court terme** – (Page 42):
Geochemical signal change that lasts about or less than a million years.

***SMOW (Standard Mean Océanique Water)* ** – (Page 42):
Reference standard for oxygen isotope ratio in waters. It is a theoric sea water whose composition is close to the present average oceanic waters.

***Spectral analysis* – Analyse harmonique** – (Page 36):
A type of mathematical analysis (such as the Fourier analysis) used to unravelling the harmonic components from a

complex stratigraphic signal recorded on a lithological cyclic succession. The harmonic components correspond to a fundamental frequency (first harmonic) characterized by other harmonics with different wavelengths and amplitudes.

Stage – **Étage** – (Page 118):
Basic chronostratigraphic unit of relatively low rank which represents an average time interval of 5 million years.

Stratigraphic classification – **Classification stratigraphique** – (Page 1):
Organization of terrains – on the basis of certain characteristics, properties or attributes – into units organized as a hierarchy according to their size and/or duration.

Stratigraphic correlations – **Corrélations stratigraphiques** – (Page 2):
Identification and definition of the relationships between stratigraphic units from different places on the base of their general characters and ages.

Stratigraphic nomenclature – **Nomenclature stratigraphique** – (Page 1):
The set of formal rules, established by national or international commissions, which is used to name stratigraphic units.

Stratigraphic terminology – **Terminologie stratigraphique** – (Page 1):
The system of words designating the different kinds of units used by the various stratigraphical methods.

Stratotype – **Stratotype** – (Page 119):
The type section of a layered stratigraphic unit that serves as the standard of reference for the definition and characterization of the unit.

Strato-volcano – **Stratovolcan (ou strato-volcan)** – (Page 113):
A polygenic volcano composed of both lava flows and pyroclastic material stratified by the accumulation during successive phases of activity. Strato-volcanoes are generally steep sided and conical and usually built over periods of tens to hundreds of thousands of years. A synonym is composite volcano.

Stratum – **Strate** – (Page 10):
Informal term used to indicate a bed of sedimentary or pyroclastic material, regardless of its thickness.

Subgroup – **Sous-groupe** – (Page 10):
Informal term used to indicate a lithostratigraphic unit of intermediate rank between the Group and the Formation.

Substage – **Sous-Etage** – (Page 118):
Subdivision of the stage generally having only a local or regional significance.

Supergroup – **Super-groupe** – (Page 10):
A lithostratigraphic unit composed of several associated groups or associated formations and groups sharing significant lithologic features.

Supersuite – **Supersuite** – (Page 102):
It is the lithodemic unit next in rank above the series. It is the equivalent, in hierarchy, to the super group (group) of the lithostratigraphic nomenclature; it comprises two or more suites vertically or laterally associated. The name is defined by the term "Superseries" and a geographic name.

Suite – **Suite** – (Page 102):
A lithodemic unit next higher in rank to lithodeme; comprises two or more lithodemes of the same class (e.g., plutonic, metamorphic). It is the equivalent, in hierarchy, to the group (formation) of the lithostratigraphic nomenclature.
The name of a suite is composed by the term "Series" followed by a descriptive adjective, such as Plutonic, Granitic, Metamorphic, etc.), and the name of geographic locality.

System – **Système** – (Page 118):
A system is a unit of major rank in the chronostratigraphic hierarchy, which groups several stages and subsystems. The duration of the systems is largely variable ranging from 25 Myr to 70 Myr.

Systems tract – **Cortège sédimentaire** – (Page 25):
Is a linkage of contemporaneous depositional systems, genetically linked by processes and environments. They represent parts of a 3rd order sequence; five types of systems tracts are generally used:
– *Falling stage systems tract, FSST:*
includes all sediments deposited during a time of progressive sea level fall during which subsidence was sufficiently high to outpace sediment supply.
– *Lowstand systems tract, LST:*
It forms during a period of relative sea-level fall and is represented by slope and basin-floor fans; lowstand wedges represented by deltaic and incised valley fill deposits develop on the continental shelf. Both prograding and aggrading stratal architectures characterize this system tract.
– *Shelf margin wedge systems tract, SMW:*
It is associated with a relative low of sea level, in a type-2 sequence. Stratal architectures are defined by prograding and/or aggrading signatures; it may be very difficult to discriminate from the highstand systems tract.
– *highstand systems tract, HST:*
It is associated to a rapid deepening of the depositional environments due to a relative sea-level rise. Sediments are trapped on the continental shelf and record aggrading or backstepping stratal stacking patterns. Basinward condensed deposits may occur within this system tract.
– *transgressive systems tract, TST:*
It occurs during decreasing rates of relative sea level rise. It is defined by rates of deposition higher than relative sea level rise; geometries are first aggrading and then prograding.

T

Taxon – **Taxon** – (Page 65):
A group of organism at any level (e.g. species, genus, family, order, etc.) in a system for classifying plants or animals (taxonomy) based on the grouping of populations having biological and space-time relationships.

Tectonic coordinates – **Coordonnées tectoniques** – (Page 61):
Direction of the magnetization of a specimen after correction for its orientation and for tectonic tilting.

Tephra – **Téphra** – (Page 110):
A generic term applied to unconsolidated materials of any type and size that are erupted from a crater or volcanic vent and deposited from the air. Tephra includes larger material like blocks and bombs, and smaller light rock debris such as scoria, pumice, reticulite, and ash. In Quaternary stratigraphy it is recommended to use this term to indicate the basic unit of ash deposite (a bed or a lamina).

Tephrochronology = Tephrostratigraphy – **Téphrochronologie = Téphrostratigraphie** – (THORARISSON, 1944) – (Page 112):
Tephrochronology refers to a dating method based on the examination of tephra. In areas where repeated activity lead to the accumulation of successive tephra layers, the latter can be used as isochronous marker beds for local correlation.

Thermal demagnetization – **Démagnétisation thermique** – (Page 59):
Demagnetization obtained by heating (and cooling) the specimen in a field free space.

TOC – **COT %** – (Page 50):
Concentration in total organic carbon expressed as weight percent in relation to dry bulk sediment.

***Total range zone* – Zone de distribution totale** – (Page 71):
Body of strata corresponding to the total known distribution of one or more taxa.

***Transgressive surface, TS* – Surface de transgression** – (Page 26):
Surface marked by the turning point from prograding to retrograding geometric configurations. It is outlined by the landward migration of marine deposits above the continental facies of the lowstand system tract; it is the equivalent of the first flooding surface on the coastal plain, and represents the lower boundary surface of the transgressive system tract.

***Type of organic matter* – Type de matière organique** – (Page 50):
The sedimentary organic matter, despite it represents a complex residue, is the product of different biomasses (land plants, marine phytoplankton, bacteria), and being so, may be differentiated. In this case, it is possible to recognize and classify different types of organic matter. The most used technique was introduced by carbon petrography. The results of the elemental analysis of the organic matter, isolated from the sediments, are plotted in a diagram where the H/C and O/C atomic ratios are represented in the y- and x-axes, respectively. In this diagram, different regions define the types of organic matter.

U

***Unconformity* – Discordance stratigraphique** – (Page 13):
A significant interruption or gap in the stratigraphic record represented by an angular discordance and/or a stratigraphic hiatus.

***Unitary association* – Association unitaire** – (Page 73):
Association of fossils taxa characterized by a real or potential overlap in their stratigraphic distribution. This kind of grouping is identified in the logical biostratigraphy under the name of "unitary association method".

***Unit Stratotype* – Stratotype d'unité chronostratigraphique** – (Page 119):
The reference section of a layered stratigraphic unit which serves as the standard for the definition and characterization of a unit.

V

***VGP (Virtual Geomagnetic Pole)* – Pole géomagnétique virtuel** – (Page 64):
Position of the geomagnetic pole calculated for a given site in the present day coordinates. Since the GAD hypothesis state that the magnetic pole must correspond to geographic pole, the displacement of the VGP from geographic pole(s) is a measure of tectonic rotations and plate motion.

***Volcanoclastic* – Volcanoclastites** – (Page 110):
A general term applied to materials entirely or partly composed by fragments of volcanic origin.

***Volcanic periodicity* – Périodicité volcanique** – (Page 113):
The alternation between eruption events and periods of quiescence. The discontinuity in the volcanic activity is recorded at different scales such as in a single volcanic edifice or a group of volcanoes in a region (volcanic province).

***Volcanogenic material* – Volcanogénique (matériau)** – (Page 109):
A deposit or a rock, entirely or partly derived from eruptive activity, not necessarily magmatic. The term refers to pyroclastic deposits essentially or totally composed of non volcanic fragments on one hand and to any other secondary vocaniclastic product (reworked or detritic).

Z

***Zijderveld plots* – Plots de Zijderveld** – (Page 60):
Projection of magnetic direction (vectors) into the horizontal and the north-south or east-west vertical planes. The choice of vertical projection usually depends on the magnitudes of N and E components.

***Zonal reference system* – Référentiel zonal** – (Page 74):
A suite of biochronozone.

***Zonule* – Zonule** – (Page 77):
It has received different meanings and is now generally used as a subdivision of a biozone or subbiozone. The use of this term is discouraged. Rehabilitated by its meaning of "3-dimensional zone" with firm ecological character, its usage would raise anyway ambiguity and cause confusion with the informal term "biohorizon".

REFERENCES

Chapter 1. – STRATIGRAPHY: FOUNDATIONS AND PERSPECTIVES

HEDBERG, H.D. (1976). – International Stratigraphic Guide. A guide to stratigraphic classification, terminology and procedure. – Wiley & Sons (eds.), New-York, 200 pp.

POMEROL, C., BABIN, C., LANCELOT, Y., LE PICHON, X., RAT, P. & RENARD, M. (1987). – Stratigraphie: principes, méthodes, applications. – Doin (ed.), Paris, 283 pp.

REY, J. (1983). – Biostratigraphie et Lithostratigraphie. Principes fondamentaux, méthodes et applications. – Technip (ed.), Paris, 18 pp.

SALVADOR, A. (1994). – International Stratigraphic Guide. A guide to stratigraphic classification, terminology and procedure (second edition). International subcommission on stratigraphic classification of I.U.G.S. International Commission on Stratigraphy. – *Geological Society of America* (eds), 214 pp.

SIGAL, J. & TINTANT, H. (1962). – Principes de classification et de nomenclature stratigraphiques. – *Com. Franç. Stratigr.*, 15 pp.

Chapter 2. – LITHOSTRATIGRAPHY

ARNAUD, H. (1981). – De la plate-forme urgonienne au bassin vocontien. – *Géologie Alpine*, mém. **12**, 1 (311pp.) et 2 (804 pp.).

ANDERSON, R.Y. & DEAN, W.B. (1988). – Lacustrine varve formation through time. – *Palaeogeogr. Palaeoclimatol. Palaeoecol.*, **62**, N1-4, 215-235.

BERGER, A.L., LOUTRE, M.F. & LASKAR, J. (1992). – Stability of the astronomical frequencies over the Earth's history for paleoclimate studies. – *Science*, **255**, 560-566.

BISTHMUTH, H., BOLTENHAGEN, C., DONZE, P., LE FEVRE, J. & SAINT-MARC, P. (1982). – Etude sédimentologique et biostratigraphique du Crétacé moyen et supérieur du Djebel Semama (Tunisie du Centre Nord). – *Cretaceous Research*, **3**, 171-185.

BOURROZ, A., SPEARS, D. & ARBEY, F. (1983). – Review of the formation and evolution of petrographic markers in coal basins. – *Mem. Soc. géol. Nord*, **16**, 115 pp.

CAVELIER, C., MEGNIEN, C., POMEROL, C. & RAT, P. (1980). – Bassin de Paris. – *In*: Lorenz, C. (ed.): Géologie des pays méditerranéens, France, Belgique, Luxembourg. – Dunod (éd.), 26ᵉ Congrès géol. int. Paris, 431-484.

COTILLON, P. (1987). – Bed-scale cyclicity of pelagic Cretaceous successions as a result of world-wide control. – *Marine Geol.*, **78**, 108-123.

COTILLON, P. (1992). – Search for eustacy record in deep-Tethyan deposits through the study of sedimentary flux variations. Application to the Upper Tithonian – Lower Aptian Series at DSDP Site 534 (Central Atlantic). – *Palaeogeogt. Palaeoclimat. Palaeoecol.*, **91**, 263-275.

COTILLON, P. & RIO, M. (1984). – Cyclicité comparée du Crétacé inférieur pélagique dans les chaines subalpine méridionales (France S.E.), l'Atlantic central (site 534D. S.D.P.) et le Golfe du Mexique (Sites 535 et 540 D.S.D.P.); implications paleoclimatiques et application aux corrélations stratigraphiques transtéthysiennes. – *Bull. soc. Géol France*, **26**, 1, 47-62.

COTILLON, P., FERRY, S., GAILLARD, C., JAUTEE, E., LATREILLE, G. & RIO, M. (1980). – Fluctuations des paramètres du milieu marin dans le domaine vocontien (France SE) au Crétacé inférieur: mise en évidence par l'étude des formations marno-calcaires alternantes. – *Bull. Soc. géol. France*, **7**, 22, 5, 733-742.

CUBAYNES, R., FAURE, P. HANTZPERGUE, P., PELISSIE, T. & REY, J. (1989). – Le Jurassique du Quercy: Unités lithostratigraphiques, stratigraphie et organization séquentielle, évolution sédimentaire. – *Géol. France*, **3**, 33-62.

DE BOER, P.L. (1983). – Aspects of middle Cretaceous pelagic sedimentation in Southern Europe. *Geologica Ultraiectina*, **31**, 112p.

DE BOER, P.L. & WONDERS, A.A.H. (1984). – Astronomically induced rhythmic bedding in Cretaceous pelagic sediments near Moria (Italy). – *In*: Berger *et al.* (eds.). – *Milankovitch and climate*, **I**, Reidel, 177-190.

DELFAUD, J. (1972). – Application de l'analyse séquentielle à l'exploration lithostratigraphique d'un bassin sédimentaire. L'exemple du Jurassique et du Crétacé inférieur de l'Aquitaine. – *Mém. Bur. Rech. géol. min.*, **77**, 593-611.

EMBRY, A. F. (1993). – Transgressive-regressive (T-R) sequence analysis of the Jurassic succession of the Sverdrup basin, Canadian Artic Archipelago. *Canadian Journal of Earth Sciences*, **30**, 301-320.

GALEOTTI, S., SPROVIERI, M., COCCIONI, R., BELLANCA, A. & NERI, R. (2003). – Orbitally modulated black shale deposition in the upper Albian Amadeus Segment (central Italy): a multi-proxy reconstruction. *Palaeogeography, Palaeoclimatology, Palaeoecology*, **190**, 441-458, Amsterdam.

FISCHER, A.G. (1986). – Climatic rhythms recorded in strata. – *Earth and Planet. Sci. Lett.*, **14**, 351-376.

GALLOWAY, W.E. (1989). – Genetic stratigraphic sequences in Basin analysis I: Architecture and genesis of flooding-surface bounded depositional units. – *Bull. amer. Assoc. Petroleum Geol.*, **73**, 2, 125-142.

GILBERT, G.K. (1985). – Sedimentary measurement of Cretaceous time. – *J. Geol.*, **3**, 121-127.

GRADSTEIN, F.M., OGG, J.G., A.G. SMITH, AGTERBERG, F.P., BLEEKER, W., COE, A., COOPER, R.A., DAVYDOV, V., GIBBARD, PH., HINOV, L. HOUSE, M.R. (†), LOURENS, L., LUTERBACHER, H.P., MCARTHUR, J., MELCHIN, M.J., ROBB, L.J., SHERGOLD, J., VILLENEUVE, M. WARDLAW, B.R. *et al.*, (2004). – A Geologic Time Scale 2004. Cambridge University Press.

HAQ, B.U., HARDENBOL, J. & VAIL, P.R. (1988). – Mesozoic and Cenozoic chronostratigraphy and cycles of sea level. *In*: Wilgus, C.K., Hastings, B.S., Kdall, C.G., Posamentier, H.W., Ross, C.A. & Van Wagoner, J.C. eds, Sea Level changes: an integrated approach. *Soc. Econ. Paleont. Mineral. sp. pub.*, **42**, 71-104, Tulsa.

HATTIN, D.E. (1971). – Widespread, synchronously deposited, burrow-mottled limestone beds in Greenhorn limestone (Upper Cretaceous) of Kansas and Central Colorado. – *Bull. amer. Assoc. Petroleum Geol.*, **55**, 412-431.

HATTIN, D.E. (1985). – Distribution and significance of widespread time-parallel pelagic limestone beds in Greenhorn limestone (Upper Cretaceous) of the Central Great Plains and Southern Rocky Mountains. – *In*: Pratte *et al.* (eds.). – S.E.P.M. field trip Guidebook, **4**, 28-37.

HECKEL, P.H. (1986). – Sea-level curve for Pennsylvanian eustatic marine transgressive-regressive depositional cycles along midcontinent outcrops belt, North America. – *Geology*, **94**, 330-334.

HEDBERG, H.D. (1976). – International Stratigraphic Guide. A guide to stratigraphic classification, terminology and procedure. – Wiley & Sons (eds.), New-York, 200 pp.

HOMEWOOD, P., GUILLOCHEAU, F., ESCHARD, R. & CROSS, T.A. (1992). – Corrélation haute résolution et stratigraphie génétique: une démarche intégrée. – *Bull. Centres Rech. Explor.- Prod. Elf Aquitaine*, **16**, 2, 357-381.

HUNT, D. & GAWTHORPE, R.L. (2000). – Responses to Forced Regressions. *Geological Society of London*, Special Publication, **172**.

HUNT, D. & TUCKER, M.E. (1992). – Sequence stratigraphy of carbonate shelves with an example from the mid-Cretaceous (Urgonian) of southeast France. *In*: Posamentier, H.W., Summerhayes, C.P., Haq, B.U. & Allen G.P. (eds) Sequence stratigraphy and facies association. Special Publication, International Association of Sedimentologists, **18**, 307-341.

HUNT, D. & TUCKER, M.E. (1993). – Stranded parasequences and the forced regressive wedge systems tract: deposition during base-level fall-reply. – *Sediment. Geol.*, **95**, 147-160.

LOMBARD, A. (1956). – Géologie sédimentaire, les séries marines. – Masson (ed.), Paris, 722 pp.

LOMBARD, A. (1972). – Séries sédimentaires. Genèse – Evolution. – Masson (ed.), Paris, 425 pp.

MUTTI, E., RICCI LUCCHI, F., SEGURET, M. & ZANZUCCHI, G. (1984). – Seismoturbidites: a new group of resedimented deposits – *Marine geology*, **55**, 103-166.

NORTH AMERICAN COMMISSION ON STRATIGRAPHIC NOMENCLATURE (1983). – North american Stratigraphic code. – *Bull. amer. Assoc. Petroleum Geol.*, Tulsa, **67**, 5, 841-87.

PAYTON, C.E. (1977). – Seismic stratigraphy: applications to hydrocarbon exploration. – *Amer. Assoc. Petroluem Geol.*, **26**, 516 pp.

PLINT, A.G. & NUMMEDAL, D. (2000). – The falling stage systems tract: Recognition and importance in sequence stratigraphic analysis. In: Sedimentary responses to forced regressions, ed by D. Hunt and R.L.G. Gawthorpe. Geological Society of London, Special Publication, **172**, 1-17.

POMAR, L. (1991). – Reef geometries, erosion surfaces and highfrequency sea-level changes, upper Miocene reef complex, Mallorca, Spain. *Sedimentology*, **38**, 243-270.

POMAR, L., WARD, W.C. & GREEN, D.G. (1996). – Upper Miocene Reef Complex of the Llucmajor area, Mallorca, Spain. *In*: Franseen, E., Esteban, M., Ward, W.C., Rouchy, J.M. (Eds.), Models for Carbonate Stratigraphy from Miocene Reef Complexes of the Mediterranean regions. Soc. Econ. Paleont. Mineral. *Concepts in Sedimentology and Paleontology*, Series 5, Tulsa, OK, pp. 191-225.

POSAMENTIER, H.W. & ALLEN, G.P. (1999). – Siliciclastic sequence stratigraphy-concepts and applications. Society of Economic Petrologists and Paleontologists, Vol. 7 Concepts in Sedimentology and Paleontology, 216 pp.

POSAMENTIER, H.W. & James, D.P. (1993). – An overview of sequence-stratigraphic concepts: uses and abuses, *in* H.W. Posamentier, C.P. Summerhayes, B.U. Haq and G.P. Allen, eds., Sequence stratigraphy and facies associations: Oxford, Blackwell, p. 3-18.

POSAMENTIER, H.W. & VAIL, P.R. (1988). – Eustatic controls on clastic deposition. II. Sequence and systems tract models. *In*: "Sea-level changes: an integrated approach". – *Soc. Econ. Paleont. Mineral. spec. Publ.*, Tulsa, **42**, 25-154.

POSAMENTIER, H.W., JERVEY, M.T. & VAIL, P.R. (1988). – eustatic controls on clastic deposition. I. Conceptual framework. – *In*: "Sea-level changes: an integrated approach". – *Soc. Econ. Paleont. Mineral. spec. Publ.*, Tulsa, **42**, 109-124.

POSAMENTIER, H.W., ALLEN, G.P., JAMES, D.P. & TESSON, M. (1992). – Forced regression in a stratigraphic framework: concepts, examples and exploration significance. – *Bull. amer. Assoc. Petroleum. Geol.*, Tulsa, **76**, 11, 1687-1709.

PURSER, B. (1972). – Subdivision et interprétation des séquences carbonatées. – *Mém. Bur. Rech. géol. min.*, **77**, 679-698.

REY, J. (1979). – Les formations bioconstruites du Crétacé inférieur d'Estramadura (Portugal). – *Géobios, mém. Spec.*, **3**, 89-99.

REY, J. (1983). – Biostratigraphie et Lithostratigraphie. Principes fondamentaux, méthodes et applications. – Technip (ed.), Paris, 181 pp.

REY, J. (2006). – Séquences de dépôt dans le Crétacé inférieur du Bassin Lusitanien. *Ciências da Terra*, Lisboa, vol. sp. VI, 120 pp., 142 fig.

RIOULT, M., DUGUÉ, O., JAN-DU-CHÊNE, R., PONSOT, C., FLY, G., MORON, J.M. & VAIL, P.R. (1991). – Outcrop sequence stratigraphy of the Anglo-Paris Basin Middle to Upper Jurassic (Normandy, Maine, Dorset). – *Bull. Centres Rech. Explor.-Prod. Elf Aquitaine*, **15**, 1, 101-194.

SALVADOR, A. (1994). – International Stratigraphic Guide. A guide to stratigraphic classification, terminology and procedure (second edition). International subcommission on stratigraphic classification of I.U.G.S. International Commission on Stratigraphy. – The international Union of Geological Sciences and the Geological Society of America Inc. (eds), 214 pp.

SCHLAGER, W. (2004). – Fractal nature of stratigraphic sequences. *Geology*, **32**(3), 185-188.

SCHWARZACHER, W. (1993). – Cyclostratigraphy and the Milankovitch theory. – *Dev. Sedimentol.*, **52**, Elsevier, 225 pp.

SERRA, O. (1979). – Diagraphies différées – bases de l'interprétation. Tome 1: Acquisition des données diagraphiques. – *Bull. Cent. Rech. Explor.-Prod. Elf- Aquitaine*, Mém. **1**, 328 pp.

SERRA, O. (1985). – Diagraphies différées – bases de l'interprétation. Tome 2: Interprétation des données diagraphiques. – *Bull. Cent. Rech. Explor.-Prod. Elf- Aquitaine*, Mém. **7**, 631 pp.

SERRA, O. (1979). – Diagraphies différées – bases de l'interprétation. Tome A: acquisition des données diagraphiques. *Bull. Cent. Rech. Explor.-prod. Elf-Aquitaine*, **1**, 328pp.

SERRA, O. (1984. – Fundamentals of Well-Log Interpretation (Vol. 1): The Acquisition of Logging Data: – *Dev. Pet. Sci.*, **15**A, Elsevier, Amsterdam.

TEN KATE, W.G. & SPRENGER, A. (1993). – Orbital cyclicities above and below the Cretaceous/Paleogene boundary at Zumaya (N. Spain), Agost and Relleu (SE Spain). – *Sediment. Geol.*, **87**, 69-101.

TESSIER, B., MONTFORT, Y., GIGOT, P. & LARSONNEUR, G. (1989). – Enregistrement des cycles tidaux en accrétion verticale, adaptation d'un outil de traitement mathématique. Exemples en baie du Mont Saint-Michel et dans la molasse marine miocène du bassin de Digne. – *Bull. Soc. géol. France*, **8**, V, 5, 1029-1041.

VAIL, P.R., AUDEMARD, F., BOWMAN, S.A., EISNER, P.N. & PEREZ-CRUZ, C. (1991). – The stratigraphic signatures of tectonics, eustacy and sedimentology – an overview. – In: "Cycles and events in stratigraphy". – Springer-Verlag (ed.), Berlin-Heidelberg, 617-659.

VAN WAGONER, J.C., POSAMENTIER, H.W., MITCHUM, R.N., VAIL, P.R., SARG, J.F., LOUTIT, T.S. & HARDENBOL, J. (1988). – An overview of the fundamentals of sequence stratigraphy and key definitions. – In: "Sea-level changes: an integrated approach". – *Soc. Econ. Paleont. Mineral. sp. Publ.*, Tulsa, **42**, 109-124.

Chapter 3. – CHEMOSTRATIGRAPHY

ACCARIE, H., RENARD, M., DECONINCK, J.F., BEAUDOIN, B. & FLEURY, J.J. (1989). – Géochimie des carbonates (Mn, Sr) et minéralogie des argiles de calcaires pélagiques sénoniens. Relations avec les variations eustatiques (Massif de la Maiella, Abruzzes, Italie). – *C.R. Acad. Sci. Paris*, (II), **309**, 1679-1685.

ACCARIE, H., RENARD, M. & JØRGENSEN, N.O. (1992). – Le manganèse dans les carbonates pélagiques: outil d'intérêt stratigraphique et paléogéographique (le Sénonien d'Italie centrale, de Tunisie et du Danemark). – *C.R. Acad. Sci. Paris*, (II), **317**, 65-72.

ALBAREDE, F. (1990). – Les anciens océans. – *Courr. CNRS*, **76**, 50-51.

ALVAREZ, L.W., ALVAREZ, W., ASARO, F. & MICHEL, H.V. (1980). – Extraterrestrial cause for Cretaceous/Tertiary boundary extinctions. – *Science*, **208**, 1095-1108.

ALVAREZ, W., ASARO, F., MICHEL, H.V. & ALVAREZ, L.W. (1982). – Iridium anomalies approximatively synchroneous with terminal Eocene extinctions. – *Science*, v. 216, p. 886.

ANDERSON, T.F. & STEINMETZ, J.C. (1981). – Isotopic and biostratigraphical records of calcareous nannofossils of a Pleistocene core. – *Nature*, **294**, 741-744.

ARRHENIUS, G. (1952). – Sediment cores from East Pacific. – *Rep. Swed. Deep Sea Exp.*, 1947-1948, **5**, 1-228.

BAKER, P.A., GIESKES, J.M. & ELDERFIELD, H. (1982). – Diagenesis of carbonates in deep-sea sediments. Evidence from Sr/Ca ratios and interstitial dissolved Sr^{2+} data. – *J. Sediment. Petrogr.*, **52**, 1, 71-82.

BARNOLA, J. M., RAYNAUD, D., KOROTKEVICH, Y. S., LORIUS, C. (1987). – Vostok ice core provides 160,000-year record of atmospheric CO_2. *Nature*, **329**(6138): 408-414.

BAUDIN, F. (1989). – Caractérisation géochimique et sédimentologique de la matière organique du Toarcien téthysien (Méditerranée, Moyen-Orient). Signification paléogéographique. – *Mem. Sci. Terre Univ. P.-et-M.-Curie Paris*, **89**, 30, 246 pp.

BENDER, M.L., KLINKHAMMER, G.P. & SPENCER, D.W. (1977). – Manganese in seawater and the marine manganese balance. – *Deep Sea Res.*, **24**, 799-812.

BERGER, W.H. & VINCENT, E. (1981). – Chemostratigraphy and biostratigraphic correlation exercices in systemic stratigraphy. – *Oceanol. Acta*, **26**, CIG, 115-127.

BERNER, R.A. (2002). – Examination of hypotheses for the Permo-Triassic boundary extinction by carbon cycle modeling). – Proceedings of the National Academy of Sciences of the United States of America. **99**(7): 4172-4177.

BICE, K.L. & MAROTZKE, J. (2002). – Could changing ocean circulation have destabilized methane hydrate at the Paleocene/Eocene boundary? *Paleoceanography*, **17**(2), Art. No. 1018.

BLUNIER, T. & BROOK, E.J. (2001). – Timing of millennial-scale climate change in Antarctica and Greenland during the last glacial period. – *Science*, **291**: 109-112.

BOCHERENS, H., FIZET, M., MARIOTTI, A., LANGE-BADRE, B., VANDERMEERSCH, B., BOREL, J.P. & BELLON, G. (1991). – Isotopic biogeochemistry (13C, 15N) of fossil vertebrate collagen: application to the study of a past food web including Neanderthal man. – *J. Human Evol.*, **20**, 481-492.

BONTE, PH., DELACOTTE, O., ROCCHIA, R., BOCLET, D. & RENARD, M. (1984). – The iridium-rich layer at the K/T boundary at Bidart (France). – *Geophys. Res. Lett.*, **11**, 473-476.

BOND, G., SHOWERS, W., CHESEBY, M., LOTTI, R., ALMASI, P., DEMENOCAL, P., PRIORE, P., CULLEN, H., HAJDAS, I. & BONANI, G. (1997). – A pervasive millennial-scale cycle in North Atlantic Holocene and Glacial Climates. – *Science*, **278**: 1257-1266.

BOND, G.C. & LOTTI, R. (1995). – Icebergs discharges into the North atlantic on millennial time scales durng the Last Glaciation. – *Science*, **267**: 1005-1010.

BOWEN, R. (1991). – Isotopes and climates. – Elsevier, London, 483 pp.

BRAND, U. & VEIZER, J. (1980). – Chemical diagenesis of a multicomponent carbonate system-1: trace elements. – *J. Sed. Petrol.*, **50**, 4, 1219-1236.

CAPO, D.C. & DE PAOLO, D.J. (1990). – Seawater strontium isotopic variation from 2.5 million years to present. – *Science*, **249**, 51-55.

CERLING, T.E. (1992). – Use of carbon isotopes in paleosols as an indicator of the $P(CO_2)$ of the Paleoatmosphere. – Global Biogeochem. *Cycles*, **6**, 3, 307-314.

CLAUSER, S. (1994). – Etudes stratigraphiques du Campanien et du Maastrichtien de l'Europe occidentale. Côte basque, Charentes (France), Limbourg. – *Doc. Bur. Rech. Géol. min.*, **235**, 243-245.

CLAYPOOL, G.E. & HOLSER, W.T., KAPLAN, I.R., SAKAI, H. & ZAK, I. (1980). – The age curves of sulfur and oxygen isotopes in marine sulfate and their mutual interpretation. – *Chem. Geol.*, **28**, 199-260.

COJAN, I., RENARD, M., COLSON, J. & EMMANUEL, L. (1994). – Essai de stratigraphie haute résolution en milieu continental. Apports et limites des variations climatiques (Provence – Crétacé supérieur/Paléocène). – 1er Congrès français Stratigraphie, Toulouse. – *Strata*, (1), **6**, 150 pp.

COJAN, I., RENARD, M. & EMMANUEL, L. (1995). – Paleoclimate changes during Maastrichtian inferred from pedogenesis and fossil geochemistry in terrestrial environment (Southern France). – 16th Regional european meeting, Aix-les-Bains, résumés, 2 pp.

CORBIN, J.C. (1994). – Evolution géochimique du Jurassique du Sud-Est de la France: Influence des variations du niveau marin et de la tectonique. – *Mém. Sci. Terre Univ. P.-et-M.-Curie*, Paris, **94**, 12, 173 pp.

CORBIN, J.C., GALBRUN, B. & RENARD, M. (1995). – La limite Campanien-Maastrichtien sur la marge N.W. australienne (leg ODP 122). Apports de la géochimie et de la magnétostratigraphie. – *C.R. Acad. Sci. Paris*, **321**, 1017-1023.

CORFIELD, R.M., CARTLIDGE, J.E., PREMOLI SILVA, I. & HOUSLEY, R.A. (1991). – Oxygen and carbon isotope stratigraphy of the Paleogene and Cretaceous limestones in the Bottacione Gorge and the Contessa Highway sections, Umbria, Italy. – *Terra nova*, **4**, 4, 414-422.

DE PAOLO, D.J. & INGRAM, B.L. (1985). – High-resolution stratigraphy with strontium isotopes. – *Science*, **227**, 938-941.

DEAN, W.E., ARTHUR, M.A. & CLAYPOOL, G. (1986). – Depletion of 13C in Cretaceous marine organic matter: source, diagenetic, or environmental signal. – *Marine Geol.*, **80**, 119-157.

EBNETH, S., SHIELDS, G.A., VEIZER, J., MILLER, J.F. & SHERGOLD, J.H. (2001). – High-resolution strontium isotope stratigraphy across the Cambrian-Ordovician transition. – *Geochimica et Cosmochimica Acta*, **65**(14): 2273-2292.

EDMOND, J.M. (1992). – Himalaya tectonics, weathering processes and the strontium isotope record in marine limestones. – *Science*, v. 258, 1594-1597.

EMILIANI, C. (1955). – Pleistocene temperatures. – *J. Geol.*, **63**, 6, 538-578.

EMMANUEL, L. (1993). – Apport de la géochimie des carbonates à la stratigraphie séquentielle. Application au Crétacé inférieur du domaine Vocontien. – *Thèse Doct. Univ. P.-et-M.-Curie*, Paris, **93**, 191 pp.

EMMANUEL, L. & RENARD, M. (1993). – Carbonate Geochemistry (Mn, d13C, d18O) of the late Tithonian-Berriasian pelagic limestones of the vocontian trough (SE France). – *Bull. Centres Rech. Explor.-Prod. elf aquitaine*, **17**, 1, 205-221.

FARRELL, J.W., CLEMENS, S.C. & GROMET, L.P. (1995. – Imroved chronostratigraphic reference curve of late Neogene seawater 87Sr/86Sr. – *Geology*, **23**(5): 403-406.

FAURE, G. (1982). – The marine strontium geochronometer. – *In*: Odin, G.S. (ed): "Numerical dating in stratigraphy", John Wiley, Chichester, 73-80.

GRADSTEIN, F.M., OGG, J.G. & SMITH, A., (2004). – A Geological time scale 2004. Cambridge University Press, 384 pp.

GALE, A.S., JENKYNS, H.C., KENNED, W.J. & CORFIELD, R.M. (1993). – Chemostratigraphy versus biostratigraphy: data from around the Cenomanian-Turonian boundary. – *J. Geol. Soc. London*, **150**, 29-32.

GODDERIS, Y. & FRANCOIS, L.M. (1995). – The Cenozoic evolution of the strontium and carbon cycles: relative importance of continental erosion and mantle exchanges. – *Chem. Geol.*, **126**, 169-190.

GRAHAM, D.W., BENDER, M.L., WILLIAMS, D.F. & KEIGWIN, L.D. (1982). – Strontium calcium ratio in Cenozoic planktonic foraminifera. – *Geochim. cosmochim. Acta*, **46**, 1281-1292.

GRANDJEAN-LECUYER, P., FEIST, R. & ALBAREDE, F. (1993). – Rare earth elements in old biogenic apatites. – *Geochim. cosmochim. Acta*, **57**, 2507-2514.

GROOTES, P.M., STEIG, E.J., STUIVER, M., WADDINGTON, E.D., MORSE, D.L. & NADEAU, M.-J. (2001). – The Taylor Dome Antarctic d18O record and globally synchronous changes in climate. – *Quaternary Research*, **56**(3): 289-298.

HADJI, S. (1991). – Stratigraphie isotopique des carbonates pélagiques (Jurassique supérieur-Crétacé inférieur) du Bassin d'Ombrie-Marches (Italie). – *Mém. Sc. Terre Univ. P.-et-M.-Curie*, Paris, **91**, 23, 160 pp.

HAYES, J.M., STRAUSS, H. & KAUFMAN, A.J. (1999). – The abundance of 13C in marine organic matter and isotopic fractionation in global biogeochemical cycle of carbon during the past 800 Ma. – *Chemical geology*, **161**: 103-125.

HESS, J., BENDER, M.L. & SCHILLING, J. (1986). – Evolution of the ratio of strontium 87 to strotium 86 in seawater from Cretaceous to Present. – *Science*, **231**, 979-984.

HILBRECHT, H. & HOEFS, J. (1986). – Geochemical and paleontological studies of the 13C anomaly in Boreal and north Tethyan Cenomanian-Turonian sediments in Germany and adjacent areas. – *Palaeogeogr. Palaeoclimatol. Palalaeoecol.*, **53**, 159-189.

HODELL, D.A. & WOODRUFF, F. (1994). – Variations in the strontium isotopic ratio of seawater during the Miocene: Stratigraphic and geochemical implications. – *Paleoceanography*, **9**(3): 405-426.

HODELL, D.A., BENSON, R.H., KENT, D.V., BOERSMA, A. & RAKIS-EL BIED, K. (1994). – Magnotostratigraphic, biostratigraphic and stable isotope statigraphy of an Upper Miocene drill core from Salé Briqueterie (northwestern Morocco): a high-resolution chronology for the Messinian stage. – *Paleoceanography*, **9**(6): 835-856.

HUGHEN, K.A., SCHRAG, D.P. & JACOBSEN, S.B. (1999). – El Niño during the last interglacial period recorded by a fossil coral from Indonesia. – *Geophysical Research Letters*, **26**(20): 3129-3132.

IATZOURA, A., COJAN I. & RENARD, M. (1991). – Géochimie des coquilles d'œufs de Dinosaures, Maastrichtien, Aix-en-Provence (France). – *C. R. Acad. Sic. Paris*, **312**, 1343-1349.

JENKYNS, H.C. (1988). – The Early Toarcian (Jurassic) anoxic event: stratigraphic, sedimentary and geochemical evidence. – *Am. J. Sci., New Haven*, **288**, 101-151.

JENKYNS, H.C., JONES, C.E., GRÖCKE, D.R., HESSELBO, S.P. & PARKINSON, D.N. (2002). – Chemostratigraphy of the Jurassic system: applications, limitations and implications for paleoceanography. – *Journal of the geological Society*, London, **159**: 351-378.

JENKYNS, H.C., GALEY, A.S. & CORFIELD, R.M. (1994). – Carbon and oxygen-isotope stratigraphy fo the english chalk and italian Scaglia and its palaeoclimatic significance. – *Geol. Mag.*, **131**, (1), 1-34.

JONES, C.E., JENKYNS, H.C. & HESSELBO, S.P. (1994). – Strontium isotopes in Early Jurassic seawater. – *Geochimica et Cosmochimica Acta*, **58**(4): 10285-1301.

KEIGWIN, L.D. (1980). – Paleoceanographic change in the Pacific at the Eocene/Oligocene Boundary. – *Nature*, **287**, 722-725.

KENNETT, J.P. (1982). – Marine Geology. – Univ. Rhode Island, Kingston. – Prentice Hall (ed.), 813 pp.

KLINKHAMMER, G.P. & BENDER, M.L. (1980). – The distribution of Manganese in the Pacific ocean. – *Earth and planet. Sci. Lett.*, **46**, 361-384.

KOCH, P.L., ZACHOS, J.C. & DETTMAN, D.L. (1995). – Stable isotope stratigraphy and paleoclimatology of the Paleocene Bighorn Basin (Wyoming, USA). – *Palaeogeogr. Palaeoclimatol. Palalaeoecol.*, **115**, 61-89.

KOEPNICK, R.B. & BURKE, W.H., DENISON, R.E., HETHERINGTON, E.A., NELSON, H.F., OTTO, J.B. & WAITE, L.E. (1985). – Construction of seawater 87Sr/86Sr curve for the Cenzoic and Cretaceous: Supporting data. – *Chem. Geol.*, 58, 55-81.

KORTE, C., KOZUR, H.W., BRUCKSCHEN, P. & VEIZER, J. (2003). – Strontiium isotope evolution of Late Permian and Triassic seawater. – *Geochimica et Cosmochimica Acta*, 67(1): 47-62.

KROOPNICK, P. & MARGOLIS, S.V. & WONG, C.S. (1977). – d13C variations in marine carbonates sediments as indicator of the C02 balance between the atmosphere and oceans. – *In*: "The fate fossil fuel CO_2 in the ocean". – *Plenum Press*, 295-321.

KROGH, T.E., KAMO, L.S., SHARPTON, V.L., MARIN, L.E. & HILDEBRAND, A.R. (1993). – U-Pb ages of single shocked zircons linking distal ejecta to the Chicxulub crater. – *Nature*, 366, 731-734.

KYTE, F.T. & BROWNLEE, D.E. (1985). – Unmelted meteoric debris in the Late Pleistocene iridium anomaly, evidence for the ocean impact of a nonchondritic asteroid. – *Geochim. cosmochim. Acta*, 49, 5, 1095-1108.

LETOLLE, R. (1979). – Oxygen 18 and C13 isotopes from bulk carbonate samples, Leg 47B. – *In*: Sibuet, J.C. *et al.* (eds.). – *init. Rep. Deep. Sea Drill. Proj.*, 47, 493-496.

LETOLLE, R. & POMEROL, B. (1980). – Mise en évidence dans les craies du Bassin de Paris d'un accident dans la répartition du sup 13 C d'âge Cénomanien terminal. – *J. Acad. Sci. Paris*, 291, 2, 133-136.

LETOLLE, R. & RENARD, M. (1980). – Evolution des teneurs en 13C des carbonates pélagiques aux limites Crétacé-Tertiaire et Paléocène-Eocène. – *C.R. Acad. Sci. Paris*, 290, 827-830.

LYLE, M. (1976). – Estimation of hydrothermal manganese input to the oceans. – *Geology*, (12), 4, 733-736.

MAGARITZ, M. (1991). – Carbon isotopes, time boundaries and evolution. – *Terra nova*, 3, 3, 251-256.

MARGOLIS, S.V., KROOPNICK, P.M., GOODNEY, D.E., DUDLEY, W.C. & MAHONEY, M. (1975). – Oxygen and carbon isotopes from calcareous nannofossils as paleoclimatic indicators. – *Science*, 189, 555-557.

MARTIN, E.E., SHACKLETON, N.J., ZACHOS, J.C. & FLOWER, B.P. (1999). – Orbitally-tuned Sr isotope chemostratigraphy for the lat middle to late Miocene. – *Paleoceanography*, 14(1): 74-83.

MCARTHUR, J.M., HOWARTH, R.J. & BALEY, T.R. (2001). – Strontium Isotope Stratigraphy: LOWESS Version 3: Best Fit to the Marine Sr-Isotope curve for the 0-509 Ma and Accompanying Look-up Table for Deriving Numerical Age. – *The Journal of Geology*, 109: 155-170.

MICHARD, G. (1969). – Contribution à l'étude du comportement du manganèse dans la sédimentation chimique. – Univ. Paris, 195 pp.

MILLER, K.G., WRIGHT, J.D. & FAIRBANKS, R.G. (1991). – Unlockng the ice house: Oligocene-Miocene oxygen isotopes, eustasy, and margin erosion. – *Journal of Geophysical Research*, 96 B4: 6829-6848.

MILODOWSKI, A.E. & ZALASIEWICZ, J.A. (1991). – Redistribution of rare earth elements during diagenesis of turbidite/hemipelagite mudrock sequences of Llandovery age from central Wales. – *Spec. Publ. geol. Soc. London*, 57, 101-124.

MONTANARI, A., ASARO, F., MICHEL, H.V. & KENNET, J.P. (1993). – Iridium anomalies of late Eocene age at Massignano (Italy) and ODP Site 698B (Maud Rise, Antarctic). – *Palaios*, 8, 420-437.

MORROWS, D.W. & MAYERS, J.R. (1977). – Simulation of limestone diagenesis. A model nased on strontium depletion. – *Canad. J. Earth Sci.*, 15, 376-396.

NORRIS, R.D. & RÖHL, U. (1999). – Carbon cycling and chronology of climate warming during the Palaecene/Eocene transition. – *Nature*, 401: 775-778.

ODIN, G.S., GALBRUN, B. & RENARD, M. (1994). – Physico-chemical tools in Jurassic stratigraphy. – *Geobios*, 17, 507-518.

ODIN, G.S., RENARD, M. & VERGNAUD-GRAZZINI, C. (1982). – Geochemical events as a mean of correlation. – *In*: "Numerical Dating in Stratigraphy", Odin, G.S. (ed). – John Wiley & Sons Publ., vol. 2, 37-71.

OSLICK, J.S., MILLER, K.G., FEIGENSON, M.D. & WRIGHT, J.D. (1994). – Oligocene-Miocene strontium isotopes: stratigraphic revisions and correlations to an inferred glacioeustatic record. – *Paleoceanography*, 9(3): 427-444.

OXBURGH, R. (1998). – Variations in the osmium isotope composition of sea water over the past 200,000 years. – *Earth and Planetary Science Letters*, 159, 183-191.

PAYTAN, A., KASTNER, M., CAMPBELL, D. & THIEMENS, M.H. (1998). – Sulfur isotopic composiion of Cenozoic seawater sulfate. – *Science*, 282(5393): 1459-1462.

PEGRAM, W.J., KRISHNASWAMI, S., RAVIZZA, G.E. & TURE-KIAN, K.K. (1992). The record of seawater 187Os/186Os variation through the Cenozoic. – *Earth Planet. Sci. Lett.*, 113, 569, 76.

PETERMAN, Z.E., HEDGE, C.E. & TOURTELOT, H.A. (1970). – Isotopic composition of strontium in seawater throughout Phanerozoic time. – *Geochim. cosmochim. Acta*, 34, 105-120.

PETIT, J.R., JOPUZEL, J., RAYNAUD, D., BARKOV, N.I., BARNOLA, J.-M., BASILE, I., BENDERS, M., CHAPPELLAZ, J., DAVIS, M., DELAYGUE, G., DELMOTTE, M., KOTLYAKOV, V.M., LEGRAND, M., LIPENKOV, V.Y., LORIUS, C., PÉPIN, L., RITZ, C., SALTZMAN, E. & STIEVENARD, M. (1999). – Climate and atmospheric history of the past 420,000 years from the Vostok ice core, Antartica. – *Nature*, 399: 429-436.

PEUCKER-EHRENBRINK, B., RAVIZZA, G. & HOFMANN, A.W. (1995). – The marine $^{187}Os/^{186}Os$ record of the past 80 million years. – *Earth Planet. Sci. Lett.*, 130, 155-167.

PIERRE, C., ROUCHY, J.M., LAUMONDAIS, A. & GROESSENS, E. (1984). – Sédimentologie et géochimie isotopique des sulfates évaporitiques givétiens et dinantiens du Nord de la France et de la Belgique, importance pour la stratigraphie et la reconstitution des paléomilieux de dépôts. – *C. R. Acad. Sci. Paris*, 299, 1, 21-26.

POMEROL, B. (1976). – Géochimie des craies du cap d'Antifer. – *Bull. Soc. Géol. France*, (7) XVIII, 4, 1051-1060.

POMEROL, B. (1984). – Géochimie des craies du bassin de Paris. – Thèse Univ. P.-et-M.-Curie, 540 pp.

PRATT, L.M., FORCE, E.R. & POMEROL, B. (1991). – Coupled manganese and carbon-isotopic events in marine carbonates at the Cenomanian-Turonian boundary. – *J. Sediment. Petrol.*, 61, 3, 370-383.

RABUSSIER-LOINTIER, D. (1980). – Variation de composition isotopique de l'oxygène et du carbone en milieu marin et coupures stratigraphiques du Cénéozoïque. – Thèse Univ. P.- et-M.-Curie, Paris, 182 pp.

RAVIZZA, G. (1993). – Variations of the 187Os/186Os ratio of seawater over the past 28 million years as inferred from metalliferous carbonates. – *Earth Planet. Sci. Lett.*, 118, 335-48.

RAVIZZA, G. & PEUCKER-EHRENBRINK, B. (2003a) – The marine $^{187}Os/^{188}Os$ record of the Eocene-Oligocene transition: the interplay of weathering and glaciation. – *Earth and Planetary Science Letters*, 210, 151-165.

Ravizza, G. & Peucker-Ehrenbrink, B. (2003b) – Chemostratigraphic evidence of Deccan Volcanism from the Marine Osmium Isotope Record. – *Science*, **302**, 1392-1395.

Reilly, T.J., Miller, K.G. & Feigenson, M.D. (2002). – Latest Eocene-earliest Miocene Sr isotopic reference section, Site 522, eastern South Atlantic. – *Paleoceanography*, **17**(3), 1046, doi: 10.1029/2001PA000745.

Renard, M. (1975). – Etude géochimique de la fraction carbonatée d'un faciès de bordure de dépôt gypseux (exemple du gypse du Bassin de Paris). – *Sediment. Geol.*, **13**, 201-231.

Renard, M. (1985a). – Géochimie des carbonates pélagiques: mises en évidence des fluctuations de la composition des eaux océaniques depuis 140 Ma. Essai de chimiostratigraphie. – *Doc. Bur. Rech. géol. min.*, **85**, 650 pp.

Renard, M. (1985b). – La chimiostratigraphie. – *Géochronique*, **13**, 16-20.

Renard, M. (1985c). – Méthodes chimiques de la stratigraphie. – *In*: "Stratigraphie: méthodes, principes, applications". – Doin (ed), Paris, 283 pp.

Renard, M. (1986). – Pelagic carbonate chemiostratigraphy (Sr, Mg, ^{18}O, ^{13}C). – *Marine Micropaleont.*, **10**, 1-3, 117-164.

Renard, M., Corbin, J.C. & Emmanuel, L. (1994). – Rôle des plates-formes carbonatées dans la régulation du système chimique océanique. – Réunion spécialisée de la Société Géologique de France "Géométrie et production des plates-formes carbonatées", Paris, volume des résumés, 1 p.

Renard, M. & Letolle, R. (1983). – Essai d'interprétation du rôle de la profondeur de dépôt dans la répartition des teneurs en manganèse et dans l'évolution du rapport isotopique du carbone des carbonates pélagiques: influence de l'oxygénation du milieu. – *C.R. Acad. Sci. Paris*, **296**, 1739-1740.

Richter, F.M. & Liang, Y. (1993). – The rate and consequences of strontium diagenesis in deep sea carbonates. – *Earth and Planet. Sci. Lett.*, **117**, 553-565.

Richter, F.M., Rowley, D.B. & De Paolo, D.J. (1992). – Sr isotope evolution of seawater: the role of tectonics. – *Earth and Planet. Sci. Lett.*, **109**, 11-23.

Robinson, L.F., Henderson, G.M. & Slowey, N.C. (2002). – U-Th dating of marine isotope stage 7 in Bahamas slope sediments. – *Earth and Planetry Science Letters*, **196** (3-4): 175-187.

Rocchia, R., Boclet, D., Bonte, P., Froget, L., Galbrun, B., Jehano, C. & Robin, E. (1992). – Iridium and others elements distribution, mineralogy and magnetostratigraphy near the Cretaceous/Tertiary boundary in Hole 761 C. – *In*: Von Rad U. H.B.U.e.a. (ed.): "*Proc. O.D.P. scientific results*", College Station TX, **122**, 753-762.

Rocchia, R., Boclet, D., Bonte, Ph., Devineau, J., Jehanno, C. & Renard, M. (1987). – Comparaison des distributions de l'Iridium observées à la limite Crétacé/Tertiaire dans divers sites européens. *In*: (ed), in "Extinctions dans l'histoire des Vertébrés". – *Mém. Soc. géol. France*, Paris, **150**, 95-103.

Rocchia, R., Boclet, D., Bonte, P., Castellarin, A. & Jehanno, C. (1986). – An iridium anomaliy in the Middle-Lower Jurassic of the Venetian Region (Northern Italy). – *J. Geophys. Res*, **91**, 13, 259-262.

Röhl, U., Wefer, G., Bralower, T.J. & Norris, R.D. (2000). – New chronology for the last Paleocene thermal maximum and its environmental implications. – *Geology*, **28**(10): 927-930.

Saito, T., Burckle, L.H. & Hays, J.D. (1975). – Late Miocene to pleistocene biostratigraphy of equatorial Pacific sediments. – *Micropaleontology spec. Publ.*, **1**, 226-244.

Scholle, P.A. & Arthur, M.A. (1980). – Carbon isotope fluctuations in Cretaceous pelagic limestones: potential stratigraphic and petroleum exploration tool. – *Bull. amer. Assoc. Petroleum. Geol.*, **64**, 67-87.

Shackleton, N.J. (1987). – The Carbon isotope record of the Cenozoic: history of organic carbon burial and of oxygen in the ocean and atmosphere. – *In*: Brooks J. & Fleet A. (eds.): Marine and Petroleum Source Rocks, Blackwell, *London Geological Society*, Special Publication, 423-434.

Shackleton, N.J. & Hall, A. (1984). – Carbon isotope data from Leg 74 sediments. – *In*: Moore T.C. Jr. et al. (eds.). – *Init. Report. Deep Sea Drill. Proj.*, Washington, **74**, 613-644.

Shackleton, N.J., Hall, M.A., Pate, D., Meynadier, L. & Valet, P. (1993). – High resolution stable isotope stratigraphy from bulk sediment. – *Paleoceanography*, **8**(2): 141-148.

Shackleton, N.J. & Kennett, J.P. (1975). – Paleotemperature history and the initiation of antarctic glaciation: oxygen and carbon isotope analysis. – *In*: DSDP site 277, 279, 280, Kennett, J.P. et al. (eds). – *Init. Report. Deep Sea Drill. Proj.*, Washigton, **29**, 743-755.

Shackleton, N.J. & Opdyke, N.D. (1973). – Oxygen isotope and paleomagnetic stratigraphy of Equatorial Pacific core V28-238. – *Quat. Res*, **3**, 39-55.

Sinha, A., Aubry, M.P., Stott, L.D., Thiry, M. & Berggren, W.A. (1995). – Chemostratigraphy of the "Lower" Sparnacian deposits (Argiles plastiques bariolées) of the Paris Basin. – *J. Earth Sci.* Israel, spec Vol. Paleocene/Eocene Boundary Events.

Stuiver, M. & Grootes, P.M. (2000). – GISP2 oxygen isotope ratios. – *Quaternary Research*, **53**(3): 277-284.

Tiedemann, R., Sarnthein, M. & Shackleton, N.J. (1994). – Astronomic timescale for the Plicene Atlantic d^{18}O and dust flux records of Ocean Drilling Program Site 659. – *Paleoceanography*, **9**(4): 619-638.

Tripati, A.K. & Elderfield, H. (2004). – Abrupt hydrography changes in the equatorial Pacific and subtropical atlantic from foraminiferal Mg/Ca indicate greenhouse origin for the thermal maximum at the Paleocene-Eocene boundary. – *Geochemistry Geophysics Geosystems*, **5**, Art. No. Q02006.

Tudhope, A.W., Chilcott, C.P., McCulloch, M.T., Cook, E.R., Chappell, J., Ellam, R.M., Lea, D.W., Lough, J.M. & Shimmield, G.B. (2001). – Variability in the El Niño-Southern Oscillation through a glacial-interglacial cycle. *Science*, **291**(5508): 1511-1517.

Turpin, L., Clauser, S., Rocchia, R., Renard, M. & Boclet, D. (1988). – ^{87}Sr/^{86}Sr and ^{143}Nd/^{144}Nd variations accross the K/T boundary at Bidart, Stevns Klint and Raton Basin. Intern. – Congress of Geochemistry and Cosmochemistry, Paris (29 Août-2 Sept. 1988), Abstr., 1 page.

Ulicny, D., Hladikova, J. & Hradecka, L. (1993). – Record of sea-level changes, oxygen depletion and d13C anomaly across the Cenomanian-Turonian boundary, Bohemian Cretaceous Basin. – *Cretaceous Res.*, **14**, 211-234.

Veizer, J. (1977). – Diagenesis of pre-quaternary carbonates as indicated by tracer studies. – *J. Sedim. Petrol.*, **47**, 2, 565-581.

Veizer, J., Holser, T. & Wilgus, C.K. (1980). – Correlation of 13C/12C and 34S/32S secular variations. – *Geochim. cosmochim. Acta*, **44**, 4, 579-588.

Vergnaud-Grazzini, C. & Oberhaensli, H. (1986). – Isotopic events at the Eocene/Oligocene boundary transition. – *In*: Pomerol, C. & Premoli-Silva, I. (eds.), "Developments in

Palaeontology and Stratigraphy". – *Elsevier Sci.*, **9**, 311-329.

VOIGT, S. & HILBRECHT, H. (1997). – Late Cretaceous carbon isotope stratigraphy in Europe: correlation and relations with sea level and sediment stability. – *Palaeogeography, Palaeoclimatology, Palaeoecology*, **134**: 39-59.

VOIGT, S. & WIESE, F. (2000). – Evidence for Late Cretaceous (Late Turonian) climate cooling from oxygen-isotope variations and palaeobiogeographic changes in Western and Central Europe. – *Journal of the Geological Society*, London, **157**: 737-743.

WANG, L. & OBA, T. (1998). – Tele-connections between East Asian Monsoon and the high-latitude climate: a comparison between the GISP 2 Ice Core record and the high resolution marine records from the Japan and the South China Sea. – *The Quaternary Research*, **37**(3): 211-219.

WANG, L., SARNTHEIN, M., ERLENKEUSER, H., GRIMALT, J., GROOTES, P., HEILIG, S., IVANOVA, E., KIENAST, M. & PFLAUMANN, U. (1999). – East Asian Monsoon climate during the Late Pleistocene: high-resolution sediment records from the South China Sea. – *Marine Geology*, **156**: 245-284.

WANG, P.X., ZHAO, Q.H., JIAN, Z.M., CHENG, X.R., HUONG, W., TIAN, J., WANG, J.L., LI, Q.Y., LI, B.H. & SU, X., (2003). – Thirty million year deep-sea records in the South China Sea. – *Chinese Science Bulletin*, **48**(23): 2524-2535.

WEISSERT, H. (1989). – C-Isotopes stratigraphy, a monitor of paleoenvironmental change: a case study from the Early Cretaceous. – *Surv. Geophys.*, **10**, 1-61.

WEISSERT, H. & CHANNELL, J.E.T. (1989). – Tethyan carbonate carbon isotope stratigraphy across the Jurassic-Cretaceous boundary: an indicator of decelerated global carbon cycling? – *Paleoceanography*, **4**, 4, 483-494.

WEISSERT, H. & ERBA, E. (2004). – Volcanism, CO_2, and palaeoclimate: a Late Jurassic-Early Cretaceous carbon and oxygen isotope record. – *Journal of the Geological Society*, London, **161**: 1-8.

Weissert, H. & LINI, A. (1991). – Ice age interludes during the time of Cretaceous greenhouse climate? – *In*: Müller *et al.* (eds.): "Controversies in modern Geology". – *Acad. Press*, 173-191.

WEISSERT, H., MC KENZIE, J. & HOCHULI, P. (1978). – Cyclic anoxic events in the Early Cretaceous Tethys Ocean. – *Geology*, **7**, 147-151.

WILLIAMS, D.F., LERCHE, I. & FULL, W.E. (1988). – Isotope Chronostratigraphy. Theory and methods. – San Diego Academic Press, Geology Series, 345 pp.

WRAY, D.S. (1995). – Origine of clay-rich beds in Turonian chalks from Lower Saxony, Germany; a rare-earth element study. – *Chem. Geol.*, **119**, 161-173.

ZACHOS, J.C., PAGANI, M., SLOAN, L., THOMAS, E. & BILLUPS, K. (2001). – Trends, rhythms, and aberrations in global climate 65 Ma to present. – *Science*, **292**: 686-693.

ZACHOS, J.C., QUINN, T.M. & SALAMY, K.A. (1996). – High-resolution (10^4 years) deep-sea foraminiferal stable isotope records.

Chapter 4. – MAGNETOSTRATIGRAPHY

ALVAREZ, W., ARTHUR, M.A., FISCHER, A.G., LOWRIE, W., NAPOLEONE, G., PREMOLI-SILVA, I. & ROGGENTHEN, W.M. (1977). – Upper Cretaceous-Paleocene magnetic stratigraphy at Gubbio, Italy. V. Type section for the late Cretaceous-Paleocene geomagnetic reversal time scale. – *Geological Society of America Bulleting*, v. 88, p. 383-389.

BAKSI, A.K. (1993). – A geomagnetic polarity time scale for the period 0-17 Ma, based on Ar plateau ages for selected field reversals. – *Geophys. Res. Lett.*, **20**, 1607-1610.

BAKSI, A.K., V. HSU, M.O. MCWILLIAMS & FARRAR, E. (1992). – Ar dating of the Brunhes-Matuyama geomagnetic field reversal. – *Science*, **256**, 356-357.

BRUNHES, B. (1906). – Recherches sur la direction d'aimentation des roches volcaniques (1). – *J. Physique*, 4^e ser., **5**, 705-724.

BUTLER, R.F. (1992). – Paleomagnetism. – Blackwell Scientific Publ., 319 pp.

CANDE, S.C. & KENT, D.V. (1992a). – Ultrahigh resolution marine magnetic anomaly profiles: A record of continuous paleointensity variations? – *Journal of Geophysical Research*, v. 97, p. 15,075-15,083.

CANDE, S.C. & KENT, D.V. (1992b). – A new geomagnetic polarity time scale for the Late Cretaceous and Cenozoic. – *J. Geophys. Res.*, **97**, B1O, 13917-13951.

CANDE, S.C. & KENT, D.V. (1995). – Revised calibration of the geomagnetic polarity time scale for the Late Cretaceous and Cenozoic. – *Journal of Geophysical Research*, v. 100, p. 6093-6095.

CANDE, S.C., LARSON, R.L. & LABRECQUE, J.L. (1978). – Magnetic lineations in the Pacific Jurassic Quiet Zone. – *Earth and Planetary Science Letters*, v. 41, p. 434-440.

COLLINSON, D.W. (1983). – Methods in Rock Magnetism and Paleomagnetism. – Chapman and Hall, London, 503 pp.

EVANS, M.E. (1976). – Test of the dipolar nature of the geomagnetic field throughout Phanerozoic time. – *Nature*, v. 262, p. 676-677.

FISHER, R.A. (1953). – Dispersion on a sphere. – Proceedings of the Royal Society of London, v. A217, p. 295-305.

GALBRUN, B. (1993). – Contribution à la connaissance de l'échelle magnétostratigraphique du Mésozoïque. – Habilitation diriger Rech., Université Paris VI, 335 pp.

GALBRUN, B., FEIST, M., COLOMBO, F., ROCCHIA, R. & TAMBAREAU, Y. (1993). – Magnetostratigraphy and biostratigraphy of Cretaceous-Tertiary continental deposits, Ager Basin, Province of Lerida, Spain. – *Palaeogeogr., Palaeoclimatol., Palaeoecol.*, **102**, 41-52.

HAILWOOD, E.A. (1989a). – Magnetostratigraphy. – *Geol. Soc. spec. Rep., Blackwell Sci. Publ.*, Oxford, **19**, 84 pp.

HAILWOOD, E.A. (1989b). – The role of magnetostratigraphy in the development of geological time scales. – *Paleoceanography*, **4**, 1, 1-18.

HARLAND, W.B., AMSTRONG, R.L., COX, A.V., SMITH, A.G. & SMITH, D.G. (1990). – A geologic time scale 1989. – Cambridge Univ. Press, 263 pp.

HARLAND, W.B., COX, A.V., LLEWELLYN, P.G., PICKTON, C.A.G., SMITH, A.G. & WALTERS, R. (1982). – A geological time scale. – Cambridge University Press, 131 p.

HEIRTZLER, J.R., DICKSON, G.O., HERRON, E.M., PITMAN, W.C. & LE PICHON, X. (1968). – Marine magnetic anomalies, geomagnetic field reversals, and motions of the ocean floor and continents. – *Journal of Geophysical Research*, v. 73, p. 2119-2136.

HILGEN, F. J. (1991). – Extension of the astronomically calibrated (polarity) time scale to the Miocene/Pliocene boundary. – *Earth Planet. Sci. Lett.*, **107**, 349-366.

KENT, D.V. (1999). – Orbital tuning of geomagnetic polarity time-scales. – *Philosophical Transactions of the Royal Society of London*, Series A, **357**, p. 1995-2007.

KENT, D.V. & OLSEN, P.E., (1999). – Astronomically tuned geomagnetic polarity time scale for the Late Triassic: *Journal of Geophysical Research*, v. 104, p. 12,831-12,841.

KENT, D.V., OLSEN, P.E. & WITTE, W.K. (1995b). – Late Triassic-earliest Jurassic geomagnetic polarity reference sequence from cyclic continental sediments of the Newark Rift Basin (eastern North America). – *Albertiana*, v. 16, p. 17-26.

KIRSCHVINK, J.L. (1980). – The least-square lines and plane and the analysis of paleomagnetic data. – *Geophys. J. R. Astron. Soc.*, **62**, 699-718.

LABRECQUE, J.L., HSU, K.J., CARMAN, M.F.J., KARPOFF, A.M., MCKENZIE, J.A., PERCIVAL, S.F.J., PETERSEN, N.P., PISCIOTTO, K.A., SCHREIBER, E., TAUXE, L., TUCKER, P., WEISSERT, H.J. & WRIGHT, R. (1983). – DSDP Leg 73, Contributions to Paleogene stratigraphy in nomenclature, chronology and sedimentation rates. – *Paleogeogr., Paleoclimatol., Paleoecol.*, **42**.

LABRECQUE, J.L., KENT, D.V. & CANDE, S.C. (1977). – Revised magnetic polarity time scale for the Late Cretaceous and Cenozoic time. – *Geology*, v. 5, p. 330-335.

LANCI, L. & LOWRIE, W. (1997). – Magnetostratigraphic evidence that "tiny wiggles" in the oceanic magnetic anomaliy record represent geomagnetic paleointensity variations. – *Earth and Planetary Science Letters*, v. 148, p. 581-592.

LANCI, L., LOWRIE, W. & MONTANARI, A. (1996). – Magnetostratigraphy of the Eocene/Oligocene boundary in a short drill-core. – *Earth and Planetary Science Letters*, v. 143, p. 37-48.

LARSON, R.L. & HILDE, T.W.C. (1975). – A revised time scale of magnetic reversals for the Early Cretaceous and Late Jurassic. – *Journal of Geophysical Research*, v. 80, p. 2586-2594.

LARSON, R.L. & PITMAN, W.C. (1972). – Worldwide correlation of Mesozoic magnetic anomalies, and its implications. – *Geological Society of America Bulletin*, v. 83, p. 3645-3662.

LOWRIE, W. (1990). – dentification of ferromagnetic minerals in a rock by coercivity and unblocking temperature properties. – *Geophysical Research Letters*, v. 17, p. 159-162.

MATUYAMA, MOTONORI (1929). – On the Direction of Magnetisation of Basalt in Japan, Tyosen and Manchuria, Japan Academy Proceedings 5, 203-5 (Reprinted 1973 by Cox Allan, ed., Plate Tectonics and Geomagnetic Reversals. – W.H. Freeman and Co., 702 pp., San Francisco, 1973.), p. 154-156.

MCELHINNY, M.W. (1964). – Statistical significance of the fold test in palaeomagnetism. – *Geophysical Journal of the Royal Astronomical Society*, v. 8, p. 338-340.

MCFADDEN, P.L. (1990). – A new fold test for palaeomagnetic studies. – *Geophysical Journal International*, v. 103, p. 163-169.

MCFADDEN, P.L. & JONES, D.L. (1981). – The fold test in palaeomagnetism. – *Geophysical Journal of the Royal Astronomical Society*, v. 67, p. 53-58.

MCFADDEN, P.L. & MCELHINNY, M.W. (1990). – Classification of the reversal test in palaeomagnetism. – *Geophysical Journal International*, v. 103, p. 725-729.

MERRILL, R.T. & MCFADDEN, P.L. (1999). – Geomagnetic polarity transitions. – *Reviews of Geophysics*, v. 37, p. 201-226.

NESS, G., LEVI, S. & COUCH, R. (1980). – Marine magnetic anomaly timescales for the Cenozoic & Late Cretaceous: A precis, critique, and synthesis. – *Reviews of Geophysics and Space Physics*, v. 18, p. 753-770.

OLSEN, P.E. & KENT, D.V. (1990). – Continental coring of the Newark Rift: Eos. – *Transactions of the American Geophysical Union*, v. 71, p. 385.

OLSEN, P.E. & KENT, D.V. (1999). – Long-period Milankovitch cycles from the Late Triassic and Early Jurassic of eastern North America and their implications for the calibration of the Early Mesozoic time-scale and the long-term behaviour of the planets. – *Philosophical Transactions of the Royal Society of London*, Series A., **357**, p. 1761-1786.

OPDYKE, N.D. & CHANNELL, J.E.T. (1996). – Magnetic Stratigraphy. – Academic Press, 346 p.

PITMAN, W.C., HERRON, E.M. & HEIRTZLER, J.R. (1968). – Magnetic anomalies in the Pacific and sea floor spreading. – *Journal of Geophysical Research*, v. 73, p. 2069-2085.

SCHNEIDER, D.A. & KENT, D.V. (1990). – Testing models of the Tertiary paleomagnetic field. – *Earth and Planetary Science Letters*, v. 101, p. 260-271.

SHACKLETON, N.J., BERGER, A. & PELTIER, W.R. (1990). – An alternative astronomical calibration of the lower Pleistocene timescale based on ODP Site 677. – *Trans. Royal Soc. of Edinburgh: Earth Sciences*, **81**, 251-261.

SPELL, T.L. & MCDOUGALL, I. (1992). – Revisions to the age of the Brunhes-Matuyama boundary and the Pleistocene geomagnetic polarity timescale. – *Geophys. Res. Lett.*, **19**, 1181-1184.

TAUXE, L. (1998). – Paleomagnetic Principles and Practice. – Kluwer Academic Publishers.

TAUXE, L. & WATSON, G.S. (1994). – The fold test: an eigen analysis approach. – *Earth and Planetary Science Letters*, v. 122, p. 331-341.

ZIJDERVELD, J.D.A. (1967). – A.C. Demagnetization of rocks: Analysis of results. – In: Collinson, D.W. *et al.* (eds.), Methods in paleomagnetism. – Elsevier, Amsterdam, 254-286.

Chapter 5. – BIOSTRATIGRAPHY

AGTERBERG, F.P. & NEL, L.D. (1982a). – Algorithms for the ranking of stratigraphic events. – *Comput. and Geosci.*, **8**, 1, 69-90.

AGTERBERG, F.P. & NEL, L.D. (1982b). – Algorithms for the scaling of stratigraphic events. – *Comput. and Geosci.*, **8**, 1, 163-189.

ALMERAS, Y., BOULLIER, A. & LAURIN, B. (1990). – Les zones de Brachiopodes du Jurassique en France. – *Ann. sci. Univ. Franche-Comté*, **4**, 10, 3-30.

ASIOLI A., TRINCARDI F., LOWE J.J. & OLDFIELD F. (1999). – Short-term climate changes during the Last Glacial – Holocene transition: comparison between Mediterranean records and the GRIP event stratigraphy. *Journal of Quaternary Science*, **14**(4): 373 – 381.

BANDY, O.L. (1963). – Cenozoic planktonic foraminiferal zonation and basinal development in the Philippines. *American Association of Petroleum Geologists Bulletin*, **47**(9), 1733–1745.

BANDY, O.L. (1964). – Cenozoic planktonic foraminiferal zonation. *Micropaleontology*, **10**(1), 1–17.

BEAUDOIN, B. & CARON, M. (1995). – Distribution of plantonic foraminifera around the Cenonian-Turonian boundary in some key sections, and the meaning of the "Archeocretacea zone". – *In*: "Symposium on the Cretaceous stage boundaries", Bruxelles, abstracts, 24-25.

BERGGRENN, W.A., KENT, D.V., SWISCHER, III, C.C. & AUBRY, M.P. (1995). – A revised cenozoic geochronology and chronostratigraphy. Geochronology, Time Scales and Global Stratigraphic Correlation. – *Spec. Publ. Soc. econ. Paleont. Mineral*, **54**, 129-212.

BERGGREN, W.A. & VAN COUVERING, J.A. (1974). – The late Neogene. Amsterdam: Elsevier Scientific Publishing Company.

BERGGREN, W.A. & VAN COUVERING, J.A. (1974). – Biochronology. *In*: Cohee, G.V., Glaessner, M.F. & Hedberg, H.D. eds. Contributions to the Geologic Time Scale. *Am. Ass. Petrol. Geol.*, Tulsa, Studies in geology, **6**, 9-32.

BLANK, R.G. (1979). – Applications of probabilistic biostratigraphy to chronostratigraphy. – *J. Geol.*, **87**, 647-670.

BLANK, R.G. & ELLIS, C.H. (1982). – The probable range concept applied to biostratigraphy of marine microfossils. – *J. Geol.*, **90**, 4, 415-533.

BOULARD, C. (1993). – Biochronologie quantitative: concepts, méthodes et validité. – *Doc. Lab. Géol. Fac. Sci. Lyon*, **128**, 259 pp.

BRACK, P. & RIEBER, H. (1993). – Towards a better definition of the Anisian-Ladinian boundary: new biostratigraphic data and correlations of boundary sections from the Southern Alps. – *Eclogae geol. Helv.*, **86**, 2, 415-527.

BROWER, J.C. (1984). – The relative biostratigraphic values of fossils. – *Comput. and Geosci.*, **10**, (1), 111-131.

BROWER, J.C. (1985). – The fossil concept and its application to quantitative biostratigraphy. – *In*: Gradstein, F.M. et al. "Quantitative stratigraphy". – Reidel, D. Publishing Co., 43-64.

BROWER, J.C., MILLENDORF, S.A. & DYMAN, T.S. (1978). – Quantification of assemblage zones based on multivariate analysis of weighted and unweighted data. – *Comput. and Geosci.*, **4**, 3, 221-227.

BUCKMAN, S.S. (1893). – The Bajocian of the Sherborne district: its relation to subjacent and superjacent strata. – *Quart. J. geol. Soc. London*, **49**, 196, 479-522.

BUCKMAN, S.S. (1898). – On the grouping of some Divisions of the so-called "Jurassic" Time. – *Quart. J. geol. Soc. London*, **54**, 215, 442-451.

BUCKMAN, S.S. (1910). – Certain Jurassic ("Inferior Oolite") species of ammonites and brachiopods. – *Quart. J. geol. Soc. London*, **66**, 90-108

CALLOMON, J.H. (1984a). – Biostratigraphy, chronostratigraphy and all that again! – *In*: Michelsen, O. & Zeiss, A. (eds.). – *Geol. Surv. Denmark*, **3**, 611-624.

CALLOMON, J.H. (1984b). – The measurement of Geological Time. – *Proc. r. Inst.* (Great Britain), **56**, 65-99.

Callomon, J.H. (1994). – Paleontological methods of Stratigraphy and Biochronology. Some introductory remarks. – *In*: 3e Symposium International de Stratigraphie du Jurassique, Poitiers, France. – *Geobios*, M.S., **17**, 1, 16-30.

CARBONNEL, G. (1986). – Ostracodes tertiaires (Paléocène à Néogène) du bassin sénégalo-guinéen. – *Bull. Bur. Rech. geol. min.*, **101**, 33-243.

CAPOTONDI, L., BORSETTI, A.M. & MORIGI, C. (1999). – Foraminiferal ecozones, a high resolution proxy for the late Quaternary biochronology in the central Mediterranean Sea, *Marine Geology*, **153**, 253-274

CARON, M., BEAUDOIN, B. TERRAB, S. & M'BAN, E. (1993). – Biozones may be diachronous: example of the Cenomanian-Turonian anoxic event and its impact on planctonic foraminifera evolution. – *In*: 1er Congrès Européen de Paléontologie, Lyon, 7-9 juillet 1993. – Rés., 27 pp.

CAVELIER, C. (1972). – L'âge Priabonien supérieur de la "zone à Ericsonia subdisticha" (Nannoplancton) en Italie et l'attribution des Latdorf Schichten allemands à l'Eocene supérieur. – *Bull. Bur. Rech. géol. min.*, 2e sér., Section IV, **1**, 15-24.

COPE, J.C.W. (1993). – High resolution biostratigraphy. – In: Hailwood, E.A. & Kidd, R.B. (eds.) "High Resolution Stratigraphy". – *Geol. Soc. Spec. Publ.*, **70**, 257-265.

COOPER R.A. (1999). – Ecostratigraphy, zonation and global correlation of earliest Ordovician planktic graptolites. – *Lethaia*, *32* (1): 1-16

COWIE, J.W., ZIEGLER, W., BOUCOT, A.J., BASSETT, M.G. & REMANE, J. (1986). – Guidelines and statuses of the International Commission on Stratigraphy. – *Courr. Fortsch. Senkenb.*, **83**, 1-14.

CUGNY, P. (1988). – Modèles paléoécologiques. Analyse quantitative des faciès dans diverses formations crétacés des marges néotéthysiennes et atlantiques; associations paléontologiques et paléoenvironnements. – *Strata*, sér. 2, **10**, 331 pp.

DE JEKHOWSKY, B. (1963). – La méthode des distances minimales, nouveau procédé quantitatif de corrélation stratigraphique; exemple d'application en palynologie. – *Rev. Inst. Franç. Pétrole*, **18**, 629-653.

DE WEVER, P. (1982). – Radiolaires du Trias et du Lias de la Téthys. Systématique, Stratigraphie. – *Mém. Soc. Géol. Nord*, **7**, 599 pp.

DE WEVER, P., GEYSSANT, J., AZEMA, J., DEVOS, I., DUEE, G. MANIVIT, H. & VRIELYNCK B. (1986). – La coupe de Santa Anna (Zone de Sciacca, Sicile): une synthèse biostratigraphique des apports des macro-, micro- et nannofossiles du Jurassique supérieur et Crétacé inférieur. – *Rev. Micropal.*, **29**, 3, 141-186.

DEBRENNE, F. & ZHURAVLEV, A. (1992). – Irregular Archaeocyaths. Morphology, Ontogeny, Systemetics, Biostratigraphy, Palaeoecology. – *Cah. Paléont.*, 212 pp.

DEBRENNE, F., ROZANOV A. & ZHURAVLEV A. (1990). – Archéocyathes réguliers. Morphologie, Systématique, Biostratigraphie, Paléogéographie, Affinités biologiques. – *Cah. Paléont.*, 218 pp.

DOMMERGUES, J.L. & MEISTER, C. (1987). – La biostratigraphie des ammonites du Carixien (Jurassique inférieur) d'Europe occidentale: un test de la méthode des associations unitaires. – *Eclogae geol. Helv.*, **80**, 3, 919-938.

DOMMERGUES, J.L. & MEISTER, C. (1991). – Fréquences de présence des horizons et mise en évidence de fluctuations de l'environnement; exemple dans le Lias nord-ouest européen. – *C. R. Acad. Sci.*, **313**, sér. II, 977-982.

EDWARDS, L.E. (1979). – Range charts and no-space graphs. – *Comput. and Geosci.*, **4**, 247-255.

EDWARDS, L.E. (1989). – Supplemented Graphic Correlation: A Powerful Tool for Paleontologists and nonpaleontologists. – *Palaios*, **4**, 127-143.

FENTON, C.L. & FENTON, M.A. (1928). – Ecological interpretation of some biostratigraphic terms. – *Amer. Midl. Natur.*, **11**, 1-40.

FISCHER, H. & GYGI, R.A. (1989). – Numerical and biochronological time scales correlated at the ammonite subzone level; K-Ar, Rb-Sr ages and Sr, Nd, and Pb seawater isotopes *in*: an Oxfordian (Late Jurassic) succession of northern Switzerland. – *Bull. Geol. Soc. Amer.*, **101**, 584-1597.

GABILLY, J. (1976). – Le Toarcien à Thouars et dans le Centre-Ouest de la France. Biostratigraphie-Evolution de la faune. – Comité Français de Stratigraphie, C.N.R.S. (ed.): Les stratotypes français, **3**, 217 pp.

GARCIA, J.P. & LAURIN, B. (1996). – Les associations de Brachiopodes du Jurassique moyen du Bassin de Paris: une échelle biochronologique ponctuée de niveaux-repères pour la contrainte des corrélations séquentielles à haute résolution. – *Bull. Soc. géol. France*, **167** (3): 435-451.

GOURINARD, Y. (1983). – Quelques vitesses d'évolution observées dans des lignées de Foraminifères néogènes. Utilisations chronologiques. – *C. R. Acad. Sci. Paris*, **297**, 269-272.

GOURINARD, Y. (1984). – Géochronologie et corrélation des séries sédimentaires. – *Géochronique*, **11**, 7.

GRADSTEIN, F.M. & AGTERBERG, F.P. (1999). – RASC and CASC – Tools for the Biostratigraphic Workstation – Users Guide. – Authors Edition, Version 17.

GRADSTEIN, F.M., KRISTIANSEN, I.L., LOEMO, L. & KAMINSKI, M.A. (1992). – Cenozoic foraminiferal and dinoflagellate cyst biostratigraphy of the central North Sea. – *Micropaleontology*, **38**, 2, 101-137.

GROUPE FRANÇAIS D'ETUDES DU JURASSIQUE (1991). – Mise à jour des zones d'Ammonites du Jurassique français. – *In*: 3rd International Symposium on Jurassic Stratigraphy, I.U.G.S. International Subcommission on Jurassic Stratigraphy. Poitiers, September 22-24, 1991. – Rés., 124-134.

GUEX, J. (1977). – Une nouvelle méthode d'analyse biochronologique. Note préliminaire. – *Bull. Lab. Géol. Univ. Lausanne*, **224**, 309-322.

GUEX, J. (1979). – Terminologie et méthodes de la stratigraphie moderne: commentaires critiques et propositions. – *Bull. Lab. Géol. Univ. Lausanne*, 355, **74**, 3, 169-216.

GUEX, J. (1987). – Corrélations biochronologiques et associations unitaires. – Presses Polytechniques Romandes Cité Universitaire, Centre midi, Lausanne, 241 pp.

GUEX, J. (1991). – Biochronological Correlations. – Springer-Verlag., Berlin. 252 pp.

HANTZPERGUE, P. (1993). – Le seuil de résolution du temps est-il atteint par les ammonites au Mésozoïque. – *In*: Gayet, M. (ed.): "Paléobiochronologie en domaines marin et continental". Réunion scientifique de l'Association paléontologique Française, Poitiers, 22 Octobre 1993. – Paleovox, 2, 31-42.

HART, M.B. (1993). – Cretaceous foraminiferal events. – In: Hailwood, E.A. & Kidd, R.B. (eds.): "High Resolution Stratigraphy". – *Geol. Soc., Spec. Publ.*, **70**, 227-240.

HAY, W.W. (1972). – Probabilistic stratigraphy. – *Eclogae geol. Helv.*, **65**, 2, 255-266.

HEDBERG, H.D. (1976). – International Stratigraphic Guide. A guide to stratigraphic classification, terminology and procedure. – Wiley & Sons (eds.), New-York, 200 pp.

HELLMANN, K.N. & LIPPOLT, H.J. (1981). – Calibration of the Middle Triassic Time Scale by Conventional K-Ar and ^{40}Ar/^{39}Ar Dating of Alkali Felspars. – *J. Geophys.*, **50**, 73-88.

HOHN, M.E. (1978). – Stratigraphic correlation by principal components: effects of missing data. – *J. Geol.*, **86**, 24-532.

HOHN, M.E. (1982). – Properties of composite sections constructud by least squares. – *In*: Cubitt, J.M. & Reyment, R.A. (eds.): "Quantitative stratigraphic correlation" – John Wiley & Sons (eds).

JORISSEN, F.J., ASIOLI, A., BORSETTI, A.M., CAPOTONDI, L., DE VISSER, J.P., HILGEN, F.J., ROHLING, E.J., VAN DER BORG, K., VERGNAUD-GRAZZINI, C. & ZACHARIASSE, W.J. (1993). – Late Quaternary central Mediterranean biochronology. – *Marine Micropaleontology*, **21**, 169-189.

KEMPER, E. & ZIMMERLEE, W. (1978). – Der Grenz-Tuff Apt/Alb von Vöhrun. – *Geol. J., Ser. A*, **45**, 125-143.

LAHM, B. (1984). – Spumellarien faunen (Radiolaria) aus den Mitteltriassischen Buchensteiner-Schichten von Recoaro (Norditalien) und den Obertriassischen Reiflingerkalken von Grossreiflieng (Osterreich) Systematik, stratigraphie. Münch. – Geowissen. – Abhandl. Reihe A, Geol. & Paläontol., 1 161 pp.

LEGRAND, P. (1996). – Étages et zones de graptolithes: définition et validité; l'exemple du Silurien inférieur. – *Bull. Soc. géol. France*, **167**, 1, 29-38.

LETHIERS, F. (1984). – Zonation du Dévonien supérieur par les Ostracodes (Ardenne et Boulonnais). – *Rev. Micropaléont.*, **27**, 1, 30-42.

LETHIERS, F. (1987). – Quelques aspects actuels de la biostratigraphie. – *Bull. Inf. Géol. Bass. Paris*, **24**, 2, 7-17.

LETHIERS, F. (1993). – La place des Ostracodes dans la biochronologie du Paléozoïque supérieur. – *In*: Gayet, M. (ed.) "Paléobiochronologie en domaines marin et continental". Réunion scientifique de l'Association paléontologique Française, Poitiers, 22 Octobre 1993. – *Paleovox*, **2**, 5-27.

LIPSON-BENITAH, S. (1992). – The retrozone: a new kind of zone for subsurface biostratigraphy and its application to Late Aptian-Early Albian succession, Israel. Isr. – *J. Earth Sci.*, **40**, 47-50.

MAGNIEZ-JANNIN, F. (1995). – Cretaceous stratigraphic scales based on benthic foraminifera in West European basins (biochronohorizons). – *Bull. Soc. géol. France*, **166**, 5, 565-572.

MC ARTHUR, KENNEDY, W.J., CHEN, M., THIRLWALL, M.F. & GALE, A.S. (1994). – Strontium isotope stratigraphy for Late Cretaceous time: Direct numerical calibration of the Sr isotope curve based on the US Western Interior. – *Palaeogeogr. Palaeoclimatol. Palaeoecol.*, **108**, 1/2, 95-119.

MC CAMMON, R.B. (1970). – On estimating the relative biostratigraphic value of fossils. – *Bull. Geol. Inst. Univ.*, Upsala, **2**, 49-57.

MILLENDORF, S.A. & HEFFNER, T. (1978). – FORTRAN program for lateral tracing of time stratigraphic units based on faunal assemblage zones. – *Comput. and Geosci.*, **4**, (3), 313-318.

MILLENDORF, S.A. & MILLENDORF, M.T. (1982). – The conceptual basis for lateral tracing of biostratigraphic units. – *In*: Cubitt, J.M. & Reyment, R.A. (eds.), "Quantitative stratigraphic correlation". – John Wiley & Sons 101-106.

MONTANARI, A., ASARO, F., MICHEL, H.V. & KENNETT, J.P. (1993). – Iridium Anomalies of Late Eocene Age at Massignano (Italy) and ODP Site 689B (Maud Rise, Antarctic). – *Palaios*, **8**, 420-437.

MOULLADE, M. (1974). – Zones de Foraminifères du Crétacé inférieur mésogéen. – *C. R. Acad. Sci. Paris*, **278**, sér. D, 1813-1816.

MOULLADE, M. (1984). – Intérêt des petits Foraminifères benthiques "profonds" pour la biostratigraphie et l'analyse des paléoenvironnements océaniques mésozoïques. – *In*: Oertli, H.J. (ed.): "Benthos' 83". 2nd Int. Symp. on Bentic Foraminifera (Pau, April 1983). – Elf Aquitaine, Esso REP and Total CFP. Pau and Bordeaux, 1984. – *Bull. Centres Rech. Explor.-Prod. Elf-Aquitaine*, Mém. **6**, 429-464.

NEAL, J.E., STEIN, J.A. & GAMBER, J.H. (1994). – Graphic correlation and sequence stratigraphy in the Palaeogene of NW Europe. – *Micropalaeontology*, **13**, 55-80.

OBRADOVICH, J.D. (1993). – A Cretaceous Time Scale. – *In*: W.G.E. Caldwell (ed.), Evolution of the Western Interior Foreland basin. – *Spec. Pap. Geol. Assoc. Canad.*, **39**, 379-396

ODIN, G.S. & ODIN, C. (1990). – Échelle numérique des temps géologiques. – *Géochronique*, **35**, 12-20.

ODIN, G.S., GILLOT, P.Y., LORDKIPANIDZE, M., HERNANDEZ, J. & DERCOUR, T.J. (1993). – Premières datations de formations à ammonites bajociennes de Caucase (Géorgie): âges K-Ar de hornblendes volcaniques. – *C. R. Acad. Sci. Paris*, **317**, sér. II, 629-638.

OPPEL, A. (1856-58). – Die Juraformation, Englands, Frankreichs und des südwestlischen Deutschlands. – *Württem. naturwiss., Jaresh.*, 14/15.

ORBIGNY, A. D' (1850). – Paléontologie française. Terrains jurassiques: I – Céphalopodes. – Masson (éd.), 2 Vol., 642 pp.

PAGE, K.N. (1995). – Biohorizons and zonules: intra-subzonal units in Jurassic Ammonite Stratigraphy. – *Paleontology*, **38**, 4, 801-814.

PARIS, F. (1990). – The Ordovician chitinozoan biozones of the Northern Gondwana Domain. – *Rev. Palaeobot. Palynol.*, **66**, 181-209.

PARIS, F., VERNIERS, J., AL-HAJRI, S. & AL-TAYYAR, H. (1995). – Biostratigraphy and palaeogeographic affinities of Early Silurian chitinozoans from central Saudi Arabia. – *Rev. Palaeobot. Palynol.*, **89**, 75-90.

PELISSIE, T., PEYBERNES, B. & REY, J. (1984). – Les grands foraminifères benthiques du Jurassique moyen-supérieur du Sud-Ouest de la France (Aquitaine, Causses, Pyrénées). Intérêt biostratigraphique, paléoécologique et paléobiogéographique). – *In*: Oertli, H.J. (ed.): "Benthos' 83". 2nd Int. Symp. on Bentic Foraminifera (Pau, April 1983). – Elf Aquitaine, Esso REP and Total CFP. Pau and Bordeaux, 1984. – *Bull. Centres Rech. Explor.-Prod. Elf-Aquitaine*, Mém. **6**, 479-489.

PERRET, M.F. & WEYANT, M. (1994). – Les biozones à Conodontes du Carbonifère des Pyrénées. Comparaison avec d'autres régions du globe. – *Geobios*, **27**, 6, 689-715.

PHELPS, M.C. (1985). – A refined ammonite biostratigraphy for the middle and upper Carixian (Ibex and Davoei zones, Lower Jurassic) in North-Western Europe and stratigraphical details on the Carixian-Domerian boundary. – *Geobios*, **18**, 3, 321-362.

REMANE, J. (1991). – From Biostratigraphy to Biochronology time correlation by fossils. – *In*: 11e Congr. Int. Stratigr. et de Géol. du Carbonifère, Beijing, 1987, C. R. 1, 187-200.

ROBASZYNSKI, F. & CARON, M. (1995). – Foraminifères planctoniques du Crétacé: commentaire de la zonation Europe-Méditerranée. – *Bull. Soc. géol. France*, **166**, 6, 681-692.

ROBASZYNSKI, F., AMEDRO, F., FOUCHER, J.C., GASPARD, D., MAGNIEZ-JANNIN, F., MANIVIT, H. & SORNAY, J. (1980). – Synthèse biostratigraphique de l'Aptien au Santonien du Boulonnais à partir de sept groupes paléontologiques: Foraminifères, Nannoplancton, Dinoflagellés et macrofaune. – *Rev. Micropaléont.*, **22**, 4, 195-321.

ROBASZYNSKI, F., CARON, M., DUPUIS, C., AMEDRO, F., GONZALEZ-DONOSO, J.M., LINARES, D., HARDENBOL, J., GARTNER, S., CALANDRA, F. & DELOFFRE, R. (1990). – A tentative integrated stratigraphy in the Turonian of Central Tunisia: Formations, Zones and Sequential Stratigraphy in the Kalaat Senan area. – *Bull. Centres Rech. Explo.-Prod. Elf-Aquitaine*, **14**,1, 213-384.

SALVADOR, A. (1994). – International Stratigraphic Guide. A guide to stratigraphic classification, terminology, and procedure (Second Edition). International Subcommission on Stratigraphic Classification of I.U.G.S. International Commission on Stratigraphy. – *Geological Society of America*, (eds.), 214 pp.

SCHAAF, A. (1985). – Un nouveau canevas biochronologique du Crétacé inférieur et moyen: les biozones à Radiolaires. – *Sci. Géo. Bull.*, **38**, 3, 227-269.

SHAW, A.B. (1964). – Time in stratigraphy. – MC Graw Hill Book Co. (ed.), 365 pp.

SIGAL, J. (1984). – La zone en biostratigraphie: quelques réflexions et conventions. – *Rev. Micropaléont.*, **27**, 1, 61-79.

SPROVIERI, R., DI STEFANO, E., INCARBONA, A. & GARGANO, M.E. (2003). – A high-resolution record of the last deglaciation in the Sicily Channel based on foraminifera and calcareous nannofossil quantitative distribution. – *Palaeogeography, Palaeoclimatology, Palaeoecology*, **202**, 119-142.

TINTANT, H. (1963). – Les Kosmocératidés du Callovien inférieur et moyen d'Europe occidentale. Essai de Paléontologie quantitative. – *Publ. Univ. Bourgogne Dijon*, **29**, 500 pp.

TINTANT, H. (1972a). – La conception biologique de l'espèce et son application en stratigraphie. – *In*: Colloque sur les méthodes et tendances de la stratigraphie; Orsay, septembre 1970. Tome 1. – *Mém. Bur. Rech. géol. min.*, **77**, 77-87.

TINTANT, H. (1972b). – Paléontologie des Invertébrés et stratigraphie. – *In*: Colloque sur les méthodes et tendances de la stratigraphie; Orsay, septembre 1970. Tome 1. – *Mém. Bur. Rech. géol. min.*, **77**, 33-44.

VERNIERS, J., NESTOR, V., PARIS, F., DUFKA, P., SUTHERLAND, S. & GROOTEL, G. VAN (1995). – A global Chitinozoa biozonation for Silurian. – *Geol. Mag.*, **132**, 6, 651-666.

VIANEY-LIAUD, M., BONIS, L. DE, BRUNET, M. & SUDRE, J. (1993). – Biochronologie mammalienne du Paléogène continental d'Europe occidentale. – *In*: "Réunion scientifique de l'Association paléontologique Française", Poitiers, 22 Octobre 1993. – *Paleovox*, 2, 45-55.

WALSH, S.L. (1998). – Fossil datum and paleobiological event terms, paleontostratigraphy, chronostratigraphy, and the definition of Land Mammal "Age" boundaries. *Journal of Vertebrate Paleontology*, **18** (1): 150-179.

WESTERMANN, G.E.G. (1984). – Gauging the Duration of Stages: A New Approach for the Jurassic. – *Episodes*, **7**, 2, 26-28.

WHITTAKER, A., COPE, J.C.W., COWIE, J.W., GIBBONS, W., HAILWOOD, E.A., HOUSE, M.R., JENKINS, D.G., RAWSON, P.F., RUSHTON, A.W.A., THOMAS, A.T. & WIMBLEDON, W.A. (1991). – A guide to stratigraphical procedure. – *J. geol. Soc.*, **148**, 813-824.

Chapter 6. – GEOCHRONOLOGY AND ISOTOPE RADIOCHRONOLOGY

BELL, K. (1985). – Geochronology of the Carswell area, Northern Saskatchewan. *In*: The Carswell structure uranium deposits, Saskatchewan, R. Lainé, Alonso D. & Svab M. Eds. – *Geol. Ass. Canada, Sp. Pap.*, **29**, 33-46.

BE MEZÈME, E., FAURE, M., COCHERIE, A. & CHEN, Y. (2005). – In situ chemical dating of superimposed magmatic and hydrothermal events in the French Variscan Belt. – *Terra Nova*, **17**, 420-426.

BROS, R., STILLE, P., GAUTHIER-LAFAYE, F., WEBER, F. & CLAUER, N. (1992). – Sm-Nd isotopic dating of Proterozoic clay materials: an example from the Francevillian sedimentary series (Gabon). – *Earth and Planet. Sci. Lett.*, **113**, 207-218.

CLAUER, N., STILLE, P., KEPPENS, E. & O'NEIL, J.R. (1992). – Le mécanisme de la glauconitisation: Apports de la géochimie isotopique du strontium, du néodyme et de l'oxygène de glauconies récentes. – *C. R. Acad. Sci., Paris*, **315**, II, 321-327.

CLAUER, N., SRODON, J., FRANCU, J. & SUCHA, V. (1997). – K-Ar dating of illite fundamental particles separated from illite/smectite. – *Clay Min.*, **32**, 181-196.

CLAUER, N. & CHAUDURI, S. (1995). – Clays in crustal environments. – Isotope dating and tracing. Springer Verlag, Berlin, Heidelberg, New York, 358 p.

CLAUER, N., HUGGETT, J.M. & HILLIER, S. (2005). – How reliable is the K-Ar glauconite chronometer? A case study of Eocene sediments from Isle of Wight. – *Clay Min.*, **40**, 167-176.

COCHERIE, A., BE MEZÈME, E., LEGENDRE, O., FANNING, M., FAURE, M. & ROSSI, PH. (2005c). – Electron microprobe dating as a tool for understanding closure of U-Th-Pb system in monazite from migmatite. – *Amer. Min.*, **90**, 607-618.

COCHERIE, A., LEGENDRE, O., PEUCAT, J-J. & KOUAMELAN, A.N. (1998). – Geochronology of polygenetic monazites constrained by in situ electron microprobe Th-U-total Pb determination: implications for lead behaviour in monazite. – *Geochim. Cosmochim. Acta*, **62**, 2475-2497.

COCHERIE, A., ROBERT, M. & GUERROT, C., (2005b). – In situ U-Pb zircon dating using LA-ICPMS and a multi-ion counting system. – Goldschmidt Conference, May 20-24, 2005, Moscow, Idaho, USA. A378.

COCHERIE, A., ROSSI, PH., FANNING, C.M. & GUERROT, C. (2005a). – Comparative use of TIMS and SHRIMP for U-Pb zircon dating of A-type granites and mafic tholeiitic layered complexes and dykes from the Corsican Batholith (France). – *Lithos*, **82**, 185-219.

COLEMAN, D.S. (2002). – U-Pb geochronologic evidence for incremental filling of the Tuolumne intrusive suite magma chamber. – *Geol. Soc. Amer.*, Annual Meeting, Denver, October 27-30, 2002, 269.

COMPSTON, W., WILLIAMS, I.S. & CLEMENT, S.W. (1982). – U-Pb ages within single zircons using a sensitive high mass-resolution ion microprobe. – *Amer. Soc. Mass Spectro. Conference*, 30th, Honolulu, 593-595.

FAURE, G. (1982). – The marine-strontium geochronometer. In: Numerical dating in stratigraphy, Odin, G.S. (ed.) – John Wiley, New York, 73-80.

FAURE, G. & MENSING, T.M. (2005). – Isotopes: Principles and applications. 3rd edition. – John Wiley & Sons Inc., Hoboken, New York, 897 p.

FOLAND, K.A., LINDER, J.S., LASKOWSKI, T.E. & GRANT, N.K. (1984). – $^{40}Ar/^{39}Ar$ dating of glauconites: measured ^{39}Ar recoil loss from well-crystallized specimens. – *Chem. Geol. Isot. Geosc. Sect.*, **2**, 241-264.

KROGH, T.E. (1973). – A low-contamination method for hydrothermal decomposition of zircon and extraction of U and Pb for isotopic age determination. – *Geochim. Cosmochim. Acta*, **37**, 485-494.

KUNK, M.J. & BUSEWIITZ, A.M. (1987). – ^{39}Ar recoil in an I/S clay from the Ordovician "Big Bentonite Bed" at Kinnekulle, Sweden. – 1st Ann. Meet. North-Central Section Geol. Soc. Amer. (Abst. With Program.), **19**, p. 230.

MATTINSON, J.M. (2005). – Zircon U-Pb chemical abrasion ("CA-TIMS") method: Combined annealing and multi-step partial dissolution analysis for improved precision and accuracy of zircon ages. – *Chem. Geol.*, **220**, 47-66.

MONTEL, J.M., FORET, S., VESCHAMBRE, M., NICOLLET, C. & PROVOST, A. (1996). – Electron microprobe dating of monazite. – *Chem. Geol.*, **131**, 37-53.

MONTEL, J.M., VESCHAMBRE, M. & NICOLLET, C. (1994). – Datation de la monazite à la microsonde électronique. – *C. R. Acad. Sci. Paris*, **318**, 1489-1495.

MOORBATH, S., TAYLOR, P.N., ORPEN, J.L., TREOLAR, P & WILSON, J.F. (1987). – First direct radiometric dating of Archean stromatolitic limestone. – *Nature*, **326**, 865-867.

ODIN, G.S. & MATTER, A. (1981). – De glauconarium origine. – *Sedimentology*, **28**, 611-641.

ODIN, G.S. (ed.) (1982). – Numerical dating in Stratigraphy. – J. Wiley & Sons Ltd, Chichester, 2 vol., 1094 p.

ODIN, G.S. & HUNZIKER, J.C. (1982). – Radiometric dating of the Albian-Cenomanian boundary. In: Numerical dating in stratigraphy, Odin G.S. (ed). – John Wiley, New York, 277-305.

ODIN, G.S. (1995). – Géochronologie stratigraphique: de l'analyse isotopique à l'âge et à l'échelle numérique. In: Odin, G.S. (ed.): Phanerozoic Time Scale. – Bull. Liais. Inform., IUGS Subcomm. Geochronol., **13**, 28-42.

PÁLFY, J., SMITH, P.L. & MORTENSEN, J.K. (2000). – A U-Pb and $^{40}Ar/^{39}Ar$ time scale for the Jurassic. – *Can. J. Earth Sci.*, **37**, 923-944.

PYLE, J.M., SPEAR, F.S., WARK, D.A, DANIEL, C.G. & STORM, L.C. (2005). – Contribution to precision and accuracy of monazite microprobe ages. – *Amer. Min.*, 90, 547-577.

ROUSSET, D. & CLAUER, N. (2003). – Discrete clay diagenesis in a very low-permeable sequence constrained by an isotopic (K-Ar and Rb-Sr) study. – *Contr. Mineral. Petrol.*, **145**, 182-198.

ROUSSET, D., LECLERC, S., CLAUER, N., LANCELOT, J., CATHELINEAU, M. & ARANYOSSY, J.F. (2004). – Age and origin of Albian glauconites and associated clay minerals inferred from a detailed geochemical analysis. – *Jour. Sedim. Res.*, **74**, 631-642.

SMITH, P.E., EVENSEN, N.M. & YORK, D. (1993). – First successful ^{40}Ar-^{39}Ar dating of glauconies: Argon recoil in single grains of cryptocrystalline material. – *Geology*, **21**, 41-44.

STACEY, J.S. & KRAMERS, J.D. (1975). – Approximation of terrestrial lead isotope evolution by a two-stage model. – *Earth Planet. Sci. Lett.*, **26**, 207-221.

STILLE, P. & CLAUER, N. (1986). – Sm-Nd isochron-age and provenance of the argillites of the Gunflint Iron Formation in Ontario, Canada. – *Geochim. Cosmochim Acta*, **50**, 1141-1146.

STILLE, P., RIGGS, S., CLAUER, N., AMES, D., CROWSON, R. & SNYDER, S. (1994). – Sr and Nd isotopic analysis of phosphorite sedimentation through one Miocene high-frequency depositional cycle on the North Carolina continental shelf. – *Mar. Geol.*, **117**, 253-273.

SUZUKI, K. & ADACHI, M. (1991). – Precambrian provenance and Silurian metamorphism of the Tsubonosawa paragneiss in the South Kitakami terrane, Northeast Japan, revealed by the chemical Th-U-total Pb isochron ages of monazite, zircon and xenotime. – *Geochem. J.*, **25**, 357-376.

TOULKERIDIS, T., GOLDSTEIN, S.L., CLAUER, N., KRÖNER, A. & LOWE, D.R. (1994). – Sm-Nd dating of Fig Tree clay minerals of the Barberton Greenstone Belt, South Africa. – *Geology*, **22**, 19-22.

TOULKERIDIS, T., CLAUER, N., CHAUDHURI, S. & GOLDSTEIN, S.L. (1998). – Multi-method (K-Ar, Rb-Sr, Sm-Nd) dating of bentonite minerals from eastern United States. – *Basin Res.*, **10**, 261-270.

WETHERILL, G.W., (1956). – Discordant uranium-lead ages 1. – Transactions of the American Geophysical Union, **37**, 320-326

WILLIAMS, M.L., JERCINOVIC, M.J. & TERRY, M.P. (1999). – Age mapping and dating of monazite on the electron microprobe: deconvoluting multistage tectonic histories. – *Geology*, **27**, 1023-1026.

YORK, D., MASLIWEC, H., HALL, C.M., KUYBIDA, P., KENYON, W.J., SPOONER, E.T.C., SCOTT, S.D. & PYE, E.G. (1981). – The direct dating of ore minerals. – *Ont. Geol. Surv., Misc. Pap.*, **98**, 334-340.

ZWINGMANN, H. (1995). – Etude des conditions de mise en place des gaz naturels dans les reservoirs gréseux (Rotliegende d'Allemagne). Aspects minéralogiques, morphologiques, géochimiques et isotopiques. – PhD Thesis, Univ. Strasbourg, France, 189 p.

Chapter 7. – SPECIFIC STRATIGRAPHIES

ALBARÈDE, F. (1985). – Age de la Terre et des météorites. – *In*: Roth, E. et Poty, B. (eds.): Méthodes de datation par les phénomènes nucléaires naturels. – Masson (éd.), Paris; 47-54.

BOND, G., BROECKER, W., JOHNSEN, S., MC MANUS, J., LABEYRIE, L., JOUZEL, J. & BONANI, G. (1993). – Correlations between climate records from North Atlantic sediments and Greenland ice. – *Nature*, London, **365**, 143-147.

BOWRING, S.A. & HOUSH, T. (1995). – The Earth's early evolution. – *Science*, **269**, 1535-1540.

B.R.G.M. (1975). – Notes d'orientation pour l'établissement de la carte géologique à 1/50 000. – Service Géologique National, Orléans, BRGM (éd.), 2ᵉ éd., 240 pp.

CAMPY, M. & MACAIRE, J.J. (1989). – Géologie des formations superficielles, Géodynamique, Faciès, Utilisation. – Masson (éd.), Paris, 433 pp.

CAS, R. & BUSBY-SPERA, C. (1991). – Volcanoclastic Sedimentation. – *Sedimentary Geology*, **74**, 1-4, 362 pp.

CAS, R. & WRIGHT, J.V. (1987). – Volcanic Successions Modern and Ancient. – Londres, Allen et Unwin, 528 pp.

CHALINE, J. (1980). – Problèmes de Stratigraphie du Quaternaire en France et dans les pays limitrophes. Actes Colloque Dijon (1978). – *Bull. Ass. Franç. étud. Quat. Paris*, n.s. **1**, 372 pp.

CLOUD, P. (1976). – Major features of crustal evolution. Alex DuToit Memorial Lecture Series, N° 14. – *Geol. Soc. South Africa*, annex., **79**, 1-32.

CLOUD, P. (1987). – Trends, transitions, and events in Cryptozoic history and their calibration: apropos recommendations by the Subcommission on Precambrian Stratigraphy. *Precambrian Res.*, **37**, 257-264.

DESNOYERS, (1829). – Observations sur un ensemble de dépôts marins. – *Annales des Sciences naturelles*, Paris, 171-214, 402-491.

DEWOLF, Y. (1965). – Intérêt et principes d'une cartographie des formations superficielles. – Faculté des Lettres et Sciences humaines Université Caen pub., 183 pp.

FAKUNDINY, R.H. & LONGACRE, S.A. (1989). – North American Commission on Stratigraphic Nomenclature, Note 57. – Aplication for amendment of North American Stratigraphic Code to provide for exclusive informal use of morphological terms such "Batholith, Intrusion, Pluton, Stock, Plug, Sill, Diapir and Body". – *Bull. amer. Assoc. Petroleum Geol.*, **73**, 11, 1452-1453.

FAURE, H. (1978). – Rapport de synthèse in colloque "Etude et cartographie des formations superficielles: Leurs applications en régions tropicales, thème 1: Formations superficielles et géologie", Sao Paulo, Brésil, **1**, 34-54.

FISHER, R.V. & SCHMINCKE, H.U. (1984). – Pyroclastic Rocks. – Springer-Verlag, Berlin, 474 pp.

FISHER, R.V. & SMITH, G. (1991). – Sedimentation in Volcanic Settings. – SEPM (Society for sedimentary geology) *Special Publ.*, **45**, 258 pp.

GERVAIS, P. (1867-1869). – Zoologie et paléontologie générale. Nouvelles recherches sur les animaux vertébrés vivants et fossiles. – Paris, 263 pp.

HATTIN, D.E. (1991). – Lithodemes, suites, supersuites, and complexes: intrusive, metamorphic, and genetically mixed assemblages of rocks now embraced by North American Code of Stratigraphic Nomenclature. – *Precambrian Res.*, **50**, 355-357.

HEDBERG, H.D. (1976). – International Stratigraphic Guide. A guide to stratigraphic classification, terminology, and procedure. – New York, Wiley and Sons (eds.), 200 pp.

HOFMANN, H.J. (1990). – Precambrian time units and nomenclature. The geon concept. – *Geology*, **18**, 340-341.

HOFMANN, H.J. (1991). – Precambrian time units. Geon or geologic unit?. – *Geology*, **19**, 958-959.

HOFMANN, H.J. (1992). – New Precambrian time scale: Comments. – *Episodes*, **15**, 2, 122-123.

IGCP 41 (1977). – Alimen, M.H.: Limite Pliocène – Quaternaire et définition du Quaternaire. – Studia Geologica Polonica, LII, working group of IGCP 41 and IUGS Stratigraphic Commission, 37-52.

IRVINE, T.N. (1982). – Terminology for layered intrusions. – *J. Petrol.*, **23**, 127-162.

IRVINE, T.N. (1987). – Glossary of terms for layered intrusions. – *In*: Parsons, I. (ed.): Origins of igneous layering, Reidel, D., Publ. Co., 641-647.

JUVIGNÉ, E. (1990). – La téphrostratigraphie et sa nomenclature de base en langue française: mise au point et suggestions. – *Ann. Soc. géol. Belgique*, **113**, 2, 295-298.

KERAUDREN, B. (1992). – Chronostratigraphie du Quaternaire méditerranéen. – *Géochronique*, **44**, 19-20.

LAAJOKI, K. (1989). – Stratigraphic classification and nomenclature of igneous and metamorphic rock bodies: Discussion. – *Bull. geol. Soc. Amer.*, **101**, 753-754.

LEBRET, P., CAMPY, M., COUTARD, J.-P., FOURNIGUET, J., ISAMBERT, M., LAUTRIDOU, J.-P., LAVILLE, P., MACAIRE, J.-J., MENILLET, F. & MEYER, M. (1993). – Cartographie des formations superficielles. Réactualisation des principes de représentation à 1/50 000. – *Géologie de la France*, Orléans, BRGM (éd.), **4**, 39-54.

LE MAITRE, R.W. (1989). – A Classification of Igneous Rocks and Glossary of Terms. Recommendations of IUGS. – Oxford, *Blackwell Scientific Publ.*, 194 pp.

LUMBERS, S.B. & CARD, K.D. (1991). – Chronometric subdivisions of the Archean. – *Geology*, **20**, 3, 56-57.

LY, M.H. (1982). – Le plateau de Perrier et la Limagne du Sud: études volcanologiques et chronologiques des produits montdoriens (Massif Central français). – Thèse 3ᵉ cycle, Univ. Clermont-Ferrand, 180 pp.

LYELL, C. (1839). – Mémoire sur les dépôts tertiaires connus sous le nom de Crag dans les comtés de Norfolk et de Suffolk. – *Mag. Hist. Nat. Publ. Spec.*, **3**, 313 pp.

MAAS, R., KINNY, P.D., WILLIAMS, I.S., FROPUDE, D. & COMPSTON, W. (1992). – The Earth's oldest known crust: A geochronological and geochemical study of 3900-4200 Ma old detrital zircons from Mt. Narryer and Jack Hills, Western Australia. – *Geochim. cosmochim. Acta*, **56**, 3, 1281-1300.

MANKINEN, E. & DALRYMPLE, G. (1979). – Revised geomagnetic time scale for the interval 0 to 5 m.y. B.P. – *J. Geophys. Res.*, **84**, 615-626.

NISBET, E.G. (1982). – Definition of "Archaean". Comment and a proposal on the recommandations of the International Subcommission on Precambrian Stratigraphy. – *Precambrian Res.*, **19**, 111-118.

NISBET, E.G. (1987). – The young Earth: an introduction to Archaean geology. – Allen & Unwin (eds.), 402 pp.

NISBET, E.G. (1991). – Of clocks and rocks – The four aeons of Earth. – *Episodes*, **14**, 4, 327-330.

NORTH AMERICAN COMMISSION ON STRATIGRAPHIC NOMENCLATURE (1983). – North American Stratigraphic Code. – *Bull. amer. Assoc. Petroleum Geol.*, **67**, 841-875.

PENK, A. & BRUCKNER, E. (1901-1905). – Die Alpen im Eiszeitalter. – Leipzig, 1-3.

PLUMB, K.A. (1991). – New Precambrian time scale. – *Episodes*, **14**, 2, 139-140.

PLUMB, K.A. (1992). – New Precambrian time scale: Reply. – *Episodes*, **15**, 2, 124-125.

PLUMB, K.A. & GEE, R.D. (1987). – Nomenclature for the Proterozoic – Recent actions by the Subcommission on Precambrian Stratigraphy. – *Precambrian Res.*, **36**, 185-187.

PLUMB, K.A. & JAMES, H.L. (1986). – "Subdivision of Precambrian time: Recommendations and suggestions by the Subcommission on Precambrian Stratigraphy". – *Precambrian Res.*, **32**, 65-92.

PLUMB, K.A. & JAMES, H.L. (1987). – Subdivision of Precambrian time: Recommendations and suggestions by the Subcommission on Precambrian Stratigraphy; Reply. – *Precambrian Res.*, **36**, 179-180.

POIDEVIN, J.L., CANTAGREL, J.M. (1984). – Un site unique du Plio-Pléistocène en Europe: le plateau de Perrier (Puy de Dôme). – Confrontation des données volcanologiques, stratigraphiques et paléontologiques. – *Rev. Sci. Nat. d'Auvergne*, **50**, 1-4, 87-95.

PREISS, W.V. (1982). – Letter to Australian Geologist. – *Austral. Geol.*, **42**, 8-9.

RAT, P. (1980). – Méthodologie stratigraphique et Quaternaire. – Problèmes de stratigraphie Quaternaire en France et dans les pays limitophes, Colloque de Dijon 1978. – *Bull. Assoc. franç. étud. Quat. Publ. Spec.*, **1**, 4-14.

SALVADOR, A. (1987). – Stratigraphic classification and nomenclature of igneous and metamorphic rock bodies: Reply. – *Bull. geol. Soc. Amer.*, **101**, 754.

SELF, S. & SPARKS, R.S. (1981). – Tephra studies. – Dordrecht, D.Reidel Publ., 482 pp.

SERRE, M. DE (1832). – Les animaux découverts dans les différentes couches des dépôts quaternaires. – *Bull. Soc. Géol. France*, **1**, 2, 450 pp.

SHACKLETON, N.J., BACKMAN, J., ZIMMERMAN, H., KENT, D.V., HALL, M.A., ROBERTS, D.G., SCHNITKER, D., BALDAUF, J.G., DESPRAIRIES, A., HOMRIGHAUSEN, R., HUDDLESTUN, P., KEENE, J.B., KALTENBACK, A.J., KRUMSIEK, K.A.D., MORTON, A.C., MURRAY, J.W. & WESTBERGSMITH, J. (1984). – Oxygen isotope calibration of the onset of ice-rafting and history of glaciations in the North Atlantic region. – *Nature*, London, **307**, 620-623.

SUC, J.-P. & ZAGWIJN, W.H. (1983). – Plio-Pleistocene in Mediterranean region and northwestern Europe according to recent biostratigraphic and paleoclimatic data. – *Boreas*, **12**, 153-166.

THORARINSSON, S. (1944). – Tefrokronologiska studier pa Island. – *Ann. Geogr. Stockholm*, **26**, 1-217.

TRENDALL, A.F. (1984). – The Archean/Proterozoic transition as a geologic event – a view from Australian evidence. – Dahlem Konferenzen 1984, Berlin, Springer-Verlag, 243-259.

VIDAL, G. (1987). – Subdivision of Precambrian time: Recommendations and suggestions by the Subcommission on Precambrian Stratigraphy; Comment. – *Precambrian Res.*, **36**, 177-178.

WINDLEY, B.F. (1984). – The Archaean-Proterozoic boundary. – *Tectonophysics*, **105**, 43-53.

WINDLEY, B.F. (1986). – The evolving continents, 2nd ed. – Wiley, J. & Sons (eds.), 399 pp.

ZAGWIJN, W.H. (1992). – The beginning of the ice age in Europe and its major subdivisions. – *Quater. Sci. Rev.*, Pergamon Press Ltd., London, **11**, 583-591.

Chapter 8. – CHRONOSTRATIGRAPHIC UNITS AND CORRELATIONS

BATES, R. & JACKSON, J.A. (1980). – Glossary on Geology, 2nd ed. – Amer. Geol. Instit. (ed.), Falls Church, 749 pp.

COMITÉ FRANÇAIS DE STRATIGRAPHIE (1962). – Principes de classification et de nomenclature stratigraphique. – *Com. franç. Strati.*, 15 pp.

COWIE, J.W. (1986). – Guidelines for boundary stratotypes. – *Episodes*, **9**, 78-82.

FOUCAULT, A. & RAOULT, J.F. (1995). – Dictionnaire de Géologie, 4ᵉ éd. – Paris, Masson (ed.), 324 pp.

EDWARDS, L.E. (1989). – Supplemented Graphic Correlation: a powerful tool for paleontologists and non paleontologists. – *Palaios*, **4**, 1, 127-143.

HEDBERG, H. (1976). – International stratigraphic guide; a guide to stratigraphic classification, terminology and procedure. – John Wiley & Sons, N.Y. (eds.), 200 pp.

MILLER, F.X. (1977). – The graphic correlation method in biostratigraphy. – *In*: Kauffman, E.G. & Hazel, J.E. (eds.): Concepts and methods in biostratigraphy. – Dowden, Hutchinson & Ross, Stroudsburg, 1681-1686.

NORTH AMERICAN COMMISSION ON STRATIGRAPHIC NOMENCLATURE (1983). – North American Stratigraphic Code. – *Bull. amer. Assoc. Petrol. Geol.*, **67**, 5, 841-875.

ODIN, G.S. & MONTANARI, A. (1989). – Age radiométrique et stratotype de la limite Eocène-Oligocène. – *C.R. Acad. Sci. Paris*, **309**, 1939-1945.

SALVADOR, A. (1994). – International Stratigraphic Guide. A guide to stratigraphic classification, terminology, and procedure. – International Subcommission on Stratigraphic Classification of I.U.G.S. – *Geological Society of America*, 2nd ed., 214 pp.

SHAW, A.B. (1964). – Time in Stratigraphy. – Mc Graw hill book Co. (ed.), 365 pp.

SIGAL, J. & TINTANT, H. (1962). – Principes de classification et de nomenclature stratigraphiques. – *Comm. Franç. Stratigr.*, 15 pp.

STEIN, J., GAMBER, J.H., KREBS, N. & LACOE, M.K. (1992). – A composite standard approach to biostratigraphic evaluation of the North Sea Paleogene. – Abstract: Petroleum Geology of North Western Europe, Aberdeen, 1992.

ZALASIEWICZ, J., SMITH, A., BRENCHLEY, P., EVANS, J., KNOX, R., RILEY, N., GALE, A., GREGORY, F.J., RUSHTON, A., GIBBARD, P., HESSELBO, S., MARSHALL, J., OATES, M., RAWSON, P. AND TREWIN, N. (2004). – Simplifying the stratigraphy of time, *Geology*, **32**, (1), 1-4.

Chapter 9. – THE GEOLOGICAL TIME SCALE

AGTERBERG, F.P. (1994). – Estimation of the Mesozoic Geological Time Scale. *Mathematical Geol.*, **26** (7), 857-876.

CHLUPÁC, I. (1993). – Geology of the Barrandian. *Senckenberg-Buch*, **69**, 163 p.

GRACIANSKY, P., HARDENBOL, J., JAQUIN, TH. & VAIL, P.R. (1998). – Mesozoic and Cenozoic Sequence Stratigraphy of European Basins. *SEPM Special Publ.*, **60**, 786 p.

GRADSTEIN, F.M. & OGG, J.G. (1996). – A Phanerozoic time scale. *Episodes*, **19** (2), 3-5.

GRADSTEIN, F. M., AGTERBERG, F. P., OGG, J. G., HARDENBOL, J., VAN VEEN, P., THIERRY, J. & HUANG, Z. (1995). – A Triassic, Jurassic, and Cretaceous time scale. *In*: Berggren *et al.*, eds., Geochronology, Time Scales and Global Stratigraphic Correlation; Tulsa, **SEPM Special Publication**, **54**, 95-126.

GRADSTEIN, F.M., OGG J.G., SMITH, A.G. AGTERBERG F.P., BLEEKER W., COOPER R.A., DAVYDOV V., GIBBARD P., HINNOV L., HOUSE M.R. (†), LOURENS L., LUTERBACHER H.P., MCARTHUR J., MELCHIN M.J., ROBB L.J., SHERGOLD J., VILLENEUVE M., WARDLAW B.R., ALI J., BRINKHUIS H., HILGEN F.J., HOOKER J., HOWARTH R.J., KNOLL A.H., LASKAR J., MONECHI S., POWELL J., PLUMB K.A., RAFFI I., RÖHL U., SADLER P., SANFILIPPO A., SCHMITZ B., SHACKLETON N.J., SHIELDS G.A., STRAUSS H., VAN DAM J., VEIZER J., VAN KOLFSCHOTEN TH. & WILSON D. (2004). – A Geologic Time Scale 2004. *Cambridge University Press*, 589p.

HAQ, B.U. & ABDUL MOTALEB AL-QAHTANI A.M. (2005). – Phanerozoic cycles of sea-level change on the Arabian Platform. *Geoarabia*, **10**, 2, 127-160, 2 charts.

HARDENBOL, J. THIERRY, J, FARLEY, M.B., JACQUIN, TH., GRACIANSKY, P.C. DE & VAIL, P.R. (1998). – Mesozoic and Cenozoic chronostratigraphy charts. *In*: Graciansky P.C. de, Hardenbol, J., Jacquin, Th., Farley, M. & Vail, P.R. eds – Mesozoic-Cenozoic Sequence Stratigraphy of European Basins; Tulsa, *SEPM Special Publication*, **60**.

HARDENBOL, J., THIERRY, J., FARLEY, M.B., JAQUIN, TH., GRACIANSKY, P.C. & VAIL, P.R., (1998). – Mesozoic and Cenozoic Sequence Chronostratigraphic Framework of European Basins. *SEPM Special Publ.*, **60**, 8 charts.

HARLAND, W. B., ARMSTRONG, R. L., COX, A. V., CRAIG, L. E., SMITH, A. G. & SMITH, D. G. (1990). – A Geologic Time Scale 1989. *Cambridge University Press*, 265 pp.

JAFFEY A. H., FLYNN K. F., GLENDENIN L. E., BENTLEY W. C. & ESSLING A. M. (1971). – Precision measurement of half-lives and specific activities of ^{235}U and ^{238}U. *Physical Review C: Nuclear Physics,* 4, 1889-1906.

KWON, J., MIN, K., BICKEL, P.J. & RENNE, P.R. (2002). – Statistical methods for jointly estimating the decay constant of ^{40}K and the age of a dating standard. *Mathematical Geol*, **34**, (4), 457-474.

MARTINSSON, A. ed. (1977). – The Silurian-Devonian Boundary: Final report of the Committee of the Siluro-Devonian Boundary within IUGS Commission on Stratigraphy and a state of the art report for Project Ecostratigraphy. Stuttgart: *Schweizerbartsche Verlagsbuchhandlung*, 347 pp.

REMANE, J., BASSETT, M.G., COWIE, J.W., GORHBANDT, K.H., LANE, H.R., MICHELSEN, O. & WANG, N. (1996). – Revised guidelines for the establishment of global chronostratigraphic standards by the International Commission on Stratigraphy (ICS). *Episodes*, **19** (3), 77-81.

REMANE, J. Ed. (2000). – Explanatory note to the International Stratigraphic Chart – Sponsored by ICS, IUGS and Unesco. 16 p. Division of Earth Sciences, *Unesco*, Paris.

RENNE, P.R., DEINO, A.L., WALTHER, R.C., THURRIN, B.D., SWISHER, C.C., BECKER, T.A., CURTIS, G.H., SHARP, W.D. & JAOUNI, A.R. (1994). – Intercalibration of astronomical and radio-isotopic time. *Geology*, **22**, 783 (1994).

ROBERTS, J., CLAOUE-LONG, J. & JONES, P.J. (1995). – Australian Early Carboniferous Time. *In*: Berggren *et al.*, eds., Geochronology, Time Scales and Global Stratigraphic Correlation; Tulsa, *SEPM Special Publication*, **54**, 24 – 40.

SALVADOR, A. Ed. (1994). – International Stratigraphic Guide-A Guide to Stratigraphic Classification, Terminology and Procedure. *Geol. Soc. of America*, 214 p.

SHARLAND, P., CASEY D.M., DAVIES R.B., SIMMONS M.B. & SUTCLIFFE O.E. (2004). – Arabian Plate Sequence Stratigraphy. *Geoarabia*, **9**, 1, 2004: 199- 214, 2 charts

STEIGER R.H. & JÄGER E. (1977). – Subcommission on geochronology; convention on the use of decay constants in geo- and cosmochronology. *Earth and Planetary Science Letters*, **36**(3), 359-362.

TUCKER, R.D. & MCKERROW, W.S. (1995). – Early Paleozoic chronology: A review in light of new U-Pb zircon ages from Newfoundland and Britain. *Can. J. Earth Sci.*, **32**, 368-379.

WILLIAMS, E.A., FRIEND, P.F. & WILLIAMS, B.P.J. (2000). – A review of Devonian time scales: databases, construction and new data. In: Friend, P.F. & Williams, B.P.J. (eds). *New Perspectives on the Old Red Sandstone.*, *Geol. Soc. London, Spec. Publ.*, **180**, 1-21.

ZALASIEWICZ, J., SMITH, A., BRENCHLEY, P., EVANS, J., KNOX, R., RILEY, N., GALE, A., GREGORY, F.J., RUSHTON, A., GIBBARD, P., HESSELBO, S., MARSHALL, J., OATES, M., RAWSON, P. & TREWIN, N. (2004). Simplifying the stratigraphy of time, *Geology*, **32**, (1), 1-4.

EDITORS

Jacques REY
Institut des Sciences de la Terre, UMR CNRS 5563, Université Paul Sabatier, 14 Avenue Edouard Belin, 314002 TOULOUSE, France.
jacques.rey3@free.fr

Simone GALEOTTI
Istituto di Geologia, Università di Urbino, Campus Scientifico, 61029, URBINO, Italy.
s.galeotti@uniurb.it

AUTHORS

François BAUDIN
Département de Géologie Sédimentaire, UMR CNRS 5143, Université Pierre et Marie Curie, 4, Place Jussieu, 75252 PARIS Cedex 05, France.
françois.baudin@upcm.fr

Nadia Kiaya BELKAALOUL
Département de Géologie, URA CNRS 723, Université de Paris Sud, 91405 ORSAY Cedex, France.

Michel CAMPY
Centre des Sciences de la Terre, UMR CNRS 5561, Université de Bourgogne, 6, boulevard Gabriel, 21000 DIJON.
michel.campy@u-bourgogne.fr

Ramon CAPDEVILA
Geosciences Rennes, UPR CNRS 4661, Université de Rennes, Campus de Beaulieu, 35042 RENNES Cedex, France.
capdevil@univ-rennes1.fr

Norbert CLAUER
Centre de Géochimie de la Surface, UPR CNRS 6251, Université Louis Pasteur, 1, rue Blessig, 67084 STRASBOURG Cedex, France.
nclauer@illite.u-strasbg.fr

Alain COCHERIE
B.R.G.M., 3, avenue Cl. Guillemin, B.P. 6009, 45060 ORLÉANS Cedex, France
a.cocherie@brgm.fr

Jean Christophe CORBIN
Cogema, 2 rue Paul Dautier, B.P. 4, 78141 VELIZY Cedex, France.
jcorbin@cogema.fr

Pierre COTILLON
Centre des Sciences de la Terre, UMR CNRS 5125, Université Claude Bernard, 69622 VILLEURBANNE Cedex, France.
Pierre.Cotillon@univ-lyon1.fr

Louis COUREL
Centre des Sciences de la Terre, UMR CNRS 5561, Université de Bourgogne, 6, boulevard Gabriel, 21000 DIJON.
louis.courel@wanadoo.fr

Valérie DAUX
UFR des Sciences de la Terre et évolution des milieux naturels, Université Pierre et Marie Curie, 4, Place Jussieu, 75252 PARIS Cedex 05, France.
valerie.daux@upcm.fr

Jean DUMAY
TOTAL E&P – Géosciences. CSTJF, Avenue Larribau, 64018 PAU Cedex, France.
jean.dumay@total.com

Laurent EMMANUEL
Laboratoire des biominéralisations et Paléoenvironnements, FR CNRS 32, Université Pierre et Marie Curie, 4, Place Jussieu, 75252 PARIS Cedex 05, France.
laurent.emmanuel@upcm.fr

Bruno GALBRUN
Centre Parisien en Géologie, FR CNRS 32, Université Pierre et Marie Curie, 4, Place Jussieu, 75252 PARIS Cedex 05, France.
bgalbrun@ccr.jussieu.fr

Simone GALEOTTI
Istituto di Geologia, Università di Urbino, Campus Scientifico, 61029, URBINO, Italy.
s.galeotti@uniurb.it

Alain de GOËR DE HERVE
Décédé

Felix M. GRADSTEIN
Geology Museum, University of Oslo, N-0318 OSLO, Norway
f.m.gradstein@nhm.uio.no

Michel ISAMBERT
I.N.R.A., Centre de Recherches d'Orléans, Service d'Etude des Sols et de la carte pédologique de la France, 45160 OLIVET (France)
infosol@orleans.inra.fr

Luca LANCI
Facoltà di Scienze Ambientali, Università di Urbino, Campus Scientifico, 61029, URBINO, Italy
and
Alpine Laboratory of Paleomagnetism, V. Madonna dei Boschi, PEVERAGNO, CN, Italy
llanci@uniurb.it

Jean-Pierre LAUTRIDOU
Laboratoire de Géomorphologie et Transferts de Surface, URA CNRS 1694, Université de Caen, 24 rue des Tilleuls, 14000 CAEN, France.
jean-pierre.lautridou@geos.unicaen.fr

Patrick LEBRET
Centre ressources minérales, B.R.G.M., Avenue de Concyr, B.P. 6009, 45060 ORLEANS Cedex, France.
p.lebret@brgm.fr

Jean-Jacques MACAIRE
Laboratoire de Géologie des Systèmes sédimentaires, EA 2100, Faculté des Sciences et Techniques de Tours, Parc de Grandmont, 32700 TOURS, France.
jean-jacques.macaire@univ-tours.fr

Pierre MAURIAUD
TOTAL E&P – Géosciences. CSTJF, Avenue Larribau, 64018 PAU Cedex, France.
pierre.mauriaud@total.com

François MENILLET
B.R.G.M.S.G.R. Alsace et 7, rue Sainte Odile, 67000 STRASBOURG, France.

Robert MEYER
280, rue Leonard de Vinci, 76960 NOTRE-DAME-DE-BONDEVILLE, France.
robert.meyer4@wanadoo.fr

Gabbi OGG
Department of Geological and atmospheric Sciences, Purdue University, WEST LAFAYETTE, in 47907, USA

James G. OGG
Department of Geological and atmospheric Sciences, Purdue University, WEST LAFAYETTE, in 47907, USA
jogg@purdue.edu

Philippe RABILLER
8, rue Jean-Baptiste Clément, 64230 LESCAR, France.

Jean François RAYNAUD
TOTAL E&P – Géosciences. CSTJF, Avenue Larribau, 64018 PAU Cedex, France.
jean-francois.raynaud@total.com

Maurice RENARD
Laboratoire des biominéralisations et Paléoenvironnements, FR CNRS 32, Université Pierre et Marie Curie, 4, Place Jussieu, 75252 PARIS Cedex 05, France.

Jacques REY
Institut des Sciences de la Terre, UMR CNRS 5563, Université Paul Sabatier, 14 Avenue Edouard Belin, 314002 TOULOUSE
jacques.rey3@free.fr

Giovanni RUSCIADELLI
Dipartimento di Scienze della Terra, Università G. d'Annunzio di Chieti, Via dei Vestini, 30, 66013 CHIETI SCALO, Italy.
grusciadelli@unich.it

Jacques THIERRY
Centre des Sciences de la Terre, UMR CNRS 5561, Université de Bourgogne, 6, boulevard Gabriel, 21000 DIJON.
jacques-thierry2@wanadoo.fr

Federica TAMBURINI
Geological Institute, ETH-Zurich, CHN, Universitaetsstrasse 16, 8092 ZURICH, Switzerland
federica.tamburini@erdw.ethz.ch

TRANSLATORS

CHAPTER 1

Catherine BREWERTON
ANGLO-FILE sarl, Pau Cité Multimédia, 45 Avenue Léon Blum, 65054 PAU Cedex9, France.
c.brewerton@anglofile.fr

CHAPTER 7

Alberto RENZULLI (paragraph 1)
Istituto di Vulcanologia e Geochimia, Università di Urbino, Campus Scientifico, 61029 URBINO, Italy
renzulli@uniurb.it

Patrizia SANTI (paragraph 5)
Istituto di Vulcanologia e Geochimia, Università di Urbino, Campus Scientifico, 61029 URBINO, Italy
santi@uniurb.it

Rowena STEAD (paragraph 2, 3, 4)
Service Eau, BROM, 3, avenue Cl. Guillemin, B.P. 6009, 45060 ORLÉANS Cedex, France
r.stead@brgm.fr

Achevé d'imprimer en mars 2008
sur les presses de EMD S.A.S
Dépôt légal : mars 2008
Numéro d'impression : 18997

Imprimé en France